Genomic Imprinting

METHODS IN MOLECULAR BIOLOGY™

John M. Walker, Series Editor

195. **Quantitative Trait Loci:** *Methods and Protocols*, edited by *Nicola J. Camp and Angela Cox, 2002*
194. **Post-translational Modification Reactions,** edited by *Christoph Kannicht, 2002*
193. **RT-PCR Protocols,** edited by *Joseph O'Connell, 2002*
192. **PCR Cloning Protocols, 2nd ed.,** edited by *Bing-Yuan Chen and Harry W. Janes, 2002*
191. **Telomeres and Telomerase:** *Methods and Protocols,* edited by *John A. Double and Michael J. Thompson, 2002*
190. **High Throughput Screening:** *Methods and Protocols,* edited by *William P. Janzen, 2002*
189. **GTPase Protocols:** *The RAS Superfamily,* edited by *Edward J. Manser and Thomas Leung, 2002*
188. **Epithelial Cell Culture Protocols,** edited by *Clare Wise, 2002*
187. **PCR Mutation Detection Protocols,** edited by *Bimal D. M. Theophilus and Ralph Rapley, 2002*
186. **Oxidative Stress and Antioxidant Protocols,** edited by *Donald Armstrong, 2002*
185. **Embryonic Stem Cells:** *Methods and Protocols,* edited by *Kursad Turksen, 2002*
184. **Biostatistical Methods,** edited by *Stephen W. Looney, 2002*
183. **Green Fluorescent Protein:** *Applications and Protocols,* edited by *Barry W. Hicks, 2002*
182. **In Vitro Mutagenesis Protocols, 2nd ed.,** edited by *Jeff Braman, 2002*
181. **Genomic Imprinting:** *Methods and Protocols,* edited by *Andrew Ward, 2002*
180. **Transgenesis Techniques, 2nd ed.:** *Principles and Protocols,* edited by *Alan R. Clarke, 2002*
179. **Gene Probes:** *Principles and Protocols,* edited by *Marilena Aquino de Muro and Ralph Rapley, 2002*
178. **Antibody Phage Display:** *Methods and Protocols,* edited by *Philippa M. O'Brien and Robert Aitken, 2001*
177. **Two-Hybrid Systems:** *Methods and Protocols,* edited by *Paul N. MacDonald, 2001*
176. **Steroid Receptor Methods:** *Protocols and Assays,* edited by *Benjamin A. Lieberman, 2001*
175. **Genomics Protocols,** edited by *Michael P. Starkey and Ramnath Elaswarapu, 2001*
174. **Epstein-Barr Virus Protocols,** edited by *Joanna B. Wilson and Gerhard H. W. May, 2001*
173. **Calcium-Binding Protein Protocols, Volume 2:** *Methods and Techniques,* edited by *Hans J. Vogel, 2001*
172. **Calcium-Binding Protein Protocols, Volume 1:** *Reviews and Case Histories,* edited by *Hans J. Vogel, 2001*
171. **Proteoglycan Protocols,** edited by *Renato V. Iozzo, 2001*
170. **DNA Arrays:** *Methods and Protocols,* edited by *Jang B. Rampal, 2001*
169. **Neurotrophin Protocols,** edited by *Robert A. Rush, 2001*
168. **Protein Structure, Stability, and Folding,** edited by *Kenneth P. Murphy, 2001*
167. **DNA Sequencing Protocols,** *Second Edition,* edited by *Colin A. Graham and Alison J. M. Hill, 2001*
166. **Immunotoxin Methods and Protocols,** edited by *Walter A. Hall, 2001*
165. **SV40 Protocols,** edited by *Leda Raptis, 2001*
164. **Kinesin Protocols,** edited by *Isabelle Vernos, 2001*
163. **Capillary Electrophoresis of Nucleic Acids, Volume 2:** *Practical Applications of Capillary Electrophoresis,* edited by *Keith R. Mitchelson and Jing Cheng, 2001*
162. **Capillary Electrophoresis of Nucleic Acids, Volume 1:** *Introduction to the Capillary Electrophoresis of Nucleic Acids,* edited by *Keith R. Mitchelson and Jing Cheng, 2001*
161. **Cytoskeleton Methods and Protocols,** edited by *Ray H. Gavin, 2001*
160. **Nuclease Methods and Protocols,** edited by *Catherine H. Schein, 2001*
159. **Amino Acid Analysis Protocols,** edited by *Catherine Cooper, Nicole Packer, and Keith Williams, 2001*
158. **Gene Knockoout Protocols,** edited by *Martin J. Tymms and Ismail Kola, 2001*
157. **Mycotoxin Protocols,** edited by *Mary W. Trucksess and Albert E. Pohland, 2001*
156. **Antigen Processing and Presentation Protocols,** edited by *Joyce C. Solheim, 2001*
155. **Adipose Tissue Protocols,** edited by *Gérard Ailhaud, 2000*
154. **Connexin Methods and Protocols,** edited by *Roberto Bruzzone and Christian Giaume, 2001*
153. **Neuropeptide Y Protocols,** edited by *Ambikaipakan Balasubramaniam, 2000*
152. **DNA Repair Protocols:** *Prokaryotic Systems,* edited by *Patrick Vaughan, 2000*
151. **Matrix Metalloproteinase Protocols,** edited by *Ian M. Clark, 2001*
150. **Complement Methods and Protocols,** edited by *B. Paul Morgan, 2000*
149. **The ELISA Guidebook,** edited by *John R. Crowther, 2000*
148. **DNA–Protein Interactions:** *Principles and Protocols* **(2nd ed.),** edited by *Tom Moss, 2001*
147. **Affinity Chromatography:** *Methods and Protocols,* edited by *Pascal Bailon, George K. Ehrlich, Wen-Jian Fung, and Wolfgang Berthold, 2000*
146. **Mass Spectrometry of Proteins and Peptides,** edited by *John R. Chapman, 2000*
145. **Bacterial Toxins:** ***Methods and Protocols,*** edited by *Otto Holst, 2000*
144. **Calpain Methods and Protocols,** edited by *John S. Elce, 2000*
143. **Protein Structure Prediction:** *Methods and Protocols,* edited by *David Webster, 2000*
142. **Transforming Growth Factor-Beta Protocols,** edited by *Philip H. Howe, 2000*
141. **Plant Hormone Protocols,** edited by *Gregory A. Tucker and Jeremy A. Roberts, 2000*
140. **Chaperonin Protocols,** edited by *Christine Schneider, 2000*
139. **Extracellular Matrix Protocols,** edited by *Charles Streuli and Michael Grant, 2000*
138. **Chemokine Protocols,** edited by *Amanda E. I. Proudfoot, Timothy N. C. Wells, and Christine Power, 2000*

METHODS IN MOLECULAR BIOLOGY™

Genomic Imprinting

Methods and Protocols

Edited by

Andrew Ward

Department of Biology and Biochemistry
University of Bath
Claverton Down, Bath, UK

Humana Press Totowa, New Jersey

999 Riverview Drive, Suite 208
Totowa, New Jersey 07512

www.humanapress.com

This publication is printed on acid-free paper. ∞

ANSI Z39.48-1984 (American National Standards Institute) Permanence of Paper for Printed Library Materials.

Cover design by Patricia F. Cleary.

Cover illustration: *In situ* hybridization showing the expression pattern of the *Igf2* gene in a mouse embryo. Artwork courtesy of Dr. Andrew Ward and Dr. Marika Charalambous.

For additional copies, pricing for bulk purchases, and/or information about other Humana titles, contact Humana at the above address or at any of the following numbers: Tel: 973-256-1699; Fax: 973-256-8341; E-mail: humana@humanapr.com or visit our website at http://humanapress.com

Printed in the United States of America. 10 9 8 7 6 5 4 3 2 1

Library of Congress Cataloging-in-Publication Data

Genomic imprinting: methods and protocols / edited by Andrew Ward.
p. cm. -- (Methods in molecular biology ; v. 181)
Includes bibliographical references and index.
ISBN 0-89603-741-X (alk. paper)
1. Genomic imprinting--Laboratory manuals. I. Ward, Andrew, 1964- II. Series.

QH 450 .G468 2001
572.8'6'028--dc21

2001024456

Preface

Genomic imprinting is the process by which gene activity is regulated according to parent of origin. Usually, this means that either the maternally inherited or the paternally inherited allele of a gene is expressed while the opposite allele is repressed. The phenomenon is largely restricted to mammals and flowering plants and was first recognized at the level of whole genomes. Nuclear transplantation experiments carried out in mice in the late 1970s established the non-equivalence of the maternal and paternal genomes in mammals, and a similar conclusion was drawn from studies of interploidy crosses of flowering plants that extend back to at least the 1930s. Further mouse genetic studies, involving animals carrying balanced translocations (reviewed in Chapter 3), indicated that imprinted genes were likely to be widely scattered and would form a minority within the mammalian genome. The first imprinted genes were identified in the early 1990s; over forty are now known in mammals and the list continues steadily to expand.

Genomic Imprinting: Methods and Protocols aims to collect protocols that have been applied to the study of imprinting or imprinted genes. Many of the protocols are based on more widely used embryology or molecular biology techniques that have been adapted for imprinting research. All of the included methods remain gainfully employed in either (or both) the discovery or analysis of imprinted genes. Chapter 1 describes the nuclear transplantation methods, first used in the 1970s, for the generation of mouse embryos with genomes of entirely maternal or entirely paternal origin. The first five chapters are specific to the mouse, though some of the principles could be applied to other species. For instance, the techniques described in Chapters 4 and 5 for generating transgenic mice using large fragments of genomic DNA have resulted in several examples of the faithful reproduction of imprinted gene expression at ectopic loci. The first few imprinted genes have recently been identified in plants and it will be interesting to know whether the imprinting of these genes can be similarly reproduced within plant transgenes.

The majority of protocols describe molecular techniques and most of these allow examination of gene structure or expression in an allele-specific manner, which is an essential aspect of most imprinting studies. Protocols are included for identifying imprinted genes (Chapters 6–8), for analyzing imprinted gene expression (Chapters 9–12), for the study of DNA methylation and methylation-sensitive DNA-binding proteins (Chapters 13–20), and for examining chromatin structure (Chapters 21–24). The final chapter is a review of genomic imprinting in plants. Although imprinting must have arisen independently in plants and animals, the available evidence suggests that the imprinting mechanisms in these species may share common features, such as the involvement of DNA methylation in distinguishing maternal and paternal alleles. Thus, the molecular methods that are already extensively used to study mammalian imprinted genes will surely find even wider employment as the genomic imprinting field continues to expand.

I thank all of the authors for their outstanding contributions to this volume. On behalf of us all I extend the hope that this effort to make these methods accessible will prove useful to genomic imprinting aficionados everywhere.

Andrew Ward

Contents

Contributors

GAIL ADAM • *Eurona Medical AB, Kungsängsvägen, Uppsala, Sweden*
ROGER L. P. ADAMS • *IBLS, University of Glasgow, Scotland, UK*
SALLY ADAMS • *Department of Biology and Biochemistry, University of Bath, Claverton Down, Bath, United Kingdom*
JUSTIN F.-X. AINSCOUGH • *Wellcome/CRC Institute of Cancer and Developmental Biology, University of Cambridge, Cambridge, United Kingdom*
MIRA ARIEL • *Department of Cellular Biochemistry and Human Genetics, Hadassah Medical School, The Hebrew University, Jerusalem, Israel*
MARISA S. BARTOLOMEI • *Howard Hughes Medical Institute and Department of Cell and Developmental Biology, University of Pennsylvania School of Medicine, Philadelphia, PA*
SHEILA C. BARTON • *Wellcome/CRC Institute of Cancer and Developmental Biology, University of Cambridge, Cambridge, United Kingdom*
DÉBORAH BOURC'HIS • *INSERM U383, Hôpital Necker-Enfants Malades, Paris Cedex, France*
GIOVANNA BRAIDOTTI • *Department of Cell Biology and Genetics, Erasmus University, Rotterdam, The Netherlands*
MARIKA CHARALAMBOUS • *Department of Biology and Biochemistry, University of Bath, Claverton Down, Bath, United Kingdom*
HENGMI CUI • *Departments of Medicine, Oncology, and Molecular Biology and Genetics, Johns Hopkins University School of Medicine, Baltimore, MD*
WENDY L. DEAN • *Laboratory of Developmental Genetics and Imprinting, Developmental Genetics Programme, The Babraham Institute, Cambridge, United Kingdom*
GHISLAINE DELL • *Department of Biology and Biochemistry, University of Bath, Claverton Down, Bath, United Kingdom*
HUGH G. DICKINSON • *Department of Plant Sciences, University of Oxford, South Parks Road, Oxford, United Kingdom*
OSMAN EL-MAARRI • *Institute of Experimental Haematology and Transfusion Medicine, Bonn, Germany*
SABINE ENGEMANN • *Max-Planck-Institut für molekulare Genetik, Ihnestr, Berlin, Germany*

ROBERT FEIL • *Institute of Molecular Genetics, CNRS-UMR5535, Montpellier, France*
ANNE C. FERGUSON-SMITH • *Department of Anatomy, University of Cambridge, Downing Street, Cambridge, United Kingdom*
CHUNGUANG GAO • *Division of Biology, Beckman Research Institute, City of Hope, Duarte, CA*
PANTELIS GEORGIADES • *Department of Anatomy, University of Cambridge, Downing Street, Cambridge, United Kingdom*
VALERIE GRANDJEAN • *Department of Anatomy, University of Cambridge, Downing Street, Cambridge, United Kingdom*
RICHARD I. GREGORY • *Fox Chase Cancer Center, Philadelphia, Pennsylvania*
PETRA HAJKOVA • *Department of Genetics, University of Saarland, Saarbrücken, Germany*
IZUHO HATADA • *Gene Research Center, Gunma University, Maebashi, Japan*
ANDREW R. HOFFMAN • *Medical Service, VA Palo Alto Health Care System and Department of Medicine, Stanford University School of Medicine, Palo Alto, CA*
JI-FAN HU • *Medical Service, VA Palo Alto Health Care System and Department of Medicine, Stanford University School of Medicine, Palo Alto, CA*
FUMITOSHI ISHINO • *Gene Research Center, Tokyo Institute of Technology, Yokohama; CREST, Japan Science and Technology Corporation (JST), Kawaguchi, Japan*
ROSALIND M. JOHN • *Wellcome/CRC Institute of Cancer and Developmental Biology, University of Cambridge, Tennis Court Road, Cambridge, United Kingdom*
PETER L. JONES • *Department of Cell and Structural Biology, University of Illinois, Urbana-Champaign, Urbana, IL*
YASUFUMI KANEDA • *Division of Gene Therapy Science, Osaka University School of Medicine, Suita, Osaka, Japan*
TOMOKO KANEKO-ISHINO • *School of Health Sciences, Tokyo University, Isehara, Japan*
STEFAN U. KASS • *Department of Genomic Technologies, Janssen Research Foundation, Beerse, Belgium*
GAVIN KELSEY • *Laboratory of Developmental Genetics and Imprinting, Developmental Genetics Programme, The Babraham Institute, Cambridge, United Kingdom*
SANJEEV KHOSLA • *Wellcome/CRC Institute of Cancer and Developmental Biology, University of Cambridge, Tennis Court Road, Cambridge, United Kingdom*
SHIN KOBAYASHI • *Gene Research Center, Tokyo Institute of Technology, Yokohama, Japan*

Takashi Kohda • *Gene Research Center, Tokyo Institute of Technology, Yokohama, Japan*
Yoshimi Kuroiwa • *Pharmaceutical Research Laboratory, Kirin Brewery Co. Ltd., Takasaki, Gunma, Japan*
Marc Lalande • *Department of Genetics and Developmental Biology, School of Medicine, University of Connecticut, Farmington, CT*
Janine LaSalle • *Medical Microbiology and Immunology, School of Medicine, University of California, Davis, CA*
Helena Malmikumpu • *Department of Animal Development & Genetics, Uppsala University, Norbyvägen, Uppsala, Sweden*
Jeffrey R Mann • *Division of Biology, Beckman Research Institute of the City of Hope, CA*
Naoki Miyoshi • *Gene Research Center, Tokyo Institute of Technology, Yokohama, Japan*
Tom Moore • *Department of Biochemistry, University College Cork, Ireland.*
Tsunehiro Mukai • *Department of Biochemistry, Saga Medical School, Saga, Japan.*
Rolf Ohlsson • *Department of Animal Development and Genetics, Uppsala University, Norbyvägen, Uppsala, Sweden*
Joachim Oswald • *Institut für Polymerforschung e.V., Dresden, Germany*
Kerstin Otte • *Wellcome/CRC Institute, Cambridge United Kingdom*
Wolf Reik • *Laboratory of Developmental Genetics and Imprinting, Developmental Genetics Programme, The Babraham Institute, Cambridge, United Kingdom*
Björn Rozell • *Unit for Morphological Phenotype Analysis and Unit for Embryology and Genetics, Clinical Research Center, Huddinge University Hospital, Stockholm, Sweden.*
Rod J. Scott • *Department of Biology and Biochemistry, University of Bath, Claverton Down, Bath, United Kingdom*
Judith Singer-Sam • *Division of Biology, Beckman Research Institute, City of Hope, Duarte, CA*
Rachel J. Smith • *Developmental Genetics Programme, The Babraham Institute, Babraham, Cambridge, United Kingdom*
Melissa Spielman • *Department of Plant Sciences, University of Oxford, South Parks Road, Oxford, United Kingdom*
Kristian Svensson • *Department of Pharmacology, Pharmacia-Upjohn, Stockholm, Sweden*
Maxine Tevendale • *Department of Anatomy, University of Cambridge, Downing Street, Cambridge, United Kingdom*
Joanne L. Thorvaldsen • *Howard Hughes Medical Institute and Department of Cell and Developmental Biology, University of Pennsylvania School of Medicine, Philadelphia, PA*

KIYOE URA • *Division of Gene Therapy Science, Osaka University School of Medicine, Suita, Osaka Japan*
ILSE VAN DEN WYNGAERT • *Department of Genomic Technologies, Janssen Research Foundation, Beerse, Belgium*
EVANI VIEGAS-PÉQUIGNOT • *INSERM U383, Hôpital Necker-Enfants Malades, Paris Cedex, France*
RINKE VINKENOOG • *Department of Biology and Biochemistry, University of Bath, Claverton Down, Bath, United Kingdom*
PAUL A. WADE • *Laboratory of Molecular Embryology, National Institute of Child Health and Human Development, National Institutes of Health, Bethesda, MD*
JOERN WALTER • *Department of Genetics, University of Saarland, Saarbrücken, Germany*
ANDREW WARD • *Department of Biology and Biochemistry, University of Bath, Claverton Down, Bath, United Kingdom*
ALAN P. WOLFFE • *Sangamo BioSciences Inc., Richmond, CA*

1

Generation of Monoparental Embryos for Investigation into Genomic Imprinting

Wendy L. Dean, Gavin Kelsey, and Wolf Reik

1. Introduction

The seminal work of McGrath and Solter *(1)* and independently of Surani et al. *(2)* in 1984 established the fundamental principle of nuclear nonequivalency; that is, chromosomes of both paternal and maternal origin are required for development to term in mammals. This was achieved through the creation of diploid reconstituted zygotes, which contained either two maternal or two paternal pronuclei. Embryos containing pronuclei exclusively of maternal or paternal origin display characteristic developmental abnormalities and fail to develop to term. This failure is partially explained by the observation that paternally and maternally derived genomes have complementary roles during embryogenesis, contributing differentially to embryonic and extraembryonic lineages *(2–5)*. These reconstitutions were accomplished by nuclear transplantation and karyoplast fusion using HVJ or Sendai virus-assisted fusion *(1)*. These experiments laid the foundation for the discovery and exploration of this unique form of non-Mendelian mammalian gene regulation whereby expression of genes and hence phenotype were dictated by the parent from whom they where inherited. This parent-of-origin phenomenon is known as genomic imprinting.

1.1. Androgenetic Embryos

Uniparental embryos possessing exclusively paternally derived chromosomes are referred to as androgenetic embryos (AG). These embryos can be produced experimentally by a limited number of routes, all involving

From: *Methods in Molecular Biology, vol. 181: Genomic Imprinting: Methods and Protocols*
Edited by: A. Ward © Humana Press Inc., Totowa, NJ

micromanipulation. Controlled bispermic fertilization of enucleated metaphase II (MII) oocytes has been used to produce AG embryos to the blastocyst stage *(6)*. However, the more usual route is to enucleate a fertilized embryo and replace the female pronucleus with a second sperm-derived pronucleus. Embryos produced by karyoplast fusion can be returned to a pseudopregnant foster mother and will develop up to day 10 of gestation. These embryos are highly variable in appearance and developmental stage. Even the most advanced AG embryos obtained are profoundly developmentally retarded, rarely achieving more than the 4- to 6-somite stage at d 10. In contrast to the poor development of the embryonic derivatives the extraembryonic tissues are highly developed and overabundant compared to the embryo *(7)*.

1.2. Parthenogenetic and Gynogenetic Embryos

Parthenogenetic embryos (PG) containing maternally derived genomic content can be derived from unfertilized oocytes by a number of methods. Parthenogenetic activation occurs spontaneously at a high frequency in the LT/SV strain and in mice deficient in *c-mos*. A variety of chemically induced methods have been described. Ethanol activation has been used extensively but results in a significant number of aneuploid activated oocytes. Exposure to Sr^{2+} has proven an effective method of activation of MII oocytes with intracellular Ca^{2+} oscillations that more closely resemble those observed following normal fertilization *(8,9)*. Alternatively, PG embryos may be produced by electroactivation of oocytes without induction of aneuploidy *(10)*.

Gynogenetic embryos (GG) containing only maternally derived genomes can also be produced through pronuclear transplantation from fertilized embryos *(5)*. However, there is not apparently any difference in development potential between pathenogenetically and gynogenetically derived embryos, and thus the simpler method of PG production is more often elected.

Clearly, the choice of PG or GG is driven by experimental questions. It remains compelling that there is equivalency of maternal pronuclei derived from potentially different circumstances, that is, following fertilization vs activation. Detailed investigation of imprinted loci may well reveal subtle differences of an epigenetic nature as predicted by the parent–offspring conflict models *(11)*.

Maternally uniparental embryos are characteristically small for gestational age, and they lack well-developed extraembryonic tissues in yolk sac and trophoblast derivatives. Development is highly variable within and between experimental groups. The most advanced maternally uniparental embryos achieve the forelimb bud stage at approximately d 10 and usually possess no more than 25 somites. This phenotype has been characterized in detail elsewhere *(12)*. The underdevelopment of trophectoderm and primitive endodermal lineages likely contributes to the variable extent of development *(13)*.

1.3. Nuclear Transplantation

1.3.1. Generation of Uniparental Embryos for Imprinting Studies

Results from nuclear transplantation studies producing uniparental mouse embryos suggested that reciprocal and complementary roles have evolved for the maternal and paternal genomes.

The generation of AG and PG embryos represents a means by which the entire repertoire of imprinted genes might be identified from the representative genomes. As relatively distinctive roles have been suggested, investigation into uniparental embryos offers the unique opportunity to study the essential gene regulation and expression of imprinted genes required for early embryonic development in isolation from those genes required for development of extraembryonic tissues and thus early placental development. Previous genetic studies have identified specific chromosomal regions within the genome known to be critical for certain aspects of development on the basis of their parent-of-origin patterns of inheritance *(14)*. Such uniparental disomic regions allow for investigation of a limited number of imprinted genes in a defined pattern on a chromosome-by-chromosome basis *(15)*.

An elegant extension of these types of experiments that makes use of the differences between AG and PG embryos and their imprinted genes has been exploited in an attempt to identify new imprinted genes by application of a cDNA subtractive hybridization strategy *(16,17)*, and *see* Chapter 7 by Ishino et al.). Methylation-based screens have been suggested as an alternative means of identifying imprinted regions. The strength of this approach is based on the observation that allele-specific methylation patterns have been found at most imprinted loci examined. One specific application encompassing uniparental embryos and methylation differences has been suggested for identification of new imprinted genes *(18)*; and *see* Chapter 8 by Smith and Kelsey). The feasibility of this approach has been demonstrated for the imprinted region of mouse chromosome 2 *(19)*.

A genome-wide survey employing this method would require the use of AG and PG embryos and their normal counterparts. The results of such a survey are eagerly awaited.

1.3.2. Technical Points

1.3.2.1. Genotype Restrictions

When generating AG or GG embryos, it is advisable to select the egg to have a different genotype from the sperm. The genetic identity of the reconstructed uniparental embryo can be identified through various means; for example, the use of isozymal variant forms of glucose-phosphate isomerase (GPI) has been employed as a means of confirming the fidelity of the manipulation. A contemporary method of identification would employ the use of microsatellite markers, which are unique for many conventionally available mouse strains ***(20)***. Identity is confirmed by a simple DNA-based polymerase chain reaction (PCR) analysis.

Perhaps more important is the genotype of the paternal genome in the generation of the AG embryos. Relatively low but consistent success rates can be obtained with outbred (MF1) or hybrid (C57Bl/6J × CBA/Ca) [B_6CB-F_1] genotypes ***(3,4)***. The 129/Sv genotype is generally preferred for AG embryos because of the relatively large proportion of androgenones that develop during pre- and postimplantation stages ***(21)*** (Dean et al., unpublished results). The overall success rates for the production of AG embryos may also be a function of the experimental method of reconstitution. Thus, the fusogenic method of choice may vary according to the genotype restrictions of the experiment.

1.3.2.2. Fusogenic Methods

The pioneering work of McGrath and Solter 1983 ***(1)*** employed inactivated Sendai virus as the fusogenic agent in their nuclear transplantation experiments. Although widely adopted, the use of Sendai virus always carries the disturbing risk of serious harm to mouse colonies in the event of failure to achieve complete inactivation of the virus. Other methods of inducing fusion of karyoplasts to enucleated donors egg membranes include polyethylene glycol (PEG) ***(22)*** and electrofusion ***(23)***.

The use of PEG as a fusogenic agent has gained wide acceptance for work in cell–cell fusion hybridization. However, its use has been limited, as deleterious effects on viability arise due to impurities in the PEG. Variable fusogenic activity has also restricted its widespread use, as well as the poor efficiency in situations of asymmetric donor-karyoplast sizes as in this case.

Electrofusion has been used widely in nuclear transplantation studies in nonmammalian systems, where the use of a viral fusogenic agent was less successful ***(24)***. In addition to the animal-welfare advantage offered by electrofusion, a further benefit is the elimination of the batch variability and titration requirement for each type of experiment when using Sendai virus. Early problems with efficient rates of fusion for asymmetric reconstructions, as encountered in nuclear transplantation studies, appear to have been overcome. A detailed and systematic analysis of the use of electrofusion in domestic animal species has been published elsewhere ***(25)***.

An application of this fusogenic method in the mouse has been described for nuclear transplantation studies ***(26,27)*** and in the nonchemical method of tetraploid embryo production ***(28,29)***.

This chapter describes the use of electrofusion for pronuclear manipulation to generate uniparental embryos. The generation of AG embryos is described as an example of the use of the technique in experimental embryology studies. This method is used routinely in our laboratory and achieves consistently high rates of karyoplast fusion, which equal and exceed the rates of fusion achieved using Sendai virus. Embryo viability is high, with good rates of development to the blastocyst stage.

2. Materials

2.1. Equipment

The generation of micromanipulated embryos of uniparental origin is a highly specialized application of experimental embryology. Any endeavor of this nature necessarily assumes that the resource of a well-equipped laboratory for routine mouse embryo work is available. Minimum requirements for equipment include an inverted fixed-stage microscope with Nomarski optics (preferably) and manipulators for the holding and enucleating instruments. A wide variety of manufacturers of the manipulators and accompanying microinjector assemblies are available. Additional equipment required to craft holding and enucleation pipets include a pipet puller (Sutter is very reliable), a microforge, and a pipet grinder (Narishige) to bevel the enucleation instruments (preferably with binocular eyepieces).

2.2. Production of Fertilized Embryos

2.2.1. Superovulation

1. 3- to 4-wk-old female B6CB-F_1 (Harlan; Charles River, U.K.) (*see* **Note 1**).
2. Phosphate-buffered saline (PBS; Sigma).
3. Pregnant mare serum gonadotropin (PMS-Folligon), 1500 IU per vial (Intervet).
4. Human chorionic gonadotropin (hCG—Chorulon) 500 IU per vial (Intervet).

2.2.2. Collection and Culture of Fertilized Zygotes

1. Dissecting tools; No. 5 watch maker's forceps; fine-pointed scissors (iris; Weiss).
2. Sterile plastic Petri dishes (50 mm).
3. Bovine serum albumin (BSA; embryo tested: Sigma A-3311).
4. Flushing and handling medium (FHM) + BSA ***(29)***: 95 m*M* NaCl, 2.5 m*M* KCl, 0.35 m*M* KH_2PO_4, 0.2 m*M* $MgSO_4$, 10 m*M* lactate, 0.2 m*M* pyruvate, 0.2 m*M* glucose, 1.0 m*M* glutamine, 1.0 mg/mL BSA, 0.01 m*M* ethylenediamine-tetraacetic acid (EDTA) (tetrasodium salt), 4.0 m*M* $NaHCO_3$, 20 m*M* HEPES, 1.71 m*M* $CaCl_2$, 100 U/mL penicillin, 50 µg/mL streptomycin.
5. Hyaluronidase (Sigma H-6254), 300 µg/mL in FHM (from stock solution).
6. 9-in. Pasteur pipets.

7. Mouth pipets (aspirator mouthpiece).
8. Simplex optimized medium (KSOM) + BSA ***(30)***: 95 m*M* NaCl, 2.5 m*M* KCl, 0.35 m*M* KH_2PO_4, 0.2 m*M* $MgSO_4$, 10 m*M* lactate, 0.2 m*M* pyruvate, 0.2 m*M* glucose, 1.0 m*M* glutamine, 1.0 mg/mL BSA, 0.01 m*M* EDTA (tetrasodium salt), 25 m*M* $NaHCO_3$, 1.71 m*M* $CaCl_2$, 100 U/mL penicillin, 50 µg/mL streptomycin (*see* **Note 2**).
9. Light paraffin oil or mineral oil (embryo tested; Sigma M-8410).
10. CO_2 incubator, set at 37°C and 5% CO_2.

2.3. Experimental Reconstruction of Monoparental Embryos

2.3.1. Generation of Instruments

The preparation of the fine glass enucleation instruments is critical for achieving consistent success in these experiments. A detailed and comprehensive outline of these instructions has been published elsewhere ***(21)***. Modifications to suit specific differences in equipment will be easily accommodated from these instructions.

2.3.2. Production of Androgenetic (AG)/Gynogenetic Embryos (GG)

2.3.2.1. Preparation of Fertilized Embryos for Pronuclear Transplantation

1. FHM (*see* **Subheading 2.2.2.**).
2. Cytochalasin B (CCB: Sigma C-6762): 5 mg/mL in dimethyl sulfoxide (DMSO) stored at –20°C.
3. Nocodazole: 3 mg/mL in DMSO stored at –20°C (Sigma: M-1404).
4. Mineral oil.
5. FNC medium: FHM + 5 µg/mL CCB + 0.5 µg/mL nocodazole.
6. Glass depression slide (BDH).
7. Inverted microscope with Nomarski optics and manipulator assemblies.

2.3.2.2. Recovery and Electrofusion

1. FHM (*see* **Subheading 2.2.2.**).
2. KSOM (*see* **Subheading 2.2.2.**), equilibrated at 37°C in 5% CO_2 in an air-humidified incubator.
3. Equilibration and fusion medium (GCM + BSA): 0.3 *M* glucose, 50 µ*M* $CaCl_2$, 100 µ*M* $MgSO_4$, 3% BSA, pH adjusted to 7.4 (*see* **Note 3**).
4. GCM: 0.3 *M* glucose, 50 µ*M* $CaCl_2$, 100 µ*M* $MgSO_4$, pH adjusted to 7.4.
5. BTX™ Electro cell manipulator with slide chamber (Genetronics, Inc.).

2.3.2.3. Embryo Culture

1. KSOM (*see* **Subheading 2.2.2.**) equilibrated at 37°C in 5% CO_2 in an air-humidified incubator.
2. Mineral oil.
3. 30-mm sterile Petri dishes.

2.3.3. Embryo Transfer to Pseudopregnant Recipient

1. Day 1 pseudopregnant female $B_6CB\text{-}F_1$ (10–15 wk old).
2. Dissection instruments.
3. Pasteur pipets: drawn (0.4 mm).
4. Wound clips (Clay Adams).
5. FHM (*see* **Subheading 2.2.2.**).
6. Mineral oil.
7. Avertin (2,2,2-tribromoethanol in *tert*-amyl alcohol) anaesthetic: Dissolve 2,2,2-tribromoethanol in *tert*-amyl alcohol to produce a 100% stock solution. Stir until a single phase is achieved. Place in a dark glass bottle and store at 4°C until use. Dilute to 1.2% by adding 120 µL to 10 mL of warmed H_2O or PBS. Shake vigorously until the oily drop mixes completely with the aqueous phase. Use at a dose of 0.4 mL/10 g body weight.

2.4. Generation of Parthenogenetic Embryos

2.4.1. Recovery of MII Oocytes

2.4.1.1. Activation of MII Oocytes

1. FHM (*see* **Subheading 2.2.2.**).
2. FHM + 300 µg/mL hyaluronidase.
3. 0.3 *M* glucose in embryo culture-grade H_2O (BDH).
4. 30-mm sterile Petri dishes.
5. 5 µg/mL CCB (*see* **Subheading 2.3.2.1.**) in KSOM (*see* **Subheading 2.2.2.**), equilibrated at 37°C in 5% CO_2 under oil.
6. KSOM (*see* **Subheading 2.2.2.**), equilibrated at 37°C in 5% CO_2 under oil.

2.4.2. Production of Tetraploid Embryos

2.4.2.1. Recovery of Two-Cell Embryos

1. Nonbeveled 33-gage flushing needles.
2. Sterile 2-mL disposable syringes.
3. Dissecting tools; No. 5 watch maker's forceps; fine-pointed scissors (iris; Weiss).
4. Mouth pipet (aspirator mouthpiece).
5. Sterile plastic Petri dishes (50 mm).
6. Drawn 9-in. Pasteur pipets (0.2 mm).

2.4.2.2. Fusion of Two-Cell Embryos

1. BTX™ Electro cell manipulator with slide chamber (Genetronics, Inc.).
2. Two dissection microscopes.
3. Sterile Petri dishes (50 mm).
4. FHM (*see* **Subheading 2.2.2.**).
5. GCM and GCM + BSA (*see* **Subheading 2.3.2.2.**).
6. Drops of KSOM (*see* **Subheading 2.2.2.**), equilibrated at 37°C in 5% CO_2 under oil.

2.4.3. Production of 4N-PG-4N Aggregation Chimeras

2.4.3.1. Preparation of Dishes for Aggregation

1. Darning needle.
2. KSOM (*see* **Subheading 2.2.2.**), equilibrated at 37°C in 5% CO_2 under oil.
3. Mineral oil.
4. Sterile plastic Petri dishes (50 mm).

2.4.3.2. Removal of Zona Pellucida

1. Acid Tyrode's (Sigma: embryo-tested T-1788).
2. FHM (*see* **Subheading 2.2.2.**)
3. Drawn 9-in. Pasteur pipets (0.2 mm; flame polished).

2.4.3.3. 4N-PG-4N "Sandwich" Aggregations

1. Petri dish with depression wells containing 4 *N* embryos with zona removed.
2. 8-cell PG embryos with zona removed.
3. Drawn Pasteur pipet (0.2-mm diameter).
4. Mouth pipet.

2.4.3.4. Embryo Transfer of Chimeric Blastocysts

1. Day 3 pseudopregnant female $B_6CB\text{-}F_1$ (10–15 wk old).
2. Dissection instruments.
3. Pasteur pipets: drawn (0.4 mm).
4. Wound clips (Clay Adams).
5. FHM (*see* **Subheading 2.2.2.**).
6. Mineral oil
7. Avertin (*see* **Subheading 2.3.3.**).
8. 1-mL sterile syringe with 27.5-gage needle.

3. Methods

3.1. Production of Embryos

3.1.1. Superovulation

Immature females $B_6CB\text{-}F_1$, 3–4 wk old, are best suited for high response to hormonal regimes. As a restricted number of strains respond to this procedure, there may be some limitations imposed where genetic rigor is paramount. Consistent high yields can be obtained from hybrid F_1 animals derived from a number of widely available mouse strains ***(31)***.

1. 5.0–7.5 IU PMS (Folligon, Intervet) is injected intraperitoneally (ip) followed after 44–48 h by 5.0–7.5 IU hCG (Chorulon, Intervet). Timing of initiation is dictated by light : dark cycles.

2. Following hCG injection, individual females are incarcerated with mature—that is, greater than 8-wk-old—129/Sv males.
3. Presence of a copulation plug is checked for the following morning. This is d 1 of gestation.

3.1.2. Collection of Fertilized Zygotes

Embryos are collected between 18 and 20 h post-hCG to ensure high rates of fertilization.

1. Sacrifice animals by cervical dislocation, excise the swollen ampullae of the oviduct intact, and place into a sterile plastic Petri dish.
2. Place oviduct in 50-μL drops of hyaluronidase solution (300 μg/mL) in FHM and tear open the swollen portion of the ampullae with a sterile 27.5-gage needle while holding the oviduct by No. 5 watch maker's forceps. Incubate at room temperature until the fertilized oocytes are denuded of cumulus cells. (This should take no more than 5 min). Wash zygotes in a series of 10 drops of FHM to clean away cumulus cells, spermatozoa, and other cellular debris. Discard obviously abnormal embryos and unfertilized oocytes (*see* **Note 4**).
4. Place one-cell embryos into equilibrated KSOM in a CO_2 incubator for approximately 1 h prior to initiation of the micromanipulation.

3.2. Production of Androgenetic/Gynogenetic Embryos

3.2.1. Preparation of Fertilized Embryos for Pronuclear Transplantation

1. Place into FNC for 20 min prior to initiating micromanipulation.
2. Prepare slide: add a small drop of FNC (10 μL) to a depression slide and cover the drop with mineral oil.
3. Place 20–30 embryos into the covered drop in the depression slide.
4. Place the prepared slide onto the microscope stage and, under low magnification (4×), place instruments carefully into the drop on the slide.

3.2.2. Micromanipulation

The following series of instructions is given to prepare AG embryos. GG embryos may be generated in an identical fashion; however, as GG and diploid PG embryos are developmentally equivalent and have no known differences in totipotency or imprinting status, it is easier to generate maternally monoparental embryonic material by derivation of PG embryos (*see* **Subheading 1.2.**).

Pronuclear micromanipulation to produce monoparental embryos requires both an enucleation step and a nuclear transfer step. These may be done as two completely separate procedures or as a single two-step operation. There are advantages and disadvantages to both. In the first case the recipients are enucleated, removing the female pronucleus. Thus the eggs have to be picked up and repositioned to locate the site of entry before introducing the male pronucleus. This can be time consuming and frustrating. The second method begins the process with removal of the male pronucleus from the donor, which must sit in the enucleation pipet until the karyoplast containing the female pronucleus is removed from the recipient oocyte and deposited on the bottom of the depression slide. The great advantage is that in the second manipulation the positioning required to introduce the donor male karyoplast is set without requiring any further adjustment. This demands a very high level of attention, but I find this strategy overall more efficient.

Regardless of the method adopted, identification of the female and male pronuclei is vital. Careful observation of the embryo for size and positioning of the pronuclei achieves this. The female pronucleus is always smaller than the male and will normally reside close to the polar body. However, embryos where this distinction is not clear should be discarded.

1. It is paramount that only embryos that have visible maternal (m), adjacent to the polar body (pb), and paternal (p) pronuclei are used (**Fig. 1A**). Introduce the transfer needle through the zona pellucida and withdraw by suction a small cytoplast to act as a buffer between the oil and FNC phase.
2. The tip of the enucleation pipet is gently inserted through the zona pellucida. The beveled opening of the tip is placed adjacent to the pronucleus and the transfer pipet pressed into the egg without breaking the membrane to make an indentation as deep as necessary to retrieve the male pronucleus (**Fig. 1B**). In the presence of cytoskeletal inhibitors, the membrane is sufficiently elastic to permit this invasive withdrawal of pronuclei without rupturing. With suction, the overlying membrane and cytoplasm surrounding the pronucleus is drawn into the pipet. Once the pronucleus is safely in the pipet, it is withdrawn through the zona and the membrane "pinched" off, thereby forming the intact karyoplast (k). The intact karyoplast should approximate the size of a polar body (pb).
3. Positioning in the first step is not critical, as the embryo only donates a male pronucleus enclosed karyoplast and is then discarded.
4. Select another fertilized embryo with clearly visible pronuclei and position it equatorially in order to remove the female pronucleus (m) (**Fig. 1C**). Repeat the process as described in **step 2**, this time removing the female pronucleus.
5. Discard the most recently removed female pronucleus (**Fig. 1C**) to the bottom of the slide. Reinsert the enucleation needle and deliver the karyoplast containing the male pronucleus (p) into the recipient zygote (**Fig. 1D**).
6. Repeat the process with the remainder of the embryos in the drop. Restricting exposure of embryos to the FNC to 1 h improves AG embryo viability. Embryos are washed and returned to the incubator in equilibrated drops of KSOM for at least 1 h prior to fusion (**Fig. 2A**).

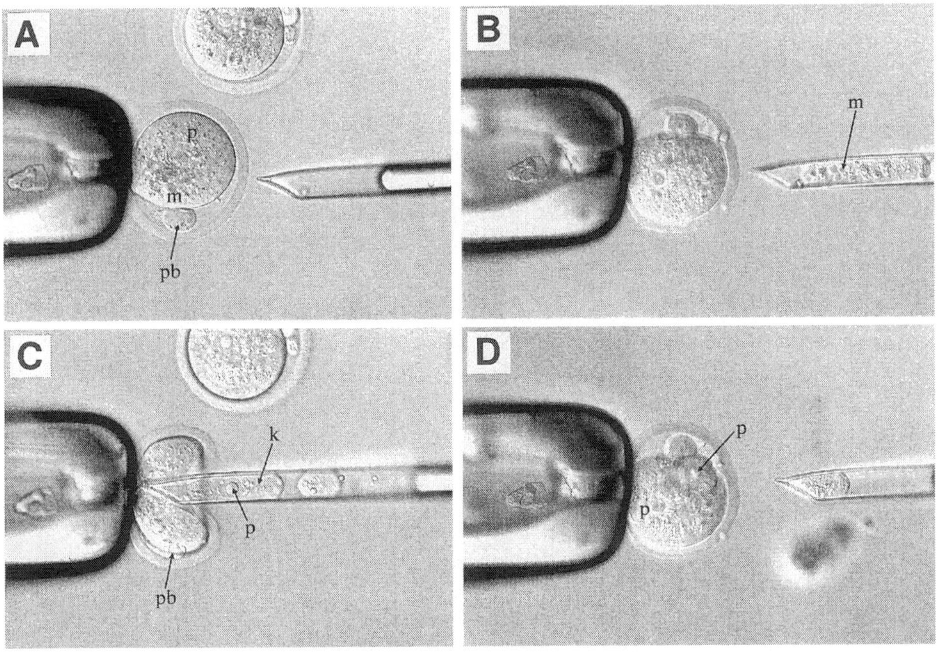

Fig. 1. Production of androgenetic embryos. **(A)** An embryo is held in the equatorial region with the polar body (pb), maternal (m), and paternal (p) pronucleus clearly visible **(B)**. The larger paternal pronucleus is removed **(C)**. The donor karyoplast (k) containing the male pronucleus is retained while the maternal pronucleus (m) is removed from a second fertilized embryo **(D)**. The maternal pronucleus is dropped to the bottom of the depression slide and the donor karyoplast containing the male pronucleus (p) is deposited in the perivitelline space.

3.2.3. Electrofusion for Diploid Androgenetic Embryo Reconstitution (Nonelectrolyte Fusion)

1. Remove embryos from the incubator and equilibrate through 3 drops of GCM + BSA (**Fig. 2A**).
2. Place embryos into the electroactivation slide chamber between the electrodes containing a large drop of GCM (without BSA), as shown schematically in **Fig. 3**.
3. Manually apply AC pulse (5 V) to affect alignment in the electric field (2–3 s). Only embryos aligned at right angles to the electrodes will be capable of undergoing fusion. Fuse by delivering a 1.5-kV/cm pulse in 2 × 70-µs intervals.
4. Wash 6 times in FHM and place into culture medium in a CO_2 incubator at 37°C.

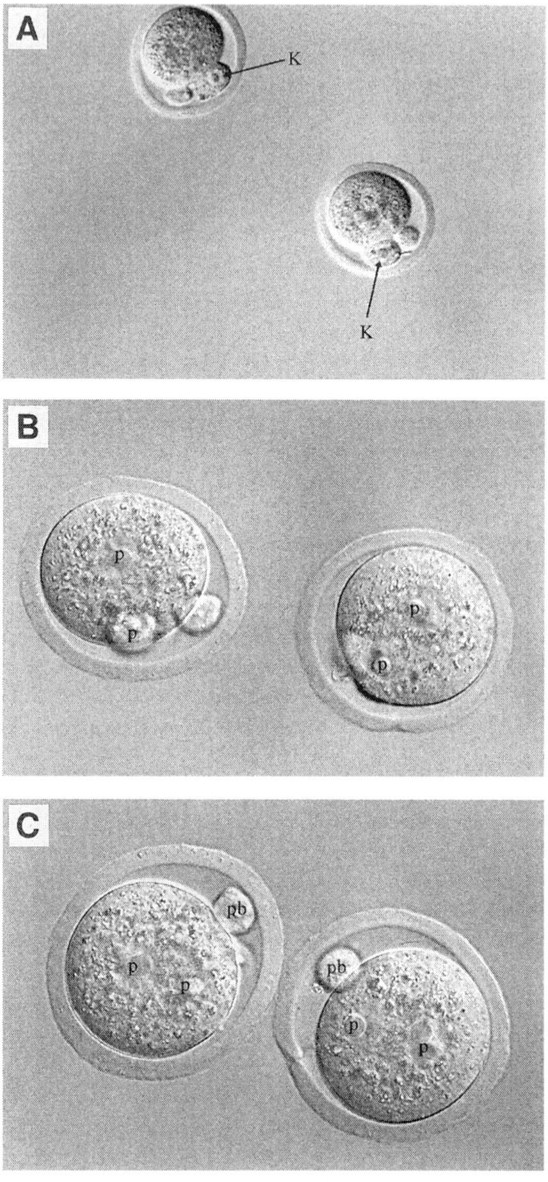

Fig. 2. Electrofusion of reconstructed embryos. **(A)** Before electrofusion, donor embryos with paternal pronuclear karyoplasts (k) are incubated to restore cytoskeletal structural integrity. **(B)** Fusion is evident in some embryos by 30 min, with the paternal pronucleus (p) clearly restored within the cytoplasm (right side). However, the kinetics of reconstruction do vary slightly within the group of manipulated embryos (left side) **(C)**. All embryos successfully undergo fusion within 45 min, as is evident by the two paternal (p) pronuclei clearly visible within the cytoplasm. The polar body (pb) remains outside, within the perivitelline space.

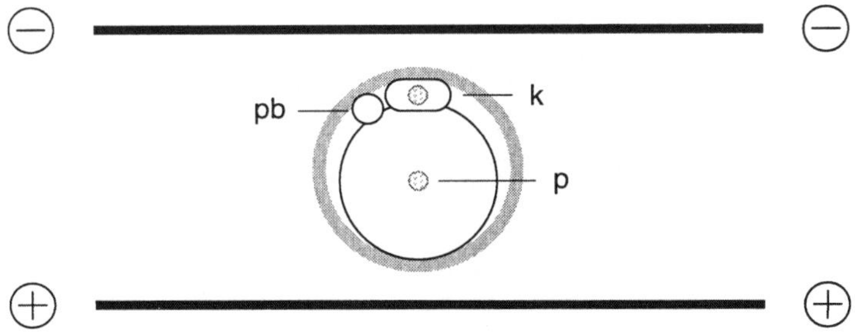

Fig. 3. Schematic representation of embryo configuration during electrofusion. Application of AC voltage in nonelectrolyte solutions permits aligment with the electric field. The apposition of membranes between the karyoplast (k) containing the male pronucleus (p) and the recipient oocyte must be parallel to ensure fusion. Note that the polar body (pb) is excluded from correct alignment and therefore does not become fused to the recipient embryo following application of the DC pulse.

5. Evidence of fusion should be apparent with 30 min. Should fusion have not been successful, repeat **steps 1–4** (**Fig. 2B,C**).
6. Culture to 4–8 cell stage. Transfer 4–8 cell embryos to d 1 pseudopregnant F1 foster mothers (*see* **Note 5** and **ref.** *32*). Dissect up to d 10 relative to recipient timing, that is, 9 d later (*see* **Note 6**).

3.3. Production of Parthenogenetic Embryos

Parthenogenetic embryos display a spectrum of phenotypes, with most dying prior to gastrulation *(12,33)*. Embryos from uniformly maternal sources are capable of relatively normal development of embryonic derivatives, but failure to produce sufficient extraembryonic structures is thought to lead to early embryonic lethality. This variability and poor rates of developments to mid-gestation can be enhanced substantially by providing PG embryos with fertilization-derived extraembryonic tissues *(13)*. This is achieved by generation of tetraploid cells, which contribute to placental structures in preference to embryonic germ layers *(28,34–36)*. In order to obtain significant quantities of maternally uniparental embryonic material for evaluation in imprinting studies, we routinely employ the "sandwich" aggregation method *(28,37)*. Tetraploid cells have been shown to be largely absent from embryonic derivatives by d 7.5 *(36)* in aggregates, leaving PG embryos of a higher frequency and degree of uniformity.

This method requires the independent production of diploid PG and tetraploid (fertilized) embryos, which are subsequently aggregated together.

3.3.1. Activation of MII Oocytes

1. Harvest unfertilized MII oocytes from superovulated F_1 juvenile females at 21 h post-hCG injection. Denude cumulus enclosed oocytes by incubation in hyaluronidase at 300 μg/mL in FHM for no more than 5 min.
2. Equilibrate oocytes × 3 in 0.3 *M* glucose (freshly made isotonic glucose works best).
3. Place oocytes into the electroactivation slide chamber and deliver 3 × 50-μs pulses DC at 1.0 kV/cm. Do not place more than 20–30 oocytes into the chamber at any given time.
4. Wash 6 times in FHM and place into 5-μg/mL CCB for 4–5 h at 37°C in a CO_2 incubator.
5. Wash embryos 9 times in CO_2-equilibrated KSOM under oil.
6. Incubate embryos in the 9th wash drop for 1 h.
7. Transfer "diploid" PG embryos to another equilibrated drop of KSOM (*see* **Note 7**).
8. Culture until the 4- to 8-cell stage or for 72 h in KSOM (*see* **Note 8**).

3.3.2. Production of Tetraploid Embryos

It is important, perhaps critical in most experiments, to be able to assess whether 4*N* cell clearance of embryonic derivatives has been achieved. To ensure that d 10 embryos are entirely PG in composition select a marker for the 4*N* embryos. We have used ROSA 26, a ubiquitously expressed *lacZ* marker derived by gene trapping ***(38)***, both to act as a marker for assessing the presence of 4*N* cells and to act as a reporter in order to evaluate and eliminate diploid ↔ PG aggregates.

3.3.3. Recovery and Electrofusion of Two-Cell Embryos

Fusion of the blastomeres of two-cell embryos to produce tetraploid embryos occurs when a square pulse (DC) is applied perpendicular to the plane of contact. The instructions that follow are designed for use with nonelectrolyte solutions.

1. Collect two-cell embryos on d 2 postcoitum (pc) by excising the oviducts and flushing the contents with a 30-gage needle containing FHM. These embryos should possess an informative marker to ensure sample fidelity ***(37)*** (*see* **Note 9**).
2. Equilibrate through 3 drops of GCM + BSA.
3. Place embryos into electrofusion chamber slide containing GCM (without BSA).
4. Set electrofusion apparatus to deliver an AC pulse of approximately 5 V for alignment at right angles to the electrodes. Fuse by delivering a 0.75 kV/cm pulse in 2 × 70-μs bursts.

5. Wash 6 times in FHM and return to equilibrated KSOM in 5% CO_2, 37°C conditions.
6. Check for fusion within 45–90 min. Embryos should appear as "one-cell zygotes".
7. Discard any two-cell embryos. Continue to culture embryos overnight in KSOM (*see* **Note 10**).

3.3.4. Production of 4N↔PG↔4N Aggregation Chimeras

3.3.4.1. Aggregation Chimeras: "Parthenogenote Sandwiches"

Embryo aggregation chimeras are produced with PG embryos at the 8-cell and the tetraploid components at the "4-cell" stage. The tetraploid embryos will "compact" after the "4-cell" stage, as this is equivalent to the diploid 8-cell stage when compaction occurs normally. Thus, the aggregates must be made early on the day after fusion.

1. Prepare aggregation depression in 50-mm bacteriological dishes using a darning needle *(28)*.
2. Remove the zona pellucida with Acid Tyrode's from both PG and 4*N* embryo populations. Remember, two 4*N* embryos are required for each PG embryo.
3. Place a single 4*N dezonaed* embryo into a microdepression in the dish. Place a single PG embryo onto this embryo and place the second 4*N* embryo onto the PG embryo, thus creating a parthenogenetic embryo sandwiched between two tetraploid embryos. Continue in a new microdepression each time until all available materials have been used.
4. Return the drops to the incubator and do not disturb them. Check after 24 h for aggregation.
5. Transfer blastocysts 48 h after aggregation to the uterus of d 3 pseudopregnant recipients (*see* **Notes 10–12**).

4. Notes

1. The genotype of the oocyte seems to have relatively little effect compared to the genotype of the sperm, that is, the paternal pronucleus. The use of 129/Sv gives considerable improvement in the yield of materials at d 10 of gestation.
2. I use this medium routinely for all applications. There is some suggestion that other embryo culture media may be particularly effective for androgenetic embryonic development *(40)*.
3. Glucose has been substituted for mannitol, as embryo viability is improved (personal communication, J. Fulka, Jr.).
4. This step is carried out under relatively low power using a dissection microscope. Thus, abnormal embryos are defined as those that are either fragmented or have grossly overrepresented perivitelline space.

5. It is assumed that the necessary manipulation skills for embryo transfer to pseudopregnant recipient females are already practised in the laboratory. A detailed account of this procedure is given in **ref.** ***32***, Protocol 7.
6. It is paramount to ensure that the material generated are authentic AG embryos. Several methods of assessing the sample integrity are available. These include the use of mouse strains that differ in their GPI isozymal variants ***(32,37)***, carry a *lacZ* gene marker ***(38,41)***, or through the use of microsatellite markers ***(20)***.
7. "Diploid" PG identifies an activated oocyte in which the extrusion of the second polar body has been suppressed.
8. We have observed that PG embryos are more acutely sensitive to culture conditions than fertilized embryos. In particular, they are very sensitive to the oxidation products in the oil used to overlay culture drops. Therefore, it is advisable to replace the mineral oil every 2–3 mo.
9. *See* **Note 6**.
10. Detailed information including practical and theoretical consideration for the use of electrofusion in tetraploid embryo production can be found in **ref.** ***29***.
11. By using a marker, for example, ROSA 26, in the 4*N* embryos, the elimination of 4*N* cells may be evaluated ***(42)***. This can also be assessed by using embryos that carry allelic variants for the tetraploid population and assessing contributions following DNA extraction from the embryo proper. All embryos containing contamination by 2*N* or residual 4*N* cells should be discarded.
12. A detailed description of the embryo transfer procedure can be found in **refs.** ***31***, ***32***, Protocol 8, and **ref.** ***39***).

References

1. McGrath, J. and Solter, D. (1983) Nuclear transplantation in the mouse embryo by microsurgery and cell fusion. *Science* **220,** 1300–1303.
2. Surani, M. A. H., Barton, S. C., and Norris, M. L. (1984) Development of reconstituted mouse eggs suggests imprinting of the genome during gametogenesis. *Nature* **308,** 548–550.
3. Barton, S. C., Surani, M. A. H., and Norris, M. L. (1984) Role of paternal and maternal genomes in mouse development. *Nature* **311,** 374–376.
4. Barton, S. C., Adams, C. A., Norris, M. L., and Surani, M. A. H. (1985) Development of gynogenetic and parthenogenetic inner cell mass and trophectoderm tissues in reconstructed blastocysts in the mouse. *J. Embryol. Exp. Morphol.* **90,** 267–285.
5. Surani, M. A. H., Barton, S. C., and Norris, M. L. (1986) Nuclear transplantation in the mouse: heritable differences between parental genomes after activation of the embryonic genome. *Cell* **45,** 127–136.
6. Kono, T., Sotomara, Y., Sato, Y., and Nakahara, T. (1993) Development of androgenetic mouse embryos produced by *in vitro* fertilisation of enucleated oocytes. *Molec. Reprod. Dev.* **34,** 43–46.
7. McGrath, J. and Solter, D. (1984) Completion of mouse embryogenesis requires both the maternal and paternal genomes. *Cell* **37,** 170–183.

8. O'Neill, Q. T., Rolfe, L. R., and Kaufman, M. H. (1991) Developmental potential and chromosome constitution of strontium-induced mouse parthenogenones. *Molec. Reprod. Dev.* **30,** 214–219.
9. Bos-Mikich, A., Swann, K., and Whittingham, D. G. (1995) Calcium oscillations and protein synthesis inhibition synergistically activate mouse oocytes. *Molec. Reprod. Dev.* **41,** 84–90.
10. Henery, C. C. and Kaufman, M. H. (1993) The incidence of aneuploidy after single pulse electroactivation of mouse oocytes. *Molec. Reprod. Dev.* **34,** 299–307.
11. Moore, T. and Haigh, D. (1991) Genomic imprinting in mammalian development: a parental tug-of-war. *Trends Genet.* **7,** 45–49.
12. Sturm, K. S., Flannery, M. L., and Pedersen, R. A. (1994) Abnormal development of embryonic and extraembryonic cell lineages in parthenogenetic mouse embryos. *Dev. Dynam.* **210,** 11–28.
13. Spindle, A., Sturm, K. S., Flannery, M., Meneses, S. J., Wu, K., and Pedersen, R. A. (1996) Defective chorioallantoic fusion in mid-gestation lethality of parthenogenone tetraploid chimeras. *Dev. Biol.* **173,** 447–458.
14. Searle, A. G. and Beechey, C. V. (1985) Non-complementation phenomena and their bearing on non-dysfunctional effects. In: *Aneuploidy: Etiology and mechanisms* (Dellarco, K. L., Voytele, P. E., and Hollaender, A., eds.), Plenum, New York, pp. 363–376.
15. Beechey, C. V. and Cattanach, B.M (1998) MRC Genetics Unit, Harwell, Oxfordshire. World Wide WebSite. Genetic and Physical Imprinting Map of the Mouse (http://www.mgu.har.mrc.ac.uk/imprinting/imprinting.html).
16. Kaneko-Ishino, T., Kuroiwa, Y., Miyoshi, N., Kohda, T., Suzuki, R., Yokoyama, M., Viville, S., Barton, S. C., Ishino, F., and Surani, M. A. (1995) Peg1/Mest imprinted gene on chromosome 6 identified by cDNA subtraction hybridization. *Nat. Genet.* **11,** 52–59.
17. Miyoshi, N., Kuriowa, Y., Kohda, T., Shitara, H., Yonekawa, H., Kaube, T., Hasegawa, H., Barton, S. C., Surani, M. A., Kaneko-Ishino, T., and Ishino, F. (1998) Identification of the Meg1/Grb10 imprinted gene on mouse proximal chromosome 11 a candidate for the Silver-Russell syndrome gene. *Proc. Natl. Acad. Sci. USA* **95,** 1102–1107.
18. Kelsey, G. and Reik, W. (1998) Analysis and identification of imprinted genes. *Methods* **14,** 211–234.
19. Peters, J., Wroe, S. F., Wells, C. A., Miller, H. J., Bodle, D., Beechey, C. V., Williamson, C. M., and Kelsey, G. (1999) A cluster of oppositely imprinted transcripts at the Gnas locus in the distal imprinting region of mouse chromosome 2. *Proc. Natl. Acad. Sci. USA* **96,** 3830–3835.
20. Dietrich, W., Katz, H., Lincoln, S. E., Shin, H. S., Friedman, J., Pracopoli, N. C., and Lander, E. S. (1992) A genetic map of the mouse suitable for typing intraspecific crosses. *Genetics* **131,** 423–447.
21. Barton, S. C. and Surani, M. A. (1993) Manipulations of genetic constitution by nuclear transplantation. *Methods Enzymol.* **225,** 732–744.

22. Eglitis, M. A. (1980) Formation of tetraploid mouse blastocysts following fusion with polyethylene glycol. *J. Exp. Zool.* **213,** 309–313.
23. Kubiak, J. Z. and Tarkowski, A. K. (1985) Electrofusion of mouse blastomeres. *Exp. Cell Res.* **157,** 561–5676.
24. Collas, P., Fissore, R., and Robl, J. M. (1993) Preparation of nuclear transplant embryos by electroporation. *Anal. Biochem.* **208,** 1–9.
25. Tatham, B. G., Giliam, K. J., and Trounson, A. O. (1996) Electrofusion parameters for nuclear transfer predicted using isofusion contours produced with bovine embryonic cells. *Molec. Reprod. Dev.* **43,** 306–312.
26. Tsuoda, Y., Kato, Y., and Shioda, Y. (1987) Electrofusion for the transplantation of mouse eggs. *Gamete Res.* **17,** 15–20.
27. Kono, T. and Tsunoda, Y. (1988) Nuclear transplantation of rat embryos. *J. Exp. Zool.* **248,** 303–305.
28. Nagy, A., Gócza, E., Diaz, E. M., Prideaux, V. R., Iványin, E., Markula, M., and Rossant, J. (1990) Embryonic stem cells alone are able to support fetal development in the mouse. *Development* **110,** 815–821.
29. McLaughlin, K. J. (1993) Production of tetraploid embryos by electrofusion. *Methods Enzymol.* **225,** 919–930.
30. Lawitts, J. A. and Biggers, J. D. (1993) Culture of preimplantation embryos. *Methods Enzymol.* **225,** 153–190.
31. Hogan, B., Costantini, F., and Lacy, E. (1986) *Manipulating the Mouse Embryo: A Laboratory Manual.* Cold Spring Harbor Laboratory Press, Cold Spring Harbor, NY.
32. Papaioannou, V. and Johnson, R. (1993) Production of chimeras and genetically defined offspring from targeted ES cells. In: *Gene Targeting: A Practical Approach* (Joyner, A. L., ed.), IRL Press at Oxford University Press, Oxford, U.K., pp. 107–146.
33. Kaufman, M. H., Barton, S. C., Surani, M. A. H. (1977) Normal postimplantation development of mouse parthenogenetic embryos to the forelimb bud stage. *Nature* **265,** 53–55.
34. Graham, C. F. (1971) Virus assisted fusion of embryonic cells. In: *Karolinska Symposia on Research Methods in Reproductive Endocrinology* (Diczfalusy, E., ed.), Karolinska Institute, Stockholm, Sweden, pp. 154–165.
35. Tarkowski, A. K., Witkowska, A., and Opas, J. (1977) Development of cytochalasin B induced tetraploid and diploid/tetraploid mosaic mouse embryos. *J. Embryol. Exp. Morphol.* **D41,** 47–64.
36. James, R. M., Klerkx, A. H. E. M., Keighnen, M., Flockhart, J. H., and West, J. D. (1995) Restricted distribution of tetraploid cells in mouse tetraploid $\leftrightarrow$ diploid chimaeras. *Dev. Biol.* **167,** 213–226.
37. Nagy, A. and Rossant, J. (1993) Production of completely ES cell derived fetuses. In: *Gene Targeting: A Practical Approach* (Joyner, A. L., ed.), IRL Press at Oxford University Press, Oxford, U.K., pp. 147–179.

38. Friedrich, G. and Soriono, P. (1991) Promoter traps in embryonic stem cells: a genetic screen to identify and mutate developmental genes in mice. *Genes Dev.* **5,** 1513–1523.
39. Stewart, C. L. (1993) Production of chimeras between embryonic stem cells and embryos. *Methods Enzymol.* **225,** 823–855.
40. Latham, K. E. and Solter, D. (1991) Effect of egg composition on the developmental capacity of androgenetic mouse embryos. *Development* **113,** 561–568.
41. Gossler, A. and Zachgo, J. (1993) Gene enhancer trap screens in ES cell chimeras. In: *Gene Targeting: A Practical Approach* (Joyner, A. L., ed.), IRL Press at Oxford University Press, Oxford, UK, pp. 181–227.
42. Dean, W., Bowden, L., Aitchson, A., Klose, J., Moore, T, Meneses, J. J., Reik, W., and Feil, R. (1998) Altered imprinted gene methylation and expression completely ES cell-derived mouse fetuses: association with aberrant phenotypes. *Development* **125,** 2273–2282.

2

Deriving and Propagating Mouse Embryonic Stem Cell Lines for Studying Genomic Imprinting

Jeffrey R. Mann

1. Introduction

Embryonic stem (ES) cells are a cell culture derivative of the blastocyst inner cell mass (ICM), the latter giving rise to the embryo, the amnion, the yolk sac, and the chrorioallantoic portion of the placenta. Blastocyst injection chimera experiments show that ES cells are similar to early-stage ICM cells in that they contribute to the primitive ectoderm and endoderm derivatives *(1)*. However, it is probably not posssible to equate these two cell types, as ES cells appear to be produced by the cell culture environment and have no exact counterpart in the blastocyst. Instead, ES cells could be thought of as being ICM cells that, instead of undergoing rapid differentiation as they would in vivo, are abnormally locked into continuing cycles of division in the undifferentiated state by virtue of the action of exogenous factors. Leukemia inhibitory factor, LIF, is one such factor *(2,3)* and is indispensable for the propagation of mouse ES cells at least when primary embryo fibroblasts (PEFs) are used as feeder layers *(4)*.

A number of features of ES cells make them useful for studying genomic imprinting. They (1) are diploid and can be derived such that they contain only maternally and paternally derived genomes, termed parthenogenetic and androgenetic cell lines, respectively *(5)*, or uniparental duplication of chromosome regions *(6)*; (2) retain imprints as assessed by the developmental potential of chimeras *(7–9)*; (3) offer a rudimentary in-vitro system of differentiation in the production of embryoid bodies *(10)* in addition to the the in-vivo system of chimera production; and (4) can provide large quantities of cellular material such as DNA for studies of chromatin structure *(11,12)*. Nevertheless, it is

From: *Methods in Molecular Biology, vol. 181: Genomic Imprinting: Methods and Protocols*
Edited by: A. Ward © Humana Press Inc., Totowa, NJ

important to realize that their derivation and unlimited capacity for division can result in epigenetic change. For example, methylation of the paternal *H19* allele in ES cells appears to resemble more the later somatic cell pattern rather than the pattern in the ICM, suggesting that this methylation is more a function of the number of cell cycles in development rather than the stage of differentiation ***(10–14)***. In addition, the allele-specific methylation and expression patterns of imprinted genes in ES cells are unstable with passage ***(10)***. While ES cells retain imprints as ascertained by the developmental potential of chimeras, they also appear to lose some. Parental-specific expression of imprinted genes is destabilized in ES cells and cannot be corrected upon differentation of the cells in chimeras ***(10,15)***, and this destabilization may contribute to the developmental abnormalities often observed ***(10,16)***. There is no substitute for the use of real embryonic material for studying genomic imprinting when possible.

2. Materials

1. 1× Dulbecco's phosphate-buffered saline; with and without calcium and magnesium ($DPBS^+$ and $DPBS^-$, respectively).
2. 1× trypsin-EDTA solution; 0.25% trypsin (1:250) and 1 m*M* ethylenediaminetetraacetic acid (EDTA) in Hank's balanced salt solution without calcium and magnesium or in $DPBS^-$. To 100 mL, add 1 mL of a 5% (w/v) solution of bovine serum albumin (BSA). To make BSA solution, dissolve 5 g of BSA (Miles, Inc., Diagnostics Division, cat. no. 82-047-3, Kankakee, IL, USA) in cell culture-grade water and filter-sterilize. Store 1-mL aliquots at –20°C.
3. 1000× 2-mercaptoethanol solution; add 70 µL of 2-mercaptoethanol to 10 mL of $DPBS^-$ to give 0.1 *M* solution. Mix, then filter-sterilize. Store at 4°C and discard after 3 wk.
4. 1× gelatin solution; add 0.5 g cell culture-grade gelatin to 500 mL $DPBS^-$ in a glass media bottle and sterilize by autoclaving. After cooling, swirl solution to mix dissolved gelatin.
5. ES cell media; to 500 mL of Dulbecco's modified Eagle's medium (DMEM) (with 4.5 g/L glucose, 2.2 g/L sodium bicarbonate, without L-glutamine, without pyruvate; *see* **Note 1**), add 75 mL fetal bovine serum (FBS), 0.6 mL of 1000× 2-mercaptoethanol, 6 mL of 100× penicillin/streptomycin solution, 6 mL of 100× L-glutamine (200 m*M*), and 6 mL of 100× nonessential amino acids solution. The latter three solutions are the standard formulations obtained from any supplier of cell culture reagents. Store media at 4°C, and if not used in 1 mo, add L-glutamine and 2-mercaptoethanol at half the initial proportion according to residual volume. Thawed FBS is also kept at 4°C and is stable for many months, although it can be refrozen if necessary.

6. Media for PEFs; as for ES cell media, except add 55 mL of FBS.
7. STO fibroblast media; as for ES cell media, except add 35 mL FBS.
8. Mineral oil (Sigma, cat. no. M 8410).
9. 100× G418 solution. Dissolve geneticin sulfate (Gibco BRL, cat. no. 11811-031) in 40 mL of $DPBS^-$ to achieve a concentration of 17.5 mg/mL active weight; proportion of active weight or microbiological potency is stated on bottle. Store as 1.5-mL aliquots in 2-mL sterile screw-capped tubes (Sarstedt, cat. no. 72.693.005) at –20°C. Thawed tubes are stable at 4°C for at least 1 mo.
10. 100× mitomycin C solution: dissolve 2 mg in 4.0 mL of $DPBS^-$. Store 0.12-mL aliquots in sterile 2-mL Sarstedt tubes. Thawed tubes are stable at 4°C for at least 2 wk.
11. Freezing solution I; Mix 12 mL of FBS with 18 mL of $DPBS^+$ in a 50-mL centrifuge tube and store at 4°C. Freezing solution II; Mix 5 mL of sterile cell culture-grade dimethyl sulfoxide (DMSO) with 20 mL of $DPBS^-$ in sterile 50-mL centrifuge tubes and store at –20°C. Thawed tubes can be kept at 4°C for a number of weeks while in use.
12. STO immortalized fibroblast cells for feeder layers. A line with suitable characteristics for ES cell culture can be obtained from the American Type Culture Collection, Mannasas, VA, USA (cat. no. CRL-2225). This line has been transfected with the *neo* and leukemia inhibitory factor, *LIF*, genes, hence is designated SNL.
13. For handling ova: Pasteur pipets pulled by hand over a flame to 0.2 and 0.3 mm in outer diameter (od) and flame polished at the tip.
14. For disaggregating blastocyst outgrowths: Borosilicate glass capillary tubing, 1 mm od and standard wall thickness, pulled and broken to 0.06 mm od and flame-polished at the tip with the aid of a pipet puller and microforge. These can be stored in aluminum blocks with holes drilled at slightly greater than 1 mm od Alternatively, Pasteur pipets can be hand-pulled.
15. For plating of disaggregated blastocyst outgrowths: 4-well tissue culture dishes (Nunc, cat. no. 176740).
16. 10× hyaluronidase solution for removal of cumulus cells from oocytes: 3 mg/mL (Sigma, cat. no. H3884) in medium M2. Stable at 4°C for months. Add 0.2 mL to approx. 1.8 mL medium M2 in a 3-cm Petri dish to give 1x solution.
17. Mouth-controlled aspiration device to control flow of medium in pipets described above ***(17)***. Thin-diameter tubing can be attached to accommodate 1-mm-od glass capillaries.
18. Ovum culture dish: Drops of medium CZB ***(18)*** of approx. 10 µL pipetted in rows onto a 3-cm Petri dish, then overlaid with mineral oil. These dishes will equilibrate in 30 min if mineral oil stock is kept permanently in an incubator. Incubator settings are 37.5°C and 6% CO_2 in air.
19. Cell digest buffer: 50 m*M* Tris-HCl pH 7.5, 20 m*M* EDTA, 0.45% each of the detergents Tween 80 and NP40, and 0.3 mg/mL proteinase K added from powder. Stable at 4°C for at least 2 mo. Do not freeze.

3. Methods

3.1. Preparation of Fibroblast Feeder Layers for ES Cells

STO cells or primary embryo fibroblasts (PEFs) are used as feeder layers for deriving and propagating ES cells (*see* **Notes 2** and **3**).

3.1.1. Propagating STO Cells

1. Thaw vial of frozen STO cells quickly in water bath and add contents directly to 10 mL of STO cell media in a 10-cm plate, then disperse the cells by crosswise agitation. Next day, replace with 10 mL of fresh medium and add 0.1 mL of 100X G418 solution.
2. Passage STO cells when they reach confluency. Rinse plate with 5 mL of DPBS$^-$, add 2 mL of trypsin-EDTA, leave in incubator for 5 min, add 3 mL of media, and pipet cells up and down in plate to produce single cell suspension. Add 4 drops of the cell suspension to a 10-cm plate containing 10 mL of media and 0.1 mL of 100× G418 solution, then disperse the cells by crosswise agitation. This is the propagating plate and is passaged for approx 50 d in this manner, at which time a new vial is plated. It is important to passage this line of plates regularly and to keep it under G418 selection so that cells retain desirable morphology and remain drug-resistant.
3. To obtain cells for making feeders, at each passage of the propagating plate, the unused portion of cells is plated at any density and grown without G418 selection for up to three passages according to the number of feeder cells that will be required. 10–15 × 10^6 cells can be obtained from one confluent 15-cm plate.

3.1.2. Deriving and Propagating PEFs

1. Some strains of PEFs grow better than others. One good combination is (C57BL × C3H)F_1 females mated to 129/Sv males.
2. Kill pregnant mouse at 12.5 d post coitum (dpc), saturate with disinfectant solution, and pull skin back with fingers to expose the body wall. With sterile instruments, remove both uterine horns containing the fetuses. Place immediately into a 10-cm plate.
3. Take plate to sterile hood. Slit uterus open and remove up to 10 fetuses. Place them into a fresh 10-cm plate with 10 mL DPBS$^+$. Using #5 watch maker's forceps, pinch off the head and remove liver from each fetus.
4. Transfer the carcasses to a fresh 10-cm plate containing 10 mL DPBS$^+$. Tilt the plate, then, with a 1-mL syringe fitted with an 18-gage needle, draw up each fetus individually into the syringe to shear it into small pieces, then expel into a fresh 10-cm plate.
5. Aspirate DPBS$^+$ from plate, add 10 mL trypsin-EDTA, then leave plate in incubator for 15 min.
6. With 5-mL pipet, transfer trypsin and fetus pieces to 50-mL centrifuge tube, then pipet vigorously up and down to create a cell suspension. Pieces should

Table 1
Plating STO and PEF Feeder Cells

	Volume required for one plate or well		
Plate type	Medium	Feeder stock[a]	Feeder stock/4[b]
3-cm	1 mL	0.5 mL	—
6-cm	2.5 mL	1 mL	—
10-cm	4 mL	3.5 mL	—
4-well	—	—	0.5 mL
12-well	—	—	1 mL
24-well	—	—	0.5 mL

[a] 1.5×10^6 cells/mL.
[b] Four times dilution of feeder stock.

dissociate almost completely. Pipette one embryo equivalent into 15 mL of PEF medium in a 15-cm plate—that is, 10 fetuses gives 10 plates.

7. After overnight culture, aspirate medium from plate and replace with 25 mL of fresh medium. Grow for 3–4 d until cells become confluent. Freeze cells (*see* **Subheading 3.1.4.**).
8. To propagate PEFs in making feeder layers, thaw one vial and transfer contents to 10 mL of PEF medium in a 10-cm plate. PEFs do not need to be grown in gelatinized plates.
9. The next day, change the medium. When the plate becomes confluent, passage the cells into up to three 15-cm plates and grow to confluency. PEFs can be passaged again, but their growth slows substantially in this and subsequent passages.

3.1.3. Making STO and PEF Feeder Layers

1. For one confluent 15-cm plate, aspirate some of the medium so that it just covers the plate. This will be approximately 12 mL. Add 0.12 mL of 1000× mitomycin C solution to achieve a concentration of approx. 5 μg/mL.
2. Leave plate for 1.5–4 h in incubator to mitotically inactivate cells, rinse plate three times with 10 mL of $DPBS^+$, add 5 mL trypsin-EDTA, then incubate plate for 5 min.
3. Add 7 mL medium, pipet up and down to disperse cells, then transfer cells to centrifuge tube. Take aliquot for cell counting, then centrifuge cells at 250*g* for 3 min.
4. Resuspend cell pellet to 1.5×10^6 cells per mL in STO or PEF medium to obtain a stock of feeder cells. Plate this feeder stock, again in STO or PEF medium, according to the guidelines given in **Table 1**. Plates must be thoroughly agitated

crosswise to achieve an even settling of cells. Feeder plates can be used from 4 h to 7 d after plating. Also, the unplated stock can be kept at 4°C for 4 d without significant loss of viability. Freezing feeder cells for plating later works well, although some viability is lost.

3.1.4. Cryopreservation

1. For one confluent 15-cm plate of STO cells or PEFs, trypsinize plate as in passaging (*see* **Subheading 3.1.1.**), transfer 12 mL of cells to centrifuge tube, then pellet cells at 250*g* for 3 min. For one confluent 6-cm plate of ES cells, trypsinize with 1 mL of trypsin-EDTA as in passaging (*see* **Subheading 3.3.6.**), then transfer cells to 3 mL of medium and pellet cells.
2. For STOs and PEFs, aspirate media, then resuspend cells in 1.25 mL of freezing solution I. Slowly add 1.25 mL of freezing solution II while gently swirling the cells, then dispense 0.5-mL aliquots of cells to five cryovials, approx. 2.5×10^6 cells per vial. Keep caps of cryovials with internal thread sterile by standing them on the inside of the 15-cm plate lid. For ES cells, aspirate media, resuspend in 2 mL solution I, then add 2 mL solution II and dispense 1-mL aliquots into four cryovials, approx. 2×10^6 cells per vial.
3. Place cryovials into a polystyrene 15-mL centrifuge tube rack "sandwich" and place at –70°C to –80°C overnight. Alternatively, use commercially available containers designed to cool at a more controlled rate. STO cells, PEFs, and ES cells are stable at –70°C for many months, but it is best to transfer to liquid nitrogen as soon as possible. In the liquid nitrogen tanks, plastic screw-capped cryovials should be stored in the vapor phase.
4. To thaw STOs and PEFs, hold the ampule in awater bath (set at 35°C) until completely thawed, then dispense contents immediately into a 10-cm plate containing 10 mL of media and agitate to disperse cells. For ES cells, thaw as above, then transfer 1 mL of cells to a 15-mL centrifuge tube, slowly add 4 mL of medium while gently swirling the tube, then pellet the cells. Resuspend in 4 mL of media and place into a 6-cm feeder plate. In this step, medium is not warmed prior to use.
5. For all types of cell, change media the next day to remove DMSO.

3.2. Obtaining Blastocysts for ES Cell Derivation

3.2.1. Parthenogenetic

1. Superovulate mice at 4–12 wk of age by injecting them in the peritoneum with 0.1 mL of saline containing 5 U of pregnant mare serum gonadotrophin followed approx. 48 h later with 5 U of human chorionic gonadotrophin (*see* **Note 4**). Eggs are ovulated approx. 12 h after the latter injection, thus 14 h after this injection, kill mice by cervical dislocation, then isolate the oviducts and place them in 2 mL of medium M2 ***(19)*** in a 3-cm Petri dish. For details on dissection, *see* **ref. *20***.

2. Transfer oviducts to 2 mL of 1× hyaluronidase solution in medium M2 and tear open the ampulla to release the "cumulus masses" of follicle cells and oocytes. Place the dish on a warm plate set at 35–37°C and leave for 10 min to release oocytes from follicle cells.
3. Activation of oocytes with ethanol ***(21)***: Using a Pasteur pipet pulled to 0.3 mm od and working with approx. 50 oocytes at a time, transfer them using the mouth-controlled aspiration device to 2 mL of 7% (v/v) ethanol in medium M2 without calcium and magnesium in a 3-cm Petri dish. Mix oocytes into medium by pipetting them up and down and around the dish.
4. After 3 min in the ethanol solution, transfer oocytes to 2 mL of standard medium M2 to wash out ethanol.
5. Transfer oocytes to culture drops of medium CZB ***(18)*** containing 5 μg/mL cytochalasin B or 1 μg/mL cytochalasin D. Culture for 4 h to inhibit extrusion of the second polar body, then transfer oocytes to standard medium CZB and culture for another 4 h.
6. Eight to 12 h after ethanol treatment, select 1-cell ova which have two pronuclei and no polar body. This should be done with a microscope under phase contrast or differential interference contrast optics. Desired ova can be set aside in a micromanipulation chamber using a holding pipet as used in micromanipulation experiments. A significant proportion of the oocytes will be unusable; some will look like 2-cell ova having undergone immediate cleavage, have fragmented—"raspberries," or may contain one pronucleus or micronuclei ***(22)***.
7. Selected ova are cultured to the blastocyst stage in medium CZB. If oocytes of inbred mice are used, it may be necessary to transfer the diploid ova into oviducts of pseudopregant recipients to enable further development.

3.2.2. Gynogenetic

1. Gynogenetic ova are produced from zygotes by pronuclear transplantation; the paternal pronucleus from a zygote is removed and replaced with the maternal pronucleus of another to produce a 1-cell ovum with two maternal pronuclei. Parthenogenetic ova are then cultured in medium CZB and should reach the blastocyst stage by 3.5 dpc or 3.5–4.5 dpc if the egg cytoplasm is from F_1 females or inbred females, respectively. For methods of pronuclear transplantation, *see* ***20***, ***23***, and ***24***. Parental-specific genetic markers are used to ensure that derived cell lines have only maternally derived genomes—for example, a convenient marker is the ubiquitously expressed glucose 6-phosphate isomerase-1 gene located on chromosome 7. Three alleles encode three electrophoretic variants detected in a simple assay ***(20)***: The A and B forms are common among laboratory strains, while the ferally derived C form is present in the 129/Sv//Tac strain (Taconic Farms, Germantown, NY).

3.2.3. Androgenetic

1. Androgenetic ova are produced by pronuclear transplantation as are gynogenetic ova, except that the maternal pronucleus from a zygote is removed and replaced

with the paternal pronucleus of another to produce a 1-cell ovum with two paternal pronuclei; zygotes for this manipulation are obtained by mating F_1 females to males of choice. Androgenetic ova are then cultured in medium CZB. Most should reach the blastocyst stage by 3.5 dpc, although some take an extra day. Parental-specific genetic markers are used to ensure that derived cell lines have only paternally derived genomes.

2. Androgenetic ova can also be produced by removing the maternal pronucleus from a zygote, then culturing the haploid ovum in cytochalasin B or D to inhibit first cleavage and make it diploid. However, only blastocysts of an XX sex-chromosome constitution can be obtained by this method, and for reasons discussed in **Note 5** it is probably best to utilize only XY cell lines.

3.2.4. Uniparental duplication

Due to the high frequencies of chromosomally unbalanced and inviable ova derived from intercrossing mice with reciprocal translocations (*see* **Note 6**), it is best to superovulate and mate females and flush out morulae at 2.5 dpc. Transfer all morula that form blastocysts to feeder layers by 4.5 dpc.

3.2.5. Wild Type

1. For inbred strains, it is best to obtain ova by natural mating. Many inbred strains do not respond well to superovulation. Often the problem lies in obtaining mating and fertilization of ovulated eggs.
2. At 2.5 dpc, dissect out the oviduct and flush out morulae in medium M2. Transfer to medium CZB and culture overnight to the blastocyst stage.

3.3. Deriving and Propagating ES Cell Lines

All culture conditions are 37.5°C and 5% CO_2 in air.

3.3.1. Culture of Blastocysts

1. Ova that have reached the blastocyst stage in medium CZB are treated with acidic Tyrode's solution to remove the zona pellucida ***(20)***. This is done as some blastocysts, especially androgenetic ones, fail to shed this coat and thereby fail to attach to the dish in subsequent steps.
2. Immediately after the zona is removed, 16 or fewer blastocysts are pipetted into one 3-cm feeder-layer plate with ES cell media and incubated.
3. 2 d after being placed in FBS culture media, the blastocysts begin to attach or "implant" to the bottom of the culture dish (**Fig. 1A**). Growth after attachment is rapid and some will already be forming outgrowths. By 3 d virtually all should be attached, and most will be forming outgrowths. By 4 d the outgrowths will be larger, and many are surrounded by trophectoderm giant cells (**Fig. 1B**).

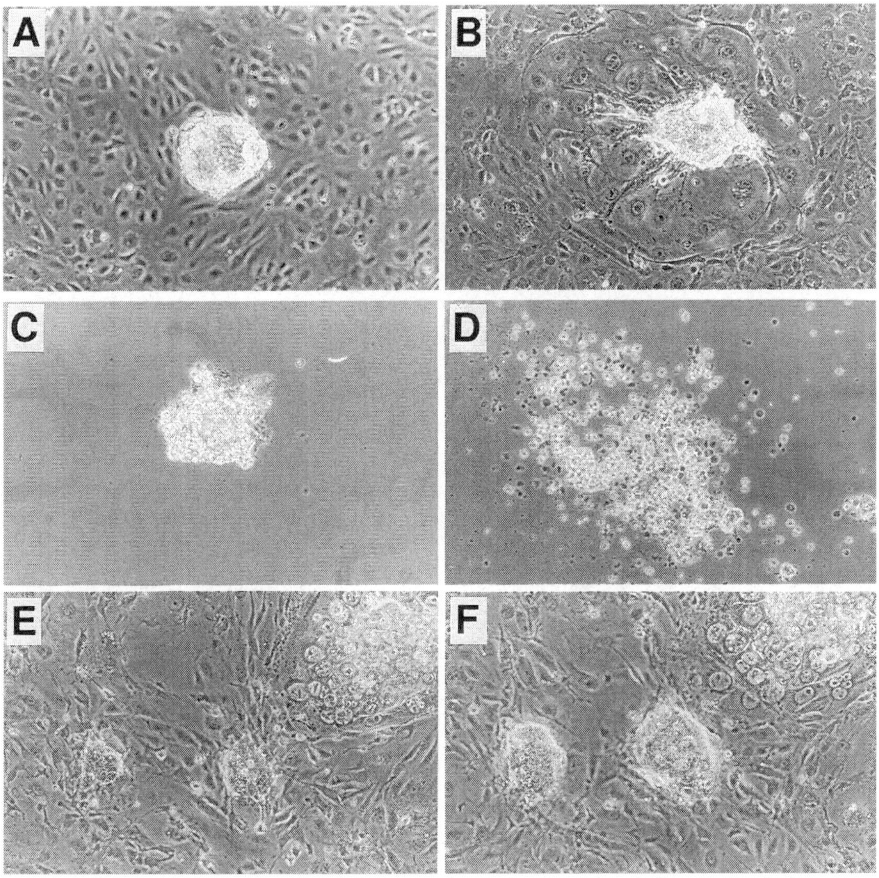

Fig. 1. Manipulation of blastocysts for deriving ES cell lines. **(A)** Blastocyst attached to STO feeder layer 3 d after transfer to feeder plate. ICM can be seen at "3-on-the-dial." **(B)** Blastocyst outgrowth 4 d after transfer to feeder plate. Note layer of trophectoderm cells surrounding outgrowth. **(C)** Picked outgrowth in trypsin-EDTA, **(D)** disaggregated outgrowth, **(E)** two 129/SvImJ ES cell colonies at passage 1 after 3 d of culture; note patch of trophectoderm cells at top right. These colonies were derived from a 4-well passaged in its entirety into a 12-well, and **(F)** colonies in **(E)** after 4 d of culture. All objects at 100× (original magnification) and under phase-contrast optics.

3.3.2. Picking and Seeding Blastocyst Outgrowths

1. In picking outgrowths, our rule of thumb is to pick no later than 1 d after attachment has begun, this attachment being depicted in **Fig. 1A**. This is usually at 4 d after 3.5-dpc blastocysts have been transferred to feeder dishes in FBS

culture media (**Fig. 1B**). Picking can be done at 2 d or more after attachment, but the frequency of cell line derivation may fall. It is difficult to obtain cell lines from outgrowths that have developed a layer of endoderm ***(25)***. Almost all blastocysts of an expected euploid chromosome constitution, that is, wild-type, parthenogenetic, gynogenetic, and androgenetic blastocysts, should attach and produce outgrowths that can be picked, although androgenetic outgrowths are often not vigorous.

2. Before picking outgrowths, a blastocyst outgrowth disaggregation dish is prepared consisting of approx. 5-µL drops of trypsin-EDTA pipetted onto a 6-cm Petri dish in *n* columns of four and overlaid with mineral oil. Also, 1–2 d in advance, 4-well feeder plates are prepared for plating of disaggregated outgrowths. The night before disaggregation, the wells are aspirated and 1 mL of ES medium is added. The next morning, half of this medium is replaced with fresh medim, and 1 drop of FBS is added from a 1-mL pipet to bring the serum concentration to approx. 16%.
3. To pick the outgrowths, rinse the dish with 2 mL of $DPBS^+$ containing 0.01% (w/v) BSA, then add 2 mL of this same solution to the dish.
4. Using a Pasteur pipet pulled to 0.2–0.3 mm od and with a stereomicroscope, nudge or pull off the outgrowth from the trophectoderm layer, then transfer it to a drop of trypsin-EDTA in the blastocyst outgrowth disaggregation dish (**Fig. 1C**). Proceed until all outgrowths in the dish are picked. Between each pick, ensure that there are no cells remaining in the pipet from the previous outgrowth by observing it briefly under the stereomicroscope and pipetting up and down to dislodge them if necessary.
5. Keep the picked outgrowths at RT, then approx. 15 min after the first outgrowth was picked, aspirate a small mount of medium into a glass capillary pulled to 0.06 mm od (**Fig. 1C**). Expel a small amount of medium over an outgrowth, aspirate it into the pipet, expel it back into the drop, then repeat this aspiration/expulsion step. A near-single-cell suspension should result (**Fig. 1D**). Aspirate all of the cells then expel them into one well of a 4-well feeder plate. Use a new capillary for each disaggregation. After all outgrowths are seeded, place the 4-well plates into the incubator, labeling them "passage 0."
6. After 3 d of culture, aspirate two-thirds of the medium from each well, then add fresh medium to 1 mL. Culture for another 2 d, then examine the wells for growth of primary ES cell colonies.

3.3.3. Selecting and Picking Primary ES Cell Colonies

1. At passage 0, emerging primary ES cell colonies will appear exactly like colonies obtained when plating an existing cell line at very low density—for example, at 200 cells per 3-cm feeder plate. In plating C57BL/6J and CBA/CaJ outgrowths, it is usual that ES cell colonies are the only type of colony that grow vigorously, thus identification is generally straightforward. However, in seeding 129/SvImJ

outgrowths, other types of cell also grow vigorously and often the primary ES cell colonies are obscured or mixed with these cells. In seeding Swiss mouse outgrowths, many colonies that appear similar to ES cells proliferate at passage 0 (*see* **ref. *25*** for a detailed discussion). In any event, at the next passage, the presence of ES cells is revealed, as generally no other cell type continues to proliferate.

2. If putative primary ES cell colonies are observed, at passage 0 and at 6 d after seeding the disaggregated outgrowths, one of two paths can be followed. (1) Individual colonies are picked as described for blastocyst outgrowths except they are placed into 0.06 mL of trypsin-EDTA in a round-bottomed well of a 96-well plate. After 10 min, with a pipetman and barrier tip, 0.1 mL of medium is added, the colony is broken up into a single cell suspension by repeated pipetting, then the well contents dispensed into a well of a 24-well feeder plate and labeled passage 1. With this method there is a high probability that the cell line will be derived from one cell of the outgrowth. (2) Passage all of the contents of the well. This may be desirable if one is uncertain that ES cell colonies have been obtained. Rinse a well with $DPBS^-$, add 0.15 mL of trypsin-EDTA, incubate for 10 min, add 0.3 mL of medium (from a 12-well feeder plate well containing 2 mL of medium), break up cell colonies by repeated pipetting, then seed all 0.45 mL back into the 12-well plate and label it passage 1. With this method, there is a significant chance that the cell line will be derived from more than one cell, and for this reason it might be expected that later clonal derivatives of such a cell line, such as are obtained in gene targeting experiments, may be more heterogeneous in their characteristics than clones obtained from cell lines derived by method 1. In any event, cell lines can be cloned at passage 1 or later.
3. With either method 1 or 2 as described in the previous step, true ES cell colonies will be clearly visible at passage 1 after 3 d (**Fig. 1E,F**). If no colonies are observed after 5 d, it is very unlikely that any will appear and such wells can be discarded. *See* **Notes 7** and **8** for the frequency of cell line derivation per blastocyst.

3.3.4. Primary Expansion of New ES Cell Lines

1. Once ES cell colonies are obtained at passage 1, a new cell line has been derived and most of them should continue to proliferate. However, in a small number of cases, significant differentiation ensues in the next couple of passages and the cell line is lost.
2. To expand the cell line, the cells are passaged in their entirety from the 24-well plate (passage 1) to a 3-cm feeder plate (passage 2), then to a 6-cm feeder plate (passage 3). An extra passage may be required to obtain confluency at this last stage. The number of days at each passage depends on the density of cells, but 2–4 d is usually required. The cells in the confluent 6-cm plate are frozen in four vials and labeled passage 4 or 5, tier I. The cell line derivation schedule is depicted in **Fig. 2**.

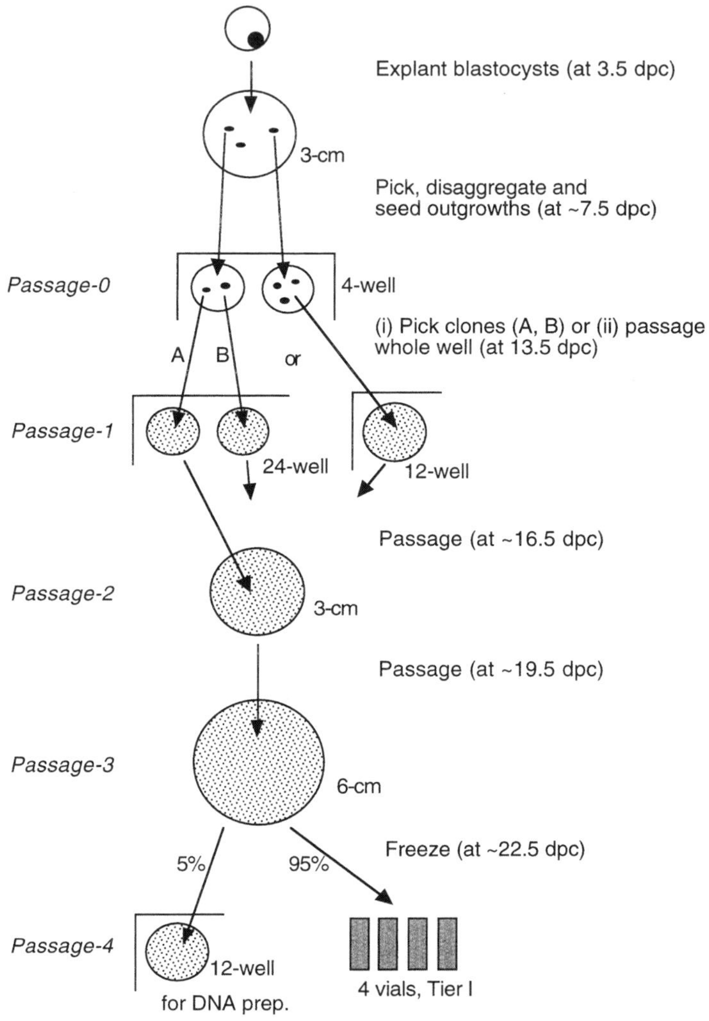

Fig. 2. Steps in the derivation of ES cell lines.

3.3.5. Characterizing New ES Cell Lines

1. Determine Y chromosome status. In freezing the cells for tier I, 5% of the cells are saved and plated into a gelatinized well of 12-well dish without feeder cells (**Fig. 2**). The ES cells should grow without differentiating and, when confluent, add 0.4 mL of cell digest buffer and incubate for at least 4 h at 37–55°C. Purify DNA and perform Southern blot. Probe pY353/B recognizes a repetitive element specific to the mouse Y chromosome ***(26)***. Cut plasmid pY353/B with EcoRI to yield a 1.5 kb-fragment for probe. Digestion of DNA with *Hin*dIII or *Eco*RI yields

strong hybridizing bands at 12 kb and at 1.5 kb, respectively. No hybridization at all is obtained with DNA from female mice. STO cells show relatively weak hybridization of a 7-kb band with *Hin*dIII digestion. Alternatively, metaphase spreads can be C-banded to reveal the Y ***(27,28)***. In deriving 129/SvImJ, C57BL/6J, and CBA/CaJ cell lines, we have obtained approximately equal numbers of pY353/B-positive and -negative lines. The remainder of the DNA can be used for further analysis of genotype.

2. Determine chromosome number. At least 80% of metaphase spreads with 40 chromosomes should be obtained with cell lines to be used for chimera production. Methods for making metaphase spreads are provided ***(25,27,28)***. There is no substitute for empirical observation in determining the efficacy of chimera formation and germ-line transmission, as a cell line or clone that is predominantly euploid may not necessarily form good chimeras.
3. Assess microbiological status ***(29)***.

3.3.6. Propagating ES Cell Lines

1. Plate a vial of cells at tier I—for example, passage 4—into one 6-cm feeder plate and grow to confluency.
2. Trypsinize cells and plate them at 0.5×10^6 cells per 6-cm feeder plate. When plates are confluent, freeze cells at four vials per plate and label them passage 6, tier II. At least 24 vials can be obtained.
3. Tier II vials can be used for further characterization of the cell line or in experiments, but it is desirable to passage tier II cells two or three more times to freeze them at passage 8 or 9, tier III, if the cell line is to be used extensively.
4. In propagating ES cells, plate them at 0.5×10^6 cells per 6-cm feeder plate in 3.5 mL of media (**Fig. 3A**). After 1 d (**Fig. 3B**), add 2 mL of fresh media, and after 2 d (**Fig. 3C**), replace all media with 6 mL of fresh media. After 3 d of growth, the plate is confluent, containing $8–10 \times 10^6$ cells (**Fig. 3D**) and is passaged again.
5. In passaging a confluent plate, if the medium is very acidic or yellow, it is replaced with 3 mL of fresh medium at least 3 h before trypsinization. This increases the viability of the cells upon passage or after cryopreservation. To passage, the plate is rinsed with 3 mL of DPBS$^-$, then 1 mL of trypsin-EDTA is added. After 10 min in the incubator, cells are broken up by vigorous pipetting in the plate with a plugged Pasteur pipet, then the 1 mL of disaggregated cells are dispensed into 4 mL of medium in a centrifuge tube and mixed immediately. Cells are counted, and 0.5×10^6 cells per 6-cm feeder plate are used for further growth (**Fig. 3A**). If the cells are plated at 1×10^6 cells or more per 6-cm plate, then passage may be required after 2 d of growth. Feeder cells and trypsin are carried over at each passage. A rule of thumb is that the trypsin-EDTA solution should not exceed 10% by volume in the culture media. If so, cells should be pelleted to remove it. At trypsinization, it is very important to create a single cell suspension, although it is impossible to avoid getting some doublets and triplets of cells (**Fig. 3A**). If undissociated clumps of ES cells are present, they

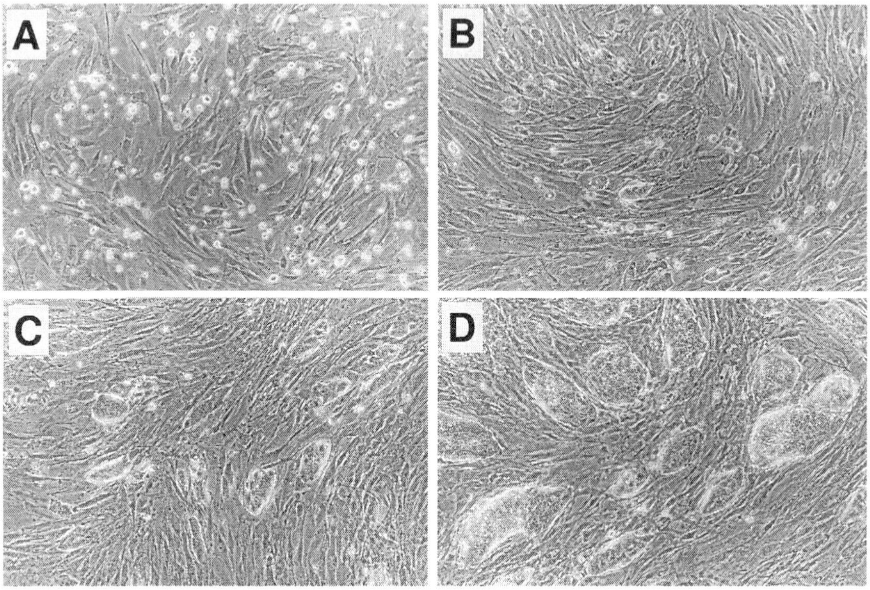

Fig. 3. One passage of a euploid XY C57BL/6J ES cell line derived from a normal blastocyst on a PEF feeder layer. (**A**) ES cells just seeded at 0.5×10^6 cells in a 6-cm plate; note single cell suspension. This same culture showing ES cell colonies at (**B**) 1 d, (**C**) 2 d, and (**D**) 3 d later. At 3 d, there are approx. 8×10^6 cells in the plate (cell doubling time approx. 18 h) and the culture is passaged again. All objects at 100X (original magnification) and under phase-contrast optics.

will become relatively large colonies and may begin to differentiate before the next passage. If this practice is continued, ultimately the cultures will deteriorate.

3.4. Testing Serum Lots

A number of commercial sources of FBS have been tested for growth of ES cells. Nevertheless, it is still a good idea to test different lots, as better or less expensive ones may often be found. To test three unknown against one control serum (control serum could be a sample of a commercial pretested lot):

1. Prepare two 12-well plates of feeder cells at least 1 d before plating ES cells.
2. Make up DMEM with all additives except serum, and dispense 1.35 mL into each well of one plate, and 1.2 mL into each well of the other plate. In the first plate, with a pipetman and barrier tip, for each serum dispense 0.15 mL into three wells (10% wells). Similarly, for the second plate, for each serum dispense 0.3 mL into three wells (20% wells).
3. Trypsinize a near-confluent 3-cm plate of ES cells, pellet cells, replate them into two nongelatinized 3-cm plates without feeders, then place in incubator.

4. After 45 min most feeder cells will have attached to the bottom of dish, while ES cells will be lightly attached at most. At this time, gently aspirate media from both dishes, then with a pipetman and barrier tip, vigorously pipet up and down over the bottom of the dishes with a total of 0.7 mL of media to dislodge ES cells, then dispense into a tube.
5. Count cells, then dilute in DPBS+ to 100 cells per 0.05 mL in a total of 10 mL.
6. Dispense 0.05 mL of cell suspension, that is, 100 cells, into each well of the two 12-well plates, then incubate.
7. Change media 3 d and 6 d after plating, then score clones, or colonies, 8 d after plating.
8. In scoring clones, first check for differentiation: Most colonies should be completely or largely undifferentiated. After this examination, aspirate media, turn plates upside down, and count the number of colonies at the bottom of each well. Choose serum which passed the differentiation test and produced the highest number of clones. An acceptable serum will give approx. 20 clones in both 10% and 20% wells. Growth of clones should have been somewhat faster in the latter wells.

4. Notes

1. DMEM used in ES cell culture typically contains approx. 2.2 g/L rather than the standard 3.7 g/L of sodium bicarbonate. This lowers the osmotic pressure of the DMEM such that it approaches that of media used widely in culture of preimplantation stage ova ***(30,31)***. For this reason it may be better suited for culturing blastocyst-derived cells, although this is anecdotal and ES cells can be cultured successfully in standard DMEM.
2. For feeder layers, investigators generally have a preference for using either PEFs or STO cells. Each type of feeder has been used with equal success, and while the choice of one over the other depends mainly on tradition, these two cell types have different minor advantages. STO cells are immortal and grow rapidly. On the other hand, ES cells can be visualized more readily on PEFs and to many the cultures are more esthetic. If one intends to use STO cells, it is very important to obtain a subline of suitable morphology and that will form a durable feeder layer—for example, the source given in the Materials section. Unsuitable batches of STOs can be subcloned by manually dispensing single cells into wells using a pulled Pasteur pipet or capillary or by seeding at very low density. The morphology of clones is examined 7 d later and desirable ones isolated and expanded. To be useful, newly derived STO cell sublines must be able to support the growth of ES cells plated at very low density—for example, 200 cells per 3-cm feeder plate.
3. Adding LIF to ES cell culture media is generally unnecessary, and this is certainly true if one is using STO feeder cells that are transfected with the *LIF* gene. Some lines of PEF feeder cells expressing NEO are not fully resistant to G418, and it may be necessary to add LIF when ES cells are placed under selection.
4. Oocytes of hybrid females, such as (C57BL female × C3H, CBA, or SLJ male)F_1 are the best for parthenogenetic activation. For inbred strains, C57BL are good,

while 129/Sv are poor. Usually the problems lie not in activation but in the viability of the ovum following the experimental procedure. A potentially gentler method than ethanol treatment for activating oocytes is to culture them for 8 h in medium CZB containing 10 m*M* strontium chloride instead of calcium chloride. Again, 5 μg/mL cytochalasin B or 1 μg/mL cytochalasin D is used to inhibit polar body extrusion ***(32)***. Alternatively, the problem can be circumvented by transplanting the two pronuclei into enucleated parthenogenetically activated haploid ova of hybrid mice ***(23)***. The parthenogenetic ova then develop to the blastocyst stage at high frequency.

5. Both X chromosomes remain active in XX ES cells, and with passage the cells appear to compensate chromosomally for an X:autosome activity ratio that is probably not compatible with long-term survival. In XX parthenogenetic lines, one X often has deletions in the distal region ***(33)***. Of 10 pY353/B-negative lines derived from wild-type 129/SvImJ blastocysts, two were XO and five had only one normal X (K. Fowler and J. Mann, unpublished data). Also, XX embryonal carcinoma cells can possess supernumerary autosomes ***(34,35)***. This chromosomal instability of XX ES cell lines obviously has implications for the production and analysis of chimeras, and careful analysis of karyotype should be carried out before they are used.
6. Mice with maternal or paternal duplication of autosome regions are produced by intercrossing mice heterozygous for reciprocal translocations (*see* Chapter 3). In deriving ES cell lines, it is important to keep in mind that in these intercrosses (1) a high frequency of chromosomally unbalanced zygotes are obtained, such that the frequency of normal postimplantation development is one-third of normal; and (2) of the mice obtained, for the chromosome regions proximal and distal to the translocation breakpoint, one-sixth and one-tenth, respectively, possess the desired maternal or paternal duplication ***(36)***. Thus, one-eighteenth and one-thirtieth of all zygotes, respectively, have the desired genotype. Because of the high frequency of chromosomal imbalance, not all morulae will develop to the blastocyst stage, and not all blastocysts will produce outgrowths. In intercrosses involving the T(7; 15)9H translocation, the frequency of outgrowths per blastocyst was similar to the frequency of 9.5 dpc embryos per implantation site ***(37)***, thus the failure of blastocysts to develop in vivo occurs also in vitro. This is an advantage, as only euploid blastocysts are likely to form new cell lines in this system. The inviability might be expected, as all noneuploid zygotes resulting from intercrosses of reciprocal translocations are partially monosomic or nullisomic for one of the two chromosomes involved. For every mouse chromosome, monosomy is lethal at the peri-implantation stage ***(38)***.
7. Using the conventional means of derivation as described in this chapter, the frequency of ES cell line derivation (per 3.5–4.5 dpc blastocyst not subjected to implantation delay) that we have obtained has varied according to the mouse strain used. With wild-type, parthenogenetic and androgenetic blastocysts of the 129/SvImJ strain, or blastocysts containing significant amounts of this genetic background, a frequency of 50% has generally been obtained. With wild-type

C57BL/6J and CBA/CaJ blastocysts, frequencies have been 10%. CBA strains have been described as resistant to ES cell line derivation when similar methods to those described here have been used ***(39,40)***. Thus, differences between CBA substrains or other undefined experimental variables determine the success of derivation, and the same may be true for many mouse strains. With hybrid wild-type blastocysts derived from crosses of the mouse subspecies *Mus musculus castaneus*, CAST/Ei, that is, (C57BL/6J or 129/SvImJ × CAST/Ei)F_1 and the reciprocal, the frequency has been 90%.

8. Regarding the type of feeder cell used at passage 0, a consistent finding has been that SNL STO cells give approximately double the frequency of cell-line derivation than standard STO cells and PEFs, and a similar phenomenon has been reported previously ***(41)***. The frequency of derivation when using these latter cell types as feeders might therefore be increased by adding LIF to the medium at passage 0.

References

1. Beddington, R. S. P. and Robertson, E. J. (1989) An assessment of the developmental potential of embryonic stem cells in the midgestation mouse embryo. *Development* **105,** 733–737.
2. Smith, A. G., Heath, J. K., Donaldson, D. D., Wong, G. G., Moreau, J., Stahl, M., and Rogers, D. (1988) Inhibition of pluripotential embryonic stem cell differentiation by purified polypeptides. *Nature* **336,** 688–690.
3. Williams, R. L., Hilton, D. J., Pease, S., Wilson, T. A., Stewart, C. L., Gearing, D. P., Wagner, E. F., Metcalf, D., Nicola, N. A., and Gough, N. M. (1988) Myeloid leukemia inhibitory factor maintains the developmental potential of embryonic stem cells. *Nature* **336,** 684–687.
4. Stewart, C. L., Kaspar, P., Brunet, L. J., Bhatt, H., Gadi, I., Köntgen, F., and Abbondanzo, S. J. (1992) Blastocyst implantation depends on maternal expression of luekemia inhibitory factor. *Nature* **359,** 76–79.
5. Mann, J. R., Gadi, I., Harbison, M. L., Abbondanzo, S. J., and Stewart, C. L. (1990) Androgenetic mouse embryonic stem cells are pluripotent and cause skeletal defects in chimeras: implications for genetic imprinting. *Cell* **62,** 251–260.
6. McLaughlin, K. J., Kochanowski, H., Solter, D., Schwarzkopf, G., Szabó, P. E., and Mann, J. R. (1997) Roles of the imprinted gene *Igf2* and paternal duplication of distal chromosome 7 in the perinatal abnormalities of androgenetic mouse chimeras. *Development* **124,** 4897–4904.
7. Mann, J. R. and Stewart, C. L. (1991) Development to term of mouse androgenetic aggregation chimeras. *Development* **113,** 1325–1333.
8. Mann, J. R. (1992) Properties of androgenetic and parthenogenetic mouse embryonic stem cell lines; are genetic imprints conserved? *Semin. Dev. Biol.* **3,** 77–85.
9. Allen, N. D., Barton, S. C., Hilton, K., Norris, M. L., and Surani, M. A. (1994) A functional analysis of imprinting in parthenogenetic embryonic stem cells. *Development* **120,** 1473–1482.

10. Szabó, P. and Mann, J. R. (1994) Expression and methylation of imprinted genes during in vitro differentiation of mouse parthenogenetic and androgenetic embryonic stem cell lines. *Development* **120,** 1651–1660.
11. Szabó, P. E., Pfeifer, G. P., and Mann, J. R. (1998) Characterization of novel parent-specific epigenetic modifications upstream of the imprinted mouse *H19* gene. *Mol. Cell Biol.* **18,** 6767–6776.
12. Khosla, S., Aitchison, A., Gregory, R., Allen, N. D., and Feil, R. (1999) Parental allele-specific chromatin configuration in a boundary-imprinting-control element upstream of the mouse *H19* gene. *Mol. Cell Biol.* **19,** 2556–2566.
13. Tremblay, K. D., Duran, K. L., and Bartolomei, M. S. (1997) A 5′ 2-kilobase-pair region of the imprinted mouse *H19* gene exhibits exclusive paternal methylation throughout development. *Mol. Cell Biol.* **17,** 4322–4329.
14. Warnecke, P. M., Mann, J. R., Frommer, M., and Clark, S. J. (1998) Bisulfite sequencing in preimplantation embryos: DNA methylation profile of the upstream region of the mouse imprinted *H19* gene. *Genomics* **51,** 182–190.
15. Dean, W., Bowden, L., Aitchison, A., Klose, J., Moore, T., Meneses, J. J., Reik, W., and Feil, R. (1998) Altered imprinted gene methylation and expression in completely ES cell-derived mouse fetuses: association with aberrant phenotypes. *Development* **125,** 2273–2282.
16. Nagy, A., Rossant, J., Nagy, R., Abramow-Newerly, W., and Roder, J. C. (1993) Derivation of completely cell culture-derived mice from early-passage embryonic stem cells. *Proc. Natl Acad. Sci. USA* **90,** 8424–8428.
17. Mann, J. R. (1993) Surgical techniques in production of transgenic mice. *Methods Enzymol.* **225,** 782–793.
18. Chatot, C. L., Ziomek, C. A., Bavister, B. D., Lewis, J. L., and Torres, I. (1989) An improved culture medium supports development of random-bred 1-cell mouse embryos *in vitro*. *J. Reprod. Fert.* **86,** 679–688.
19. Wood, M. J., Whittingham, D. G., and Rall, W. F. (1987) The low temperature preservation of mouse oocytes and embryos, in *Mammalian Development: A Practical Approach* (Monk, M., ed.), IRL, Oxford, U.K./Washington, DC, pp. 255–280.
20. Hogan, B., Beddington, R., Costantini, F., and Lacy, E. (1994) *Manipulating the Mouse Embryo: A Laboratory Manual*, 2d ed., Cold Spring Harbor Laboratory, Cold Spring Harbor, NY.
21. Cuthbertson, K. S. R. (1983) Parthenogenetic activation of mouse oocytes in vitro with ethanol and benzyl alcohol. *J. Exp. Zool.* **226,** 311–314.
22. Kaufman, M. H. (1983) in *Early Mammalian Development: Parthenogenetic Studies*, Cambridge, UK, Cambridge University Press, pp. 9–16.
23. McGrath, J. and Solter, D. (1983) Nuclear transplantation in the mouse embryo by microsurgery and cell fusion. *Science* **220,** 1300–1302.
24. Stewart, C. L. (1993) Genomic imprinting in the regulation of mammalian development. *Adv. Dev. Biol.* **2,** 73–118.
25. Robertson, E. J. (1987) Embryo-derived stem cell lines. In *Teratocarcinomas and Embryonic Stem Cells. A Practical Approach* (Robertson, E. J., ed.), IRL, Oxford, U.K./Washington, DC, pp. 71–112.

26. Bishop, C. E., Boursot, P., Baron, B., Bonhomme, F., and Hatat, D. (1985) Most classical *Mus musculus domesticus* laboratory mouse strains carry a *Mus musculus musculus* Y chromosome. *Nature* **315,** 70–72.
27. Evans, E. P. (1987) Karyotyping and sexing of gametes, embryos and fetuses and in situ hybridization to chromsomes. In *Mammalian Development: A Practical Approach* (Monk, M., ed.), IRL, Oxford, U.K./Washington DC, pp. 93–114.
28. Abbondanzo, S. J., Gadi, I., and Stewart, C. L. (1993) Derivation of embryonic stem cell lines. *Methods Enzymol.* **225,** 803–823.
29. Lincoln, C. K. and Gabridge, M. G. (1998) Cell culture contamination: sources, consequences, prevention, and elimination. *Methods Cell Biol.* **57,** 49–65.
30. Biggers, J. D., Whitten, W. K., and Whittingham, D. G. (1971) The culture of mouse embryos *in vitro*. In *Methods in Mammalian Embryology* (Daniels, J. C. J., ed.), Freeman, San Francisco, pp. 86–116.
31. Lawitts, J. A. and Biggers, J. D. (1993) Culture of preimplantation embryos. *Methods Enzymol.* **225,** 153–164.
32. Bos-Mikich, A., Whittingham, D. G., and Jones, K. T. (1997) Meiotic and mitotic Ca^{2+} oscillations affect cell composition in resulting blastocysts. *Dev. Biol.* **182,** 172–179.
33. Robertson, E. J., Evans, M. J., and Kaufman, M. H. (1983) X-chromosome instability in pluripotential stem cell lines derived from parthenogenetic embryos. *J. Embryol. Exp. Morphol.* **74,** 297–309.
34. McBurney, M. W. and Adamson, E. D. (1976) Studies on the activity of the X-chromosomes in female teratocarcinoma cells in culture. *Cell* **9,** 57–70.
35. Martin, G. R., Epstein, C. J., Travis, B., Tucker, G., Yatsiv, S., Martin, D. W., Clift, S., and Cohen, S. (1978) X-chromosome inactivation during differentiation of female teratocarcinoma stem cells *in vitro*. *Nature* **271,** 329–333.
36. Searle, A. G. and Beechey, C. V. (1990) Genome imprinting phenomena on mouse chromosome 7. *Genet. Res., Camb.* **56,** 237–244.
37. McLaughlin, K. J., Szabó, P. E., Haegel, H., and Mann, J. R. (1996) Mouse embryos with paternal duplication of an imprinted chromosome 7 region die at midgestation and lack placental spongiotrophoblast. *Development* **122,** 265–270.
38. Epstein, C. J. (1986) Developmental genetics. *Experientia* **42,** 1117–1128.
39. McWhir, J., Schnieke, A. E., Ansell, R., Wallace, H., Colman, A., Scott, A. R., and Kind, A. J. (1996) Selective ablation of differentiated cells permits isolation of embryonic stem cell lines from murine embryos with a non-permissive genetic background. *Nature Genet.* **14,** 223–226.
40. Brook, F. A. and Gardner, R. L. (1997) The origin and efficient derivation of embryonic stem cells in the mouse. *Proc. Natl. Acad. Sci. USA* **94,** 5709–5712.
41. Ledermann, B. and Bürki, K. (1991) Establishment of a germ-line competent C57BL/6 embryonic stem cell line. *Exp. Cell Res.* **197,** 254–258.

3

Balanced Translocations for the Analysis of Imprinted Regions of the Mouse Genome

Anne C. Ferguson-Smith, Maxine Tevendale, Pantelis Georgiades, and Valerie Grandjean

1. Introduction

Experimental studies that investigate the functional and mechanistic properties of an imprinted locus require material in which the two parental chromosome homologs can be easily distinguished. The use of animals with uniparental duplications and deficiencies of imprinted regions of interest is one powerful approach. This material not only allows the successful analysis of the monoallelic expression and genome modifications associated with imprinting, but also is useful for studying the developmental roles of imprinted genes through the analysis of conceptuses in which the dosage of imprinted genes has been perturbed *(1–3)*.

Robertsonian and reciprocal translocation heterozygote intercrosses have been used to generate mice with uniparental disomies (UPD) and uniparental duplications/deficiencies of whole or selected chromosomal regions, respectively. This genetic approach was pioneered for imprinting studies by Bruce Cattanach and his colleagues at the MRC Mammalian Genetics Unit at Harwell in Oxfordshire. A mouse carrying a UPD or uniparental duplication/deficiency (hereafter to referred to as a partial disomy) has a normal diploid chromosome content; however, the parental origin of a subset of that genome is perturbed such that the animal has inherited both copies of a chromosome, or chromosomal region, from one parent and none from the other. This is shown schematically for UPD in **Fig. 1**. Genetic analysis has shown that there are abnormal consequences for development if the parental origin of particular mouse autosomal regions is perturbed *(4)*. Isolation of these defective embryos

From: *Methods in Molecular Biology, vol. 181: Genomic Imprinting: Methods and Protocols*
Edited by: A. Ward © Humana Press Inc., Totowa, NJ

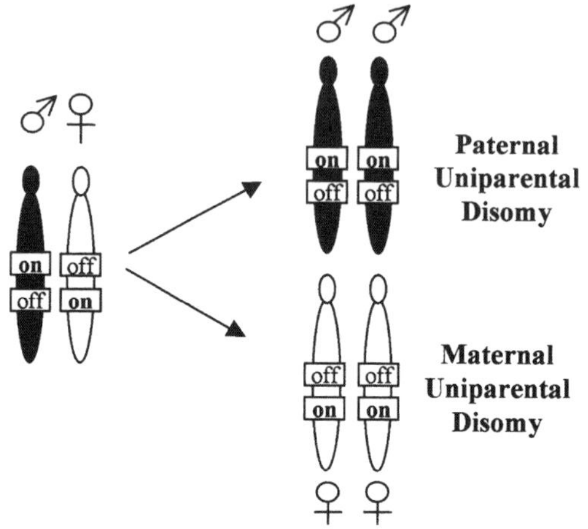

Fig. 1. Schematic representation of uniparental disomy (UPD) chromosomes, which by definition involve the whole chromosome. In this example, maternal chromosomes are in white and the paternal chromosomes in black. Robertsonian translocation heterozygous intercrosses are used to generate UPD. For simplicity, the other chromosome involved in the translocation is not shown. Also included are two reciprocally imprinted genes (on = expressed allele; off = silent allele), and their dosages are perturbed with maternal and paternal uniparental disomy.

and their normal littermates results in a valuable source of material for the molecular and developmental analysis of the imprinted region.

In the mouse it is relatively easy to generate conceptuses with imbalances in the parental origin of imprinted domains. In humans this occurs on rare occasions and has resulted in patients with imprinted disorders ***(5,6)***. These patients have been a very valuable resource for the study of human imprinting but are sometimes limited through availability of material and difficulties/inconsistencies with clinical diagnoses. In the mouse, these problems are not encountered.

1.1. The Harwell Studies and Imprinting Map to Date

Over a 20-year period, the Harwell team has generated an "imprinting map" (previously known as a "noncomplementation map") of mouse autosomal chromosomal regions that exhibit gross phenotypic abnormalities when their parental origin is perturbed. In this analysis, abnormal phenotypes including lethality, growth anomalies, and obvious behavioral defects were noted. Ten imprinted chromosomal regions have been identified from these genetic studies, and most of the imprinted genes identified to date map to these regions. They are located on proximal and distal chr2; proximal chr6; proximal, central, and distal chr7; proximal chr11; distal chr12; and proximal and distal chr17 ***(7)***.

Reports from Oakey and colleagues suggest an imprinted domain on chr18 ***(8)***. Imprinted regions causing more subtle defects or containing imprinted genes with functional redundancy would not have been identified by this method. A minority of the imprinted genes identified to date map outside these genetic regions. The map of imprinted regions and locations of known imprinted genes are kept up to date and can be found at *http://www.mgu.har.mrc.ac.uk/anomaly/anomaly.html*. A table of all the known imprinted genes and the key imprinting references can also be found at this site.

2. Reciprocal and Robertsonian Translocations

The ability to generate conceptuses with partial and whole chromosome disomy depends on the use of balanced reciprocal or Robertsonian translocation heterozygote animals, respectively, which, during gametogenesis, undergo meiotic segregation events that result in unbalanced gametes. Zygotes formed from gametes with complementary duplications and deficiencies of most chromosome regions usually give rise to normal, viable mice. If an imprinted chromosomal region is involved in the translocation, the perturbed parental origin of the region can cause developmental defects associated with the abnormal dosage of the imprinted gene or genes.

A catalog of chromosomal variants, including reciprocal and Robertsonian translocations and some information on their origin, fertility, and frequency of nondisjunction, has been documented by Beechey and Evans ***(9)***. Most mouse stocks carrying chromosomal anomalies are available from Harwell (UK), *http://www.mgu.har.mrc.ac.uk/ anomaly/ anomaly.html*, and the Jackson Laboratory (USA), *http://jaxmice.jax.org/index.shtml*.

2.1. Nondisjunction in Robertsonian Translocation Heterozygotes

A Robertsonian translocation is characterized by fusion at the centromeres of two nonhomologous chromosomes, resulting in the translocation of a whole murine chromosome. During meiosis in animals heterozygous for a Robertsonian translocation, nondisjunction events occur. This results in the formation of unbalanced gametes. When a nullisomic egg is fertilized by a sperm harboring a disomy of that same region (or vice versa), a balanced zygote is formed. The zygote is diploid with a balanced chromosomal constitution; however, for the regions involved in the translocation, there are two copies from one parent and no copies from the other (**Fig. 1**). Although the frequency of nondisjunction can be up to 30% in each parent, the actual frequency of maternal or paternal UPD in the progeny can be very low. This is especially true for single heterozygote intercrosses. Higher frequencies of nondisjunction occur in animals that are heterozygous for two Robertsonian translocations with both translocations having a chromosome in common—known as monobrachial homology. For example, in a recent report using Robertsonian translocation double heterozygote intercrosses, with the translocations sharing monobrachial homology for chromosome 12, UPD12 conceptuses were obtained at a total frequency of 6% of the total number of implantations *(10)*.

2.2. Adjacent-1 and Adjacent-2 Disjunction in Reciprocal Translocation Heterozygotes

A balanced reciprocal translocation results from breakage in two nonhomologous chromosomes followed by symmetrical exchange, thus producing four different chromosomes rather than two homologous chromosome pairs. Because there are two regions associated with the breakpoint, one proximal and one distal, meiotic segregation events result in gametes with either proximal uniparental partial disomy (after adjacent-2 segregation) or distal uniparental partial disomy (after adjacent-1 segregation). This is illustrated schematically in **Fig. 2**. The frequency of adjacent-2 segregation is lower than that of adjacent-1 segregation. This results in a greater frequency of distal partial disomies and makes the study of more proximal regions more difficult. Imprinting effects associated with proximal or distal partial disomy for the 10 autosomal regions are reviewed in **ref.** ***4***.

When using translocation heterozygote intercrosses, in addition to the normal littermates and those obtained with balanced partial or complete uniparental disomies, many of the conceptions will have whole chromosome or segmental aneuploidies, including trisomies (Ts) and monosomies (Ms). Ts conceptuses can be distinguished from those with UPD using genetic markers, and most Ts progeny die prenatally. Their phenotypes have been described in the literature *(11,12)*. Monosomies die before or around the time of implantation *(13)*.

2.3. Nomenclature

In human and mouse whole chromosome uniparental duplications associated with a balanced diploid karyotype have been referred to in the literature as uniparental disomies and abbreviated to mUPDn or pUPDn, where m and p refer to either maternal or paternal origin, respectively, and n represents the chromosome involved. Partial disomy (uniparental duplication/deficiency), which by definition involves part of a chromosome, is not a UPD. The term "uniparental duplication/deficiency" is a cumbersome one, especially if it is extended to include key details of the particular duplicated/deficient chromosomal region, such as its proximal or distal nature or its parental origin. For uniparental duplications/deficiencies for part of a chromosome generated by reciprocal translocation heterozygote intercrosses, there is nomenclature incon-

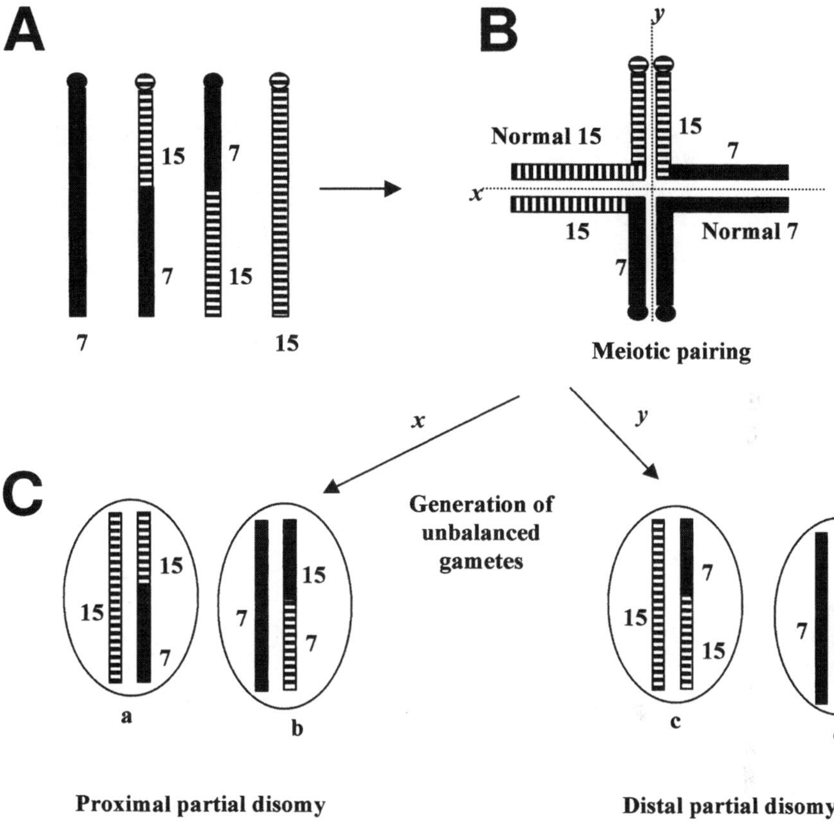

Fig. 2. Mice heterozygous for the reciprocal translocation T(7;15)9H will produce normal and unbalanced gametes during meiosis. **(A)** Schematic representation of the chromosomes 7 and 15 in a translocation heterozygote. **(B)** Pairing during meosis results in quadrivalent rather than bivalent formation due to the presence of both normal and derivative chromosomes. **(C)** The unbalanced gametes produced after adjacent-2 (*x*) and adjacent-1 (*y*) segregation are illustrated. Normal gametes are also produced. In (a) and (b), the proximal partial disomies and partial nullisomies are represented; (c) and (d) are distal partial disomies and partial nullisomies are shown. Complementation via fertilization of a partially nullisomic egg by a sperm with the corresponding partial disomy (or vice versa) will result in a balanced zygote with a partial disomy. Fertilization between a normal gamete and an unbalanced gamete will result in an aneuploid conceptus.

sistency in the literature. Several different abbreviated forms have been used. For example, MatDi7 has been used to indicate a maternal duplication/paternal deficiency of the distal part of chromosome 7 *(2)*. The same genotype has also been described by McLaughlin and colleagues as MatDup.d7 *(3)*, and MatDi has also been more recently used to describe mUPD *(11)*. A distal maternal duplication/paternal deficiency of chromosome 2 has been called MatDp2 ***(14)***. In all these studies, the translocations involved are noted and the abbreviations, though inconsistent, are reasonably clear. More recently, terminology has been approved by the Committee on Standard Genetic Nomenclature for Mice (2000) and can be found on the Harwell website (*see* **Subheading 2.**).

2.4. Choosing Translocation Stocks

Several factors should be considered when choosing the translocations to be used for experimental analysis of imprinted regions:

1. To generate partial disomies, because two chromosomal regions are involved in the translocation, it is prudent to choose a translocation in which only one of the chromosomes involved exhibits evidence of imprinting. For example, with T(7;15)9H, there is no current evidence for imprinting on chromosome 15 ***(15)***, hence imprinting phenotypes can be attributed to chromosome 7 imprinting rather than to the proximal or distal chromosome 15 partial disomies that will also arise from the cross (**Fig. 2**).
2. Some translocations are unsuitable for a variety of reasons such as sterility in one sex, or they may not be available as homozygous stocks due to lethality or poor fertility. If they can only be maintained as heterozygote stocks, then extensive cytogenetic expertise is needed to characterize the outcomes associated with stock maintenance.
3. Experiments that require the analysis of regions proximal to a reciprocal translocation breakpoint may be impractical due to the low frequency of partial disomies recovered.
4. To facilitate the molecular genotyping of progeny, it is preferable that translocations be maintained on two separate genetic backgrounds if possible. Not all translocations are available on multiple genetic backgrounds. Alternatively, visible genetic markers may be used, and again, suitable stocks may not be available or phenotypes associated with these genetic markers may not be visible at prenatal stages.

Information regarding these factors can be obtained from the Harwell and/or Jackson Laboratory websites and some can also be found in **ref. *9***. Researchers requiring further information are encouraged to consult with experts at the Jackson Laboratory or at the MRC Mammalian Genetics Unit at Harwell.

3. Breeding Strategy

Here, three sets of crosses are described: two for the maintenance and generation of the required stocks and parent animals, and one for generating the experimental animals:

1. a. Maintenance of the translocation in the homozygous state.
   b. Generation of heterozygotes for the subsequent intercross.

2. The heterozygous intercross for generating the uniparental duplication/deficiency conceptuses.

3.1. Maintainance as Homozygotes

Translocation lines are best maintained as homozygous stocks. An animal that is homozygous for a reciprocal or Robertsonian translocation has both chromosome homologs associated with the translocation. Because of the homozygosity, normal pairing will occur during meiosis and segregation will be balanced. A minimum number of breeding trios (one male and two females) is maintained and replaced on average every 6 mo to 1 yr. The number of trios and length of reproductive success depends primarily on the genetic background and environmental conditions.

3.2. Generation of Heterozygous Stocks for Subsequent Experimental Crosses

3.2.1. Single Heterozygotes

Single heterozygotes are generated when males and females that are heterozygous for reciprocal translocations are required. These are made by crossing male or female translocation homozygotes with cytogenetically normal inbred laboratory strains. The choice of laboratory strain depends on the experimental use and the molecular markers required (see below). Usually, strains with the same genetic background as the translocation homozygote are chosen to maximize genomic homogeneity, thus minimizing genetic background effects—for example, Cross 1 = T(7;15)9H/T(7;15)9H (BALB/c) × BALB/c; Cross 2 = T(7;15)9H/T(7;15)9H (129/Sv) × 129/Sv. However, if there is a need to identify the grandparental origin of each of the four homologs in subsequent intercrosses (for example, to identify the parental origin of each chromosome in partial Ts or Ts offspring), then using four strains can be useful—for example, Cross 1 = T(7;15)9H/T(7;15)9H (BALB/c) × C57BL/6J; Cross 2 = T(7;15)9H/T(7;15)9H (129/Sv) × AKR.

Translocation heterozygote breeders can be set up in trios. Initially the translocation is transmitted through either male or female homozygotes to compare the fertility between laboratory strains and translocation males. Subsequently, it is useful to revert to the more productive cross for generating the male and female heterozygotes for the experimental stock. The number of heterozygous breeding trios varies considerably depending on the numbers required and the fertility of the breeders.

3.2.2. Double Heterozygotes

Double heterozygote crosses are preferred when using Robertsonian translocations because the frequency of meiotic nondisjunction with Robertsonian translocations is increased in double heterozygote parents. To generate double heterozygotes, two different Robertsonian translocation homozygotes with monobrachial homology for the chromosome of interest are intercrossed.

For example, in recent work where UPD12 conceptuses were generated and analyzed *(10)*, Rb(8,12)5Bnr/Rb(8,12)5Bnr animals were intercrossed with Rb(4,12)9Bnr/Rb(4,12)9Bnr (both C57BL/6J) parents to generate Rb5/Rb9 double heterozygote animals. For the second parental cross, Rb(8,12)5Bnr homozygotes were mated with Rb(6,12)3Sic homozygotes (both BALB/c) to generate Rb5/Rb3 heterozygote females. In this latter cross the majority of males produced are sterile, so Rb5/Rb9 males are mated to Rb5/Rb3 females in the final experimental cross such as that described below.

The gametes arising in the double heterozygote parents used above are described in **ref. *10***. Predicted zygotic outcomes are also illustrated schematically in this paper. Many aneuploid conceptuses are predicted, but because of their unbalanced chromosomal constitution are peri-implantation lethal.

3.3. Translocation Heterozygote Intercrosses and the Generation of Conceptuses with Uniparental Duplications

This is the cross for generating the experimental material and is housed as stock cages of translocation heterozygous females that are mated to translocation heterozygous stud males of a different genetic background. As fertility is sometimes reduced in translocation heterozygous animals *(9)*, stud males are used for 6 mo to 2 yr, and information regarding plug frequency is kept for each male. In addition, males that regularly give rise to UPD or partial UPD offspring are noted and preferred as studs, because frequencies of meiotic nondisjunction can vary among individuals. As indicated above, single heterozygous intercrosses are used for reciprocal translocations, and double heterozygous intercrosses are preferred with Robertsonian translocations. Translocation heterozygote females are plug-checked on the morning after mating. Pregnant females are sacrificed at the required day after plugging and conceptuses are dissected for further analysis. Before genotyping, the dissected material is washed carefully through several changes of sterile phosphate-buffered saline to minimize contamination from maternal cells that might interfere with polymerase chain reaction (PCR)-based genotyping assays.

4. Genetic Analysis

4.1. Identifying Parental Chromosomes

4.1.1. Genetic Markers

A successful strategy for generating animals containing alterations in the parental origin of imprinted chromosomes depends on the ability to ensure that the two parental chromosomes can easily be distinguished and that disomy or partial disomy animals can be identified from their normal littermates and from the aneuploid progeny generated. Before the advent of molecular technology, recessive alleles with obvious phenotypes were very useful genetic markers to determine uniparental inheritance of chromosomal regions of interest. Examples of these include *c*, the recessive tyrosinase mutation for albinism located on distal chromosome 7, and *vt*, vestigial tail, on chromosome 11 ***(9)***. By having one of the parents homozygous for the recessive allele and the other homozygous for the wild-type allele, any offspring inheriting both copies of the recessive allele would be easily identified as having uniparental inheritance. One disadvantage of this method is that some of the marker phenotypes are only evident later in gestation (eye pigment) or after birth (coat phenotypes), and this has limited the earlier gestation identification and analysis of imprinted phenotypes.

4.1.2. Chromosome-Specific Molecular Markers

Nowadays, the parental origin of chromosomes is identified through the use of strain-specific DNA sequence polymorphisms, especially those detected by PCR amplification of mouse microsatellite markers. These are located in several databases, including *http://www.hgmp.mrc.ac.uk/*. For a simple analysis, translocation heterozygotes of one strain are intercrossed with those of another, and hence offspring that are heterozygous for a marker carry both parental chromosomes. Those that are homozygous for the marker have contribution from only one parent and are likely to have uniparental disomy or partial disomy. Sometimes it is useful to be able to characterize the transmission of each of the four grandparental chromosomes in offspring of a translocation heterozygote intercross. This is useful (1) for determining isodisomy (homozygous) or heterodisomy (heterozygous)—see below; (2) in identifying trisomic or monosomic conceptuses in addition to the uniparental duplication/deficiency embryos; or (3) assessing the contribution of genetic background effects. The four parental strains used to generate the translocation heterozygotes must be different, and strain-specific chromosome-specific markers of all the chromosomal regions involved in the translocation must be resolvable by gel electrophoresis.

For genotyping, a small amount of tail or yolk sac DNA can be isolated from the conceptus and subjected to PCR amplification using primers specific for the parent-specific polymorphic microsatellite markers (Research Genetics, Inc., Huntsville, AL). Both paternal and maternal control DNAs are always run alongside progeny for comparison. Amplification conditions depend on the primers used. The quickest and easiest approach to resolve genotype-specific bands is on a high-percentage agarose gel (3/1 Nusieve/agarose) as shown in **Fig. 3**. Polyacrylamide-urea gels can also be used if bands are too close in size to be resolved on agarose gels.

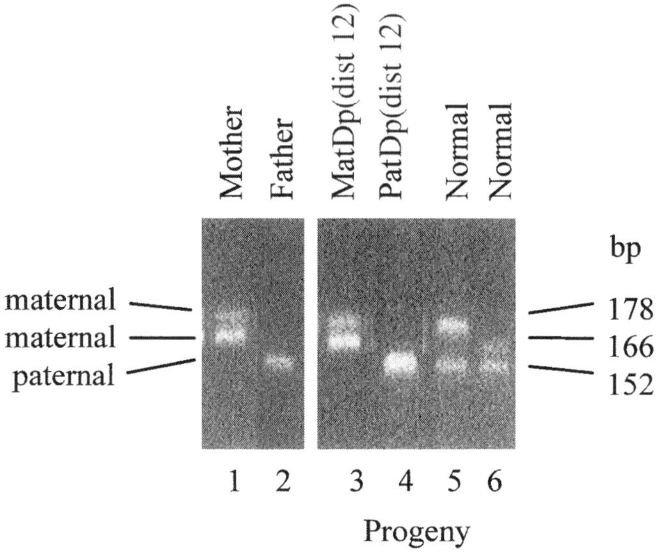

Fig. 3. Genotyping of maternal (MatDp(dist 12)) and paternal (PatDp(dist 12)) partial disomy 12 embryos. Genetic background-specific amplification products of different lengths can be generated by PCR using primers that span microsatellite and other short sequence repeats. Here, primers that map to the distal portion of chromosome 12 are used to differentiate between maternally and paternally inherited chromosomes. Maternal alleles measure 178 and 166bp (*lane 1*) and the paternal alleles both measure 152bp (*lane 2*). Normal littermates who have inherited one maternal and one paternal allele are shown (*lanes 5* and *6*). MatDp(dist 12) lack the paternal allele and PatDp(dist 12) lack either of the maternal alleles. (These data are part of an ongoing project conducted in collaboration with Dr. Bruce Cattanach, Harwell, UK).

An example of the use of PCR amplification of strain-specific microsatellite repeat polymorphisms to genotype offspring in a cross used to generate maternal and paternal partial disomy for chromosome 12 offspring is shown in **Fig. 3**. A 101/129Sv mother heterozygous for the T(4;12)47H translocation (the translocation is on 101 and the normal chr12 on a 129/Sv genetic background) has been crossed with a male who is C3H (from both his parents) and heterozygous for the translocation. Two normal littermates who have inherited paternal and maternal alleles are shown in lanes 5 and 6. Lane 3 shows amplification from a maternal partial distal disomy conceptus (MatDp(dist 12)) who is heterodisomic for the distal region (i.e., both maternal grandparental alleles are represented). Lane 4 shows a paternal partial distal disomy embryo that is isodisomic for the paternal C3H alleles. All partial disomy embryos should also be genotyped for the distal chromosome 4 alleles that are also involved in the translocation. This

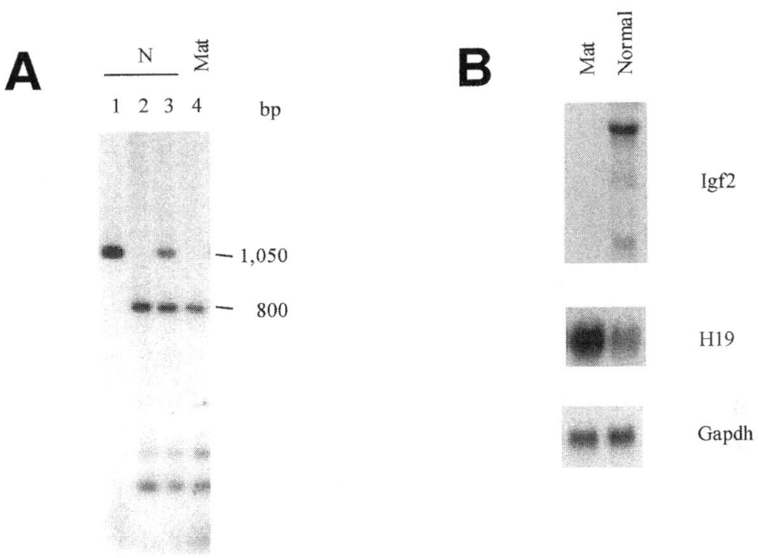

Fig. 4. Southern and Northern hybridization using DNA and RNA isolated from embryos with mat maternal partial disomy (uniparental duplication/paternal deficiency) for the region distal to the T(7,15)9H translocation breakpoint. Hybridization signals are compared with those of normal littermates (N). Techniques and probes used were as described previously ***(16)***. (**A**) Using this DNA, parental-origin-specific DNA methylation is evident for the promoter of the H19 gene. DNA was digested with *Apa*I alone (*lane 1*) or double-digested with either *Msp*I (*lane 2*) or *Hpa*II (*lanes 3* and *4*). (**B**) Total RNA was isolated from MatDp(dist 7) and normal embryos and hybridized to radiolabeled coding sequences from *Igf2*, *H19*, and *Gapdh*. MatDp(dist 17) embryos do not express *Igf2*, express a double dose of *H19*, and equivalent levels of *Gapdh* to that expressed in normal embryos.

confirms the genotype because embryos with partial disomy for the imprinted region distal to the translocation breakpoint (chromosome 12) will also carry a partial disomy of the distal region of the other chromosome (chromosome 4) inherited from the other parent (*see also* legend for **Fig. 2**).

4.2. Further Analysis

An example of the use of partial disomy material for further molecular study is shown in **Fig. 4**. Imprinted genes, such as *H19* and *Igf2* on distal chromosome 7, can be assayed for allele-specific DNA methylation and mono- or bi-allelic expression, which is evident when DNA and RNA from embryos with a maternal partial disomy of distal chromosome 7 (MatDp(dist 7)) are compared with material from a normal littermate. For the methylation analysis shown in **Fig. 4A**, a probe from the promoter of the maternally expressed, paternally repressed *H19* gene is hybridized to a Southern blot of DNA digested with methylation-sensitive and insensitive restriction enzyme isoschizomers, *Hpa*II and *Msp*I, respectively. The promoter of this gene is unmethylated on the active maternal allele, hence complete digestion is evident with HpaII in MatDi7 embryos (lane 4). In the adjacent lane 3, the normal embryonic DNA shows 50% digestion. Comparison of lanes 3 and 4 indicates that this digestion is allele-specific, with the paternal allele being methylated and uncut and the maternal cutting to completion. The use of the MatDp(dist 7) embryo allows for clean analysis with hybridization signals that are easier to interpret than alternative approaches that depend on the occurrence and characterization of allele-specific DNA polymorphisms. **Figure 2B** illustrates how the MatDp(dist 7) material can be used for allele-specific expression analysis. The genes shown here are *Igf2*, which is expressed from the paternal allele, and the reciprocally imprinted *H19*. In the absence of a paternally inherited *Igf2* allele in the MatDp(dist 7) embryo, there is no activity compared to normal. For the maternally expressed *H19*, we see a measurable double dose in the MatDp(dist 7) compared to normal. Both are quantitated against the nonimprinted *Gapdh* control. Thus, using the MatDp(dist 7) material, differences can be attributed to the missing paternal chromosome and hence methylation and expression can be shown for each parental allele. The standard protocols used for these experiments are as described previously ***(16)***. Quantitation is easily carried out through measurement of band intensities. In addition, the analysis of expression and methylation can be done without using PCR-based approaches because in most instances, adequate amounts of material can be generated. An exception to this, pertinent for the example shown, is the PatDp(dist 7) embryo. These conceptuses, which have inherited two copies of the region from the father and none from their mother, die at e9.5 of gestation and thus do not provide adequate amounts of material for molecular studies of the kind described in **Fig. 2** ***(3)***.

5. Conclusion

In addition to the use of dissected animal tissue, cell cultures can be made from uniparental disomy or partial disomy conceptuses. Primary embryonic fibroblasts usually retain their imprints in culture and have been used successfully to assess and manipulate the epigenetic state of imprinted loci ***(17,18)***. Embryonic stem (ES) cells harboring a disomy or partial disomy seem to maintain their imprints at early passage but do not usually maintain them after longer passage in vitro. ES cells have been useful for the analysis of imprinting in partial paternal distal disomy for chromosome 7. For example, they have been successfully incorporated into chimearas with normal cells to study the later gestation defects associated with paternal uniparental duplication of distal chr 7 ***(19)***.

The are several advantages associated with using balanced translocations to generate conceptuses with uniparental duplications/deficiencies for the study of imprinting. (1) One can cleanly observe the properties of an imprinted region from one parental source in the absence of "contamination" by the chromosome from the other parent. (2) Imprinted genes can be studied in an undisturbed chromatin environment, unlike situations in which imprinting has been disrupted by natural or targeted mutation. This is a relevant consideration for mechanistic studies. (3) It is not necessary to identify and characterize polymorphisms specific to the locus or expressed sequence of interest in order to identify DNA or expression associated with parental alleles. However, there are two assumptions to be borne in mind when using this approach. First, it is currently assumed that the chromosomal region being investigated behaves appropriately in the absence of a homolog from the other parent. To date there is no evidence to suggest that this is not the case. Second, and this is more relevant for expression studies, usually the region involved in the translocation can be quite large, so anomalies in gene dosage that might be attributed to imprinting at that locus could potentially be secondary effects associated with imprinting at another locus. To date there is no evidence of an imprinted gene affecting the expression of a neighboring nonimprinted gene.

Acknowledgment

We are indebted to Colin Beechey at the MRC Mammalian Genetics Unit, Harwell, for extensive comments and advice associated with an earlier draft of this chapter.

References

1. Cattanach, B. M. and Kirk, M. (1985) Differential activity of maternally and paternally derived chromosome regions in mice. *Nature* **315,** 496–498.
2. Ferguson-Smith, A. C., Cattanach, B. M., Barton, S. C., Beechey, C. V., and Surani, M. A. (1991) Embryological and molecular investigations of parental imprinting on mouse chromosome 7. *Nature* **351,** 667–670 (1991).
3. McLaughlin, K. J., Szabo, P., Haegel, H., and Mann, J. R. (1996) Mouse embryos with paternal duplication of an imprinted chromosome 7 region die at midgestation and lack placental spongiotrophoblast. *Development* **122,** 265–270.
4. Cattanach, B. M. (1986) Parental origin effects in mice. *J. Embryol. Exp. Morphol.* **97** (Suppl.), 137–150.
5. Ledbetter, D. H. and Engel, E. (1995) Uniparental disomy in humans: development of an imprinting map and its implications for prenatal diagnosis. *Human Mol. Genet.* **4,** 1757–1764.
6. Joyce, J. and Ferguson-Smith, A. C. (1999) Genomic imprinting in development and disease, in *Developmental Genetics, Epigenetics and Environmental Regula-*

tion (V. Russo, D. Cove, L. Edgar, R. Jaenisch, and F. Salamani, eds.), Springer-Verlag, Berlin, pp. 421–434.

7. Cattanach, B. M. and Beechey, C. V. (1997) Genomic imprinting in the mouse: possible final analysis, in *Genomic Imprinting: Frontiers in Molecular Biology* (Reik, W. and Surani, M. A., eds.), IRL, Oxford, U.K., pp. 118–141.
8. Oakey, R. J., Matteson, P. G., Litwin, S., Tilghman, S. M., and Nussbaum, R. L. (1995) Nondisjunction rates and abnormal embryonic development in a mouse cross between heterozygotes carrying a (7, 18) Robertsonian translocation chromosome. *Genetics* **141,** 667–674.
9. Beechey, C. V. and Evans, E. P (1996) Chromosomal variants—numerical variants and structural rearrangements, in *Genetic Variants and Strains of the Laboratory Mouse, Vol. 2* (Lyon, M. F., Rastan, S., and Brown, S. D. M., eds.), Oxford University Press, Oxford, pp. 1452–1506.
10. Georgiades, P., Watkins, M., Surani, M. A., and Ferguson-Smith, A. C. (2000). Parental origin specific developmental defects in mice with uniparental disomy for chromosome 12. *Development* **127,** 4719–4728.
11. Epstein, C. J. (1986) Generation and properties of mouse aneuploids, in *The Consequences of Chromosome Imbalance*, Cambridge University Press, Cambridge, chapter 10.
12. Gropp, A., Giers, D., and Kolbus, U. (1974). Trisomy in the fetal backcross progeny of male and female metacentric heterozygotes of the mouse. I. *Cytogenet. Cell Genet.* **13,** 511–535.
13. Magnuson, T., Debrot, S., Dimpfl, J., Zweig, A., Zamora, T., and Epstein, C. J. (1985) The early lethality of autosomal monosomy in the mouse. *J. Exp. Zool.* **236,** 353–360.
14. Williamson, C. M., Schofield, J., Dutton, E. R., Seymour, A., Beechey, C. V., Edwards, Y. H., and Peters, J. (1996) Glomerular-specific imprinting of the mouse gsalpha gene: how does this relate to hormone resistance in albright hereditary osteodystrophy? *Genomics* **36,** 280–287.
15. Searle, A. G. and Beechey, C. V. (1990) Genome imprinting phenomena on mouse chromosome 7. *Genet. Res.* **56,** 237–244.
16. Ferguson-Smith, A. C., Sasaki, H., Cattanach, B. M., and Surani, M. A. (1993) Parental origin specific epigenetic modification of the mouse H19 gene. *Nature* **362,** 751–755.
17. Eversole-Cire, P., Ferguson-Smith, A. C., Sasaki, H., Brown, K., Cattanach, B. M., Gonzales, F. A., Surani, M. A., and Jones, P. A. (1993) Induced activation of an imprinted Igf2 gene in mouse somatic cell cultures. *Mol. Cell. Biol.* **13,** 4928–4938.
18. Eversole-Cire P., Ferguson-Smith, A. C., Surani, M. A., and Jones, P. A. (1995) Coordinate regulation of Igf2 and H19 in cultured cells. *Cell Growth Differentiation* **6,** 337–345.
19. McLaughlin, K. J., Kochanowski, H., Solter, D., Schwarzkopf, G., Szabo, P. E., and Mann, J. R. (1997) Roles of the imprinted gene Igf2 and paternal duplication of distal chromosome 7 in the perinatal abnormalities of androgenetic mouse chimeras. *Development* **124,** 4897–4904.

4

Production of YAC Transgenic Mice by Pronuclear Injection

Justin F.-X. Ainscough, Rosalind M. John, and Sheila C. Barton

1. Introduction

1.1. Suitability of Transgenes for Imprinting Studies

The production of transgenic mice using small DNA constructs has been widely used for many years to investigate the regulation of gene activity. Small plasmid-based constructs (less than 20 kb) have been favored for a number of reasons, particularly the ease with which they can be manipulated and purified in large quantities. While this approach is powerful, there are some problems associated with the size of these transgenes. In particular, many of these small transgenes do not reproduce accurately the expression seen from the endogenous gene. For some genes the regulatory elements that control activity are located at a distance from the promoter and can be omitted from the transgene. These may be enhancers, repressors, boundary elements, or even locus control regions (LCRs), which are responsible for maintaining the correct spatial and temporal expression patterns of a number of genes, such as the globin clusters in mouse and humans *(1)*. More important, small transgenes are susceptible to position effects from the chromatin environment in which they integrate, which often results in either ectopic expression (from trapping of nearby enhancers for other genes) or suppression of gene activity. Finally, small transgenes usually integrate in a multicopy tandem arrangement that does not accurately reflect the situation seen at the endogenous locus.

There is growing evidence from studies in mouse and humans that the regulatory elements for many imprinted genes may be widely dispersed within "imprinted domains," which may span hundreds of kilobases *(2,3)*. Therefore,

From: *Methods in Molecular Biology, vol. 181: Genomic Imprinting: Methods and Protocols*
Edited by: A. Ward © Humana Press Inc., Totowa, NJ

it is unlikely that analysis of small transgenes will provide much useful information concerning the expression or mechanism of imprinting for the majority of this unusual class of genes.

1.2. Availability of Larger Transgenes

In recent years technologies have been developed to make transgenic mice using large DNA constructs based on artificial chromosomes from a number of sources, including bacteria (BACs; **ref.** ***4***) and yeast (YACs; for examples, *see* **refs.** ***5–8***). These transgenes can first be manipulated in the host organism due to efficient mechanisms of homologous recombination, before purification and transfer into mammalian cells. Numerous constructs are available commercially for YACs, and methods of modification have been well documented that can be used for generating point mutations, deletions, replacements, and insertions *(9)*. Therefore it is possible, for example, to introduce reporter genes into genes of interest, or to modify the size of the YAC construct under investigation, with relative ease. The methods available for modifying BACs are described by John et al. (Chapter 5).

The focus of this chapter is on the purification of YAC DNA and subsequent transfer into fertilized mouse eggs. As mentioned earlier, one of the major advantages that YAC-based transgenes have over smaller plasmid-based transgenes is that they are more likely to possess the full array of regulatory elements for the gene of interest, and thus are more likely to behave in a manner similar to that seen at the endogenous locus. There is also an increased likelihood of obtaining low-copy-number (or even single-copy) transgenic animals using YAC DNA, because fewer molecules are introduced, so these transgenes should provide a more accurate representation of both the patterns of expression and also the levels of expression of genes located therein.

1.3. Study of Imprinted Genes

To date we and others have investigated two regions containing imprinted genes using the YAC transgenic approach. These are a 130-kb region from mouse distal chromosome 7, which contains the reciprocally imprinted *Igf2* and *H19* genes *(5)*, and a 300-kb region from mouse chromosome 17, which contains the maternally expressed *Igf2r* gene *(8)*. In both cases the YACs demonstrated appropriate imprinting of the genes at a number of ectopic genomic locations. For the 130-kb *Igf2/H19* YAC, little evidence was found for position effects such as the trapping of nearby enhancer elements. These analyses therefore demonstrate the power of using this technique as a first step toward defining the minimal critical regions for imprinting of particular genes, with the ultimate aim of identifying and characterizing the regulatory elements involved.

The *Igf2r* YAC transgene was further modified by targeted deletion of a differentially methylated intronic CpG island, which led to loss of imprinting, thus defining a specific role for this region. We have recently modified the 130-kb *Igf2/H19* YAC transgene by targeting *loxP* recombination sites around a putative control element of unknown function, which had allowed us to delete this element from the YAC after it had been integrated into the mouse genome. We were therefore able to compare directly the activity of the YAC transgene with and without the targeted element, at the same location in the mouse genome, providing rapid and conclusive data on the function of the deleted region ***(10)***.

It is therefore evident that the use of large transgenes has many advantages over smaller, plasmid-based transgenes for initial investigations into dissecting imprinted domains, and is a powerful tool that is underexploited at the present time.

2. Materials

2.1. Preparation of High-Density Yeast Plugs

1. Synthetic minimal (SD) medium. 0.17% yeast nitrogen base, 0.5% ammonium sulfate, 2% dextrose. Add appropriate amino acids (nonselective). For plates, adjust pH to 5.8 with NaOH and add 2% bacto-agar.
2. Hemocytometer (Weber Scientific).
3. 50 m*M* ethylenediaminetetraacetic acid (EDTA), pH 8.0.
4. Yeast lytic enzyme (ICN Biochemicals), 50 mg/mL in water.
5. 1.5% low-melting-point (LMP) agarose (Seaplaque-FMC) in 125 m*M* EDTA, pH 8.0. Melt in microwave and equilibrate to 40°C.
6. Agarose plug molds (Pharmacia), sealed on one side with tape and cooled on ice.
7. LET buffer: 0.5 *M* EDTA, 10 m*M* Tris-HCl, pH 7.5. Just before use, add yeast lytic enzyme to a final concentration of 2 mg/mL.
8. Yeast lysis buffer (YLB): 100 m*M* EDTA, 10 m*M* Tris-HCl pH 8.0, 1% lithium dodecyl sulfate (LDS).
9. 100 m*M* EDTA, pH 8.0.

2.2. Purification of YAC DNA by Pulsed-Field Gel Electrophoresis

1. Pulsed-field gel electrophoresis (PFGE) apparatus (Pharmacia Gene Navigator System).
2. Running buffer (0.5x TBE).
3. 1.2% LMP agarose (Seaplaque-FMC) in 0.5x TBE.
4. 0.5% LMP agarose (Seaplaque-FMC) in 0.5x TBE.
5. Ethidium bromide solution: 5 µL of 10 mg/mL per 100 mL 0.5x TBE.
6. Agarose digestion buffer: 10 m*M* Tris-HCl pH 7.5, 1.0 m*M* EDTA pH 8.0, 100 m*M* NaCl, 30 µ*M* spermine, 70 µ*M* spermidine.

7. β-Agarase (NEB).
8. Injection buffer: 10 m*M* Tris-HCl pH 7.5, 0.1 m*M* EDTA pH 8.0, 100 m*M* NaCl, 30 μ*M* spermine, 70 μ*M* spermidine. Filter-sterilize and store at 4°C. Make fresh each week.
9. Dialysis membrane (Millipore 0.05 μm, VMWP02500).

2.3. Preperation of Fertilized Mouse Eggs

1. F1 hybrid mice in which the maternal component is C57BL/6, paternal can be any vigorous inbred strain, e.g., CBA, C3H, DBA, 129/Sv. We generally use (C57BL/6 × CBA) F1 mice; females for superovulation at 5–6 wk old, males mature and fertile.
2. Pregnant mare serum (PMS), "Folligon"; Intervet UK) 50 IU/mL sterile phosphate-buffered saline (PBS). Store at 4°C.
3. Human chorionic gonadotrophin (HCG), "Chorulon"; Intervet UK), 50 IU/mL sterile PBS. Store at 4°C.
4. 27G needles.
5. Phosphate-buffered medium (PB1): NaCl (5.97 g/L), KCl (0.2 g/L), KH_2PO_4 (0.19 g/L), $CaCl_2 \cdot 2H_2O$ (0.14 g/L), $MgCl_2 \cdot 6H_2O$ (0.1 g/L), glucose (1.0 g/L), sodium pyruvate (0.04 g/L), penicillin G (0.06 g/L), streptomycin sulfate (0.05 g/L), phenol red (0.01 g/L), Na_2HPO_4 (anhydrous, 1.14 g/L—dissolved seperately and added last). Filter-sterilize and store in aliquots at 4°C.
6. Bicarbonate-buffered medium (T6): NaCl (4.72 g/L), KCl (0.11 g/L), $NaH_2PO_4 \cdot 2H_2O$ (0.06 g/L), $CaCl_2 \cdot 2H_2O$ (0.26 g/L), $MgCl_2 \cdot 6H_2O$ (0.1 g/L), glucose (1.0 g/L), sodium pyruvate (0.03 g/L), penicillin G (0.06 g/L), streptomycin sulfate (0.05 g/L), phenol red (0.01 g/L), 3.4 mL sodium lactate (60% syrup), $NaHCO_3$ (2.1 g/L—dissolved seperately and added last). Filter-sterilize and store in aliquots, with minimum air pocket above medium, at 4°C.
7. Bovine serum albumin (BSA), fraction V; Sigma). Add to PB1 and T6 media at 4 mg/mL before use.
8. Hyaluronidase solution (6 mg/mL sterile water, stored at –20°C in 50-μL aliquots). Make up to 1 mL with PB1 + BSA before use (final concentration 300 IU/mL).
9. Light mineral oil, embryo tested (Sigma, M8410).
10. Bacteriological Petri dishes (30 and 50 mm, Sterilin).
11. Dissection microscope, e.g., Wild M3C.
12. Fine forceps (e.g., Dumont no. 5).
13. Pasteur pipets, drawn to internal diameter of approximately 100 μm, for egg transfer by mouth pipetting system.
14. Humidified CO_2 incubator at 38°C (5% CO_2 in air).

2.4. Injection of DNA into Pro-Nuclei

1. Standard micromanipulation setup. We use a Leitz micromanipulator and Zeiss Ergaval upright fixed-stage microscope equipped with Nomarski optics, and work in hanging drops in a Puliv chamber filled with heavy mineral oil (Sigma).

2. DNA delivery is by means of a 50-mL ground-glass syringe connected by an airline to the instrument holder.
3. Holding pipet, controlled by an oil-filled pressure device (e.g., Narashige IM-6 or Eppendorf Cell Tram Oil).
4. Equipment for glass instrument making: microforge (Beaudouin), needle puller (Sutter Brown-Flaming), Bunsen burner.
5. Thin-walled capillary with filament (Clark Electromedical Instruments, GC100TF-15) for injection pipets, prepared in house as required.
6. Thick-walled capillary without filament (Clark Electromedical Instruments, GC100-15) for holding pipets, prepared as required.
7. PB1 + BSA (4 mg/mL), prepared as in **Subheading 2.3.**

2.5. Preparation of Recipients for Embryo Transfer

1. (C57BL/6 × CBA) F1 females (6–8 wk old), mated naturally with vasectomized males of proven sterility. Day of vaginal plug counts as d 1 of "pregnancy."

2.6. Screening for Transgenic Offspring

1. Tail lysis buffer (TLB): 50 m*M* EDTA, 10 m*M* Tris-HCl pH 8.0, 1% sodium dodecyl sulfate (SDS).
2. 1 Phenol/1 [chloroform ***(24)*** *iso*-amyl alcohol ***(1)***].
3. 3 *M* NaCl.
4. 100% isopropanol.
5. TE: 10 m*M* Tris-HCl, 1 m*M* EDTA pH 8.0.

3. Methods

3.1. Prepare High-Density Yeast Plugs

1. Inoculate 100 mL of selection media (SD without selective amino acids) with a few colonies from a freshly grown selective SD plate. Grow with rapid shaking at 30°C for 16–24 h (*see* **Note 1**).
2. Count cell density using a hemocytometer, and determine the final volume that will give a density of 4×10^9 yeast/mL.
3. Spin cells at 1500*g* for 5 min and pour off media. Resuspend cells in an equivalent volume of 50 m*M* EDTA, spin, and repeat once. Briefly spin cells once more and remove residual liquid with pipet tip.
4. Using a wide-bore pipet tip, thoroughly resuspend cells in 50 m*M* EDTA to a concentration of 1.2×10^{10} yeast/mL (*see* **Note 2**).
5. Warm briefly to 37°C (30 s) before adding 5 µL yeast lytic enzyme/100 µL. Vortex briefly.
6. Add 2 vol of 1.5% LMP agarose. Vortex briefly and quickly transfer 100 µL to each PFGE plug mold using a wide-bore pipet tip (*see* **Note 3**).
7. Leave 15 min to set and transfer plugs into LET buffer (20 mL/mL of plug), prewarmed to 37 °C (*see* **Note 2**).
8. Incubate for 16–24 h at 37 °C, with periodic shaking (*see* **Note 4**).

9. Replace LET buffer with equal volume of YLB, prewarmed to 50°C. Change for fresh YLB after 1 h and incubate at 50 °C for 16–24 h (*see* **Note 4**).
10. Equilibrate the plugs in 100 m*M* EDTA at room temperature. Replace 3–4 times to remove residual traces of LDS before storing at room temperature or 4°C (*see* **Note 5**).

3.2. Purification of YAC DNA

3.2.1. PFGE Separation of YAC from Yeast Chromosomes

1. Equilibrate enough plugs for a single PFGE run (10–11) in 0.5× TBE for 30 min. Repeat two times.
2. Load plugs onto 1.2% LMP agarose gel. Marker lanes can be included at both ends of the gel. Seal the lanes with 0.5% LMP agarose.
3. Run gel under conditions appropriate for adequate seperation of yeast chromosomes, depending on size of YAC of interest (*see* **Note 6**).
4. Cut off both edges of the gel including approximately one plug width at each side, and stain in ethidium bromide solution for 30 min. Store the central part of the gel in running buffer (*see* **Note 7** and **Fig. 1**).
5. Rinse the stained gel edges and, under UV illumination, cut notches in the gel to mark the position of the YAC band (*see* **Fig. 1**).
6. Reassemble the gel using Saran wrap to keep the stained portions seperated from the clean gel containing the YAC DNA to be purified. Slice out the unstained YAC band between the notches and transfer to 0.5x TBE at 4°C until ready to agarase treat (*see* **Note 8**).
7. Stain the remaining parts of the gel and reassemble under UV illumination to ensure adequate excision of the YAC band (*see* **Fig. 1**).

3.2.2. Agarase Treatment of Purified YAC Band

1. Equilibrate the gel slice in 20 mL of agarose digestion buffer at least three times for 30 min each.
2. Remove all the buffer with a pipet tip and transfer the gel slice to a 2-mL microfuge tube. Determine the weight of the gel by comparing to an empty tube. Use approximately 500 mg of gel/tube.
3. Melt the gel at 68°C for 3 min, spin briefly, then melt for a further 5 min.
4. Transfer to 42°C and equilibrate for 5 min.
5. Equilibrate 2 units of β-agarase per 100 mg of gel in a 2-mL microfuge tube at 42°C for 30 s, then transfer the melted gel into the β-agarase-containing tube with a prewarmed wide-bore 1-mL tip, and mix very gently, but quickly, 3–4 times.
6. Incubate at 42°C for 3–4 h, then transfer to ice for 10 min.
7. Spin down any undigested agarose for 15 min at room temperature, and transfer the digested solution to a new tube.
8. Store at 4°C (*see* **Note 9**). Check intactness of the YAC DNA by running 15 µL on a pulsed-field gel, and determine the concentration by comparing serial

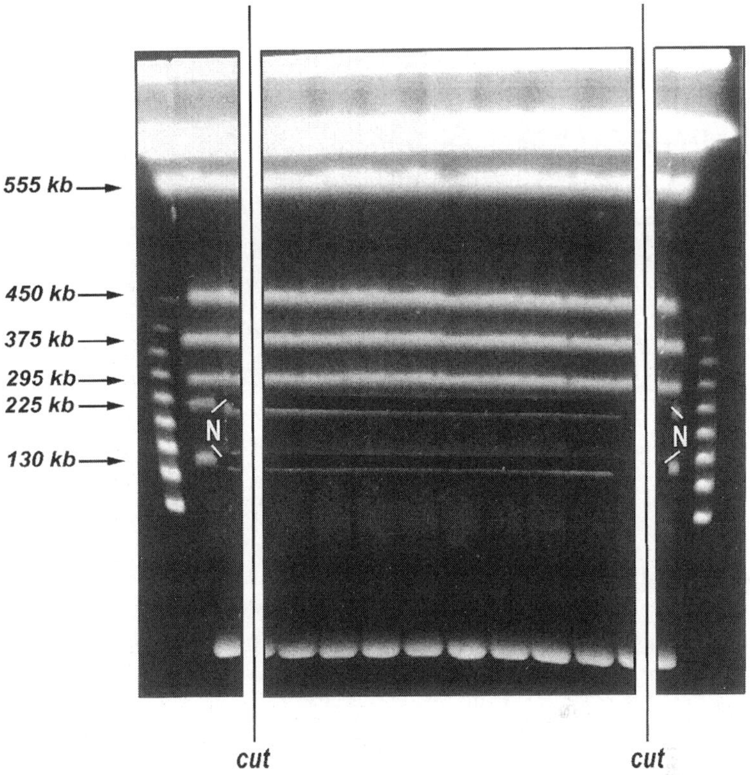

Fig. 1. Stained pulsed-field gel run for 24 h at 195 V with a pulse time of 40 s at 9°C, on a Pharmacia Gene Navigator system, containing yeast DNA plugs with a 130-kb YAC. Positions of where to cut off the edges of the gel for staining to reveal the location of the YAC band are indicated (cut). Notches cut into the stained gel edges at the position(s) of the band of interest are also shown (N). The gel is shown reassembled after slicing out the YAC band and the 225-kb yeast chromosome, demonstrating that the DNA has been excised efficiently before staining.

dilutions to known concentrations of standard marker DNA. An approximate concentration of 0.5–1 ng/µL is ideal for injection to generate low-copy-number transgenic mice (*see* **Note 10**).

3.2.3. Prepare YAC DNA for Injection

1. Dialyze 100 µL of YAC DNA solution against injection buffer *immediately* before injecting. Transfer the DNA, using a wide-bore tip, onto dialysis membrane floating on top of 20 mL of injection buffer in a Petri dish at 4°C for 1–2 h.
2. Transfer the dialyzed DNA into an microfuge tube and spin down any debris for 5 min before injecting. Keep on ice.

The protocol of **Subheading 3.2.3.** should be repeated every day on which injections are done. Repeat as in **Subheading 3.2.2.** every few days. Repeat as in **Subheading 3.2.1.** when required.

3.3. Preparation of Fertilized Mouse Eggs

1. Using a 27G needle, superovulate F1 females by intraperitoneal injection of 150 µL of PMS (*see* **Note 11**).
2. After 48 h, similarly inject 150 µL of HCG.
3. Immediately transfer each female to a cage containing a fertile male.
4. Check females for vaginal plugs the following morning.
5. At 20 h after HCG injection, sacrifice the females and excise the oviducts with care to avoid bursting the swollen ampullae. Transfer to a Petri dish.
6. Cover each pair of oviducts with a drop of hyaluronidase solution and open the ampullae with forceps under the dissecting scope.
7. A mass of eggs embedded in a cloud of cumulus cells should be released.
8. When most of the cumulus cells have detached (3–5 min), collect the eggs with a mouth-operated Pasteur pipet and wash through 3–4 large drops of T6 + BSA.
9. Check that the eggs have been fertilized.
10. Incubate in a drop of T6 + BSA covered with light mineral oil in the CO_2 incubator until ready to proceed with YAC DNA injection (*see* **Note 12**).

3.4. Injection of YAC DNA into Pro-Nuclei and Transfer into Recipients

Since the procedures for setting up the micromanipulator depend on the type of microscope and micromanipulator used, we will assume a working system is available to the user and provide only a brief outline of particular procedures we use in this section.

1. Prepare injection pipets immediately before use to avoid atmospheric contamination of the tip.
2. Transfer eggs in batches of 20–30 to drop(s) of PB1 + BSA in the injection chamber (*see* **Note 13**).
3. Load the injection pipet from the rear by standing it in the DNA solution for a few minutes, tapping if necessary. Alternatively, place a drop of DNA solution within the pipet against the glass filament using a fine-drawn capillary.
4. Check that the piston of the syringe can move freely and without catching, then mount the injection pipet.
5. Open the tip of the injection pipet in the injection chamber by gently knocking it against the holding pipet until the outside diameter is approximately 0.75 µm, preferably with a slight bevel.
6. Inject DNA into the pro-nucleus in a gentle and controlled manner to avoid shearing the DNA. Injection can be into either the male or female pro-nucleus, although the larger size of the male pro-nucleus makes the procedure easier.

Inject until the pro-nucleus is observed to swell to about twice its normal volume. Withdraw the pipet gently.

7. If the tip is found to block after several injections, repeat the knocking procedure as many times as the needle shape will allow before the outside diameter gets too large (approximately 2 µm).
8. After injection, transfer the eggs to a drop of T6 + BSA, and remove any that have lysed. Incubate overnight in the CO_2 incubator.
9. Select eggs that have not blocked (now at the two-cell stage) to transfer to oviducts of d 1 psuedo-pregnant recipient females, following standard procedures (*see* **Notes 14** and **15**).
10. Monitor the pregnancy by weighing at d 15 and d 18 after transfer (d 1). If the female has not given birth by mid-day of d 20, deliver the pups by Caesarian section and foster them to suitable foster mothers. This applies to any litter size, since birth at a later stage will result in increased perinatal lethality.

3.5. Screening for Transgenic Offspring

1. Take 0.5-cm tail snips from 10- to 14-d-old pups and incubate in 1 mL of TLB in a 2-mL microfuge tube at 37°C for at least 16 h.
2. Add 0.8 mL of phenol/chloroform and shake vigorously, repeated several times over a minimum of 15 min.
3. Spin at maximum speed in a microcentrifuge for 15 min at room temperature.
4. Transfer 0.8 mL of the supernatant into a 2-mL microfuge tube using a wide-bore pipet tip, taking care not to disturb the interphase.
5. Add 80 µL of 3 *M* NaCl.
6. Add 0.88 mL of isopropanol and mix gently until DNA strands form.
7. Leave at room temperature for 30 min for full precipitation, with occasional mixing.
8. Pellet the DNA in the microcentrifuge at maximum speed for 15 min at room temperature.
9. Pour off the supernatant and rinse the DNA pellet once with 70% ethanol.
10. Briefly spin the tubes once more and remove residual ethanol with a pipet tip.
11. Resuspend the DNA pellet in 150 µL of TE. Vortex briefly, then stand at room temperature for at least 30 min. To ensure homogeneity of the DNA solution before use, it is reccomended to pass the solution through a narrow-bore pipet tip at least 10 times (*see* **Note 16**).

4. Notes

1. Longer incubation times of up to 48 h may be required for some strains.
2. β-Mercaptoethanol can be added to a final concentration of 14 µ*M* to increase the quality of the DNA preparation, although we have not generally found this step to be essential.
3. Ensure that the yeast cells are well mixed with the agarose solution, since homogeneity of the sample is very important for the preparation of good DNA

plugs. We find that brief vortexing at this step provides efficient mixing and is not detrimental to the final DNA quality.

4. These incubation times can each be reduced to 4–6 h by changing the buffers regularly.
5. Plugs can be stored for many months in 100 m*M* EDTA at room temperature with little or no degradation. If storage at 4 °C is preferred, it is essential that all traces of LDS have been removed, as this will precipitate as fine crystals.
6. We find conditions of 195 V, 40-s pulse time, 9°C for 24 h is adequate to separate chromosomes up to 550 kb.
7. It is very important that no ethidium bromide comes into contact with the YAC DNA to be purified at any stage, as this will enhance degradation.
8. It is best to go on to the next step as soon as possible, although we have stored the gel slices at 4 °C in 0.5x TBE overnight with no detrimental results.
9. Immediate use for injection is preferable, but we have stored DNA in this way for up to 1 wk with little degradation.
10. In our hands the quality and quantity of DNA that can be purified by this method is suitable for the efficient generation of low-copy-number transgenic mice. It has been suggested, however, that more concentrated DNA samples can be advantageous, especially for very large transgenes (over 300 kb). Additional steps that can be taken to increase the final concentration of YAC DNA are described in **ref. *11***.
11. The aim is to produce about 200 fertilized eggs, from which 100–120 can be selected for injection. Yield varies from 20–40 eggs/female.
12. Eggs of this genotype enter the ideal stage for injection (enlarging pro-nuclei migrating toward the center of the egg) at around 23 h post-hCG injection, and are harvested 2–3 h before this. Harvesting earlier than this will increase the number of unfertilized eggs present.
13. For convenience, we generally produce a line of drops of PB1 + BSA in the injection chamber and transfer 3–4 eggs into each one. Including a different number in the last drop provides an indicator of when the procedure is complete.
14. If more than 15% of the eggs have blocked, try using a more dilute DNA solution next time.
15. Do not overcrowd the uterus; a single healthy embryo can develop to term (although it will probably require delivery by Caesarian section), whereas more than 8 surviving embryos will result in less vigorous pups at birth. We generally transfer up to 15 two-cell eggs into each recipient female, which results in a litter of between 1 and 6 pups on a good day.
16. DNA prepared in this way is suitable for use in both Southern analysis (use 10–15 µL per lane for identifying a single-copy fragment) and PCR (use 0.25 µL in a 50-µL reaction).

References

1. Dillon, N. and Grosveld, F. (1993), Transcriptional regulation of multigene loci: multilevel control. *TIG* **9,** 134–137.

2. Ainscough, J. F.-X., John, R., and Surani, M. A. (1998), Mechanism of imprinting on mouse distal chromosome 7. *Gen. Res.* **72,** 237–245.
3. Buiting, K., Saitoh, S., Gross, S., Dittrich, B., Schwartz, S., Nicholls, R.D. and Horsthemke, B. (1995) Inherited microdeletions in the Angelman and Prader-Willi syndromes define an imprinting centre on human chromosome 15. *Nature Genet.* **9,** 395–400.
4. Yang, X. W., Model, P., and Heintz, N. (1997) Homologous recombination based modification in *Esherichia coli* and germline transmission in transgenic mice of a bacterial artificial chromosome. *Nature Biotechnol.* **15,** 1–7.
5. Ainscough, J. F.-X., Koide, T,, Tada, M., Barton, S., and Surani, M. A. (1997) Imprinting of *Igf2* and *H19* from a 130 kb YAC transgene. *Development* **124,** 3621–3632.
6. Lee, J. T., Strauss, W. M., Dausman, J. A., and Jaenisch, R. (1996) A 450 kb transgene displays properties of the mammalian X-inactivation center. *Cell* **86,** 83–94.
7. Schedl, A., Ross, A., Lee, M., Engelkamp, D., Rashbass, P., van Heyningen, V., and Hastie, N. D. (1996) Influence of *PAX6* gene dosage on development: overexpression causes severe eye abnormalities. *Cell* **86,** 71–82.
8. Wutz, A., Smrzka, O. W., Wagner, E. F., and Barlow, D. P. (1997) Imprinted expression of the *Igf2r* gene depends on an intronic CpG island. *Nature* **389,** 745–749.
9. Rothstein, R. (1991) Targeting, disruption, replacement and allele rescue: integrative DNA transformation in yeast, in *Guide to Yeast Genetics and Molecular Biology* (Guthrie, C., and Fink, G., eds.), Academic, San Diego, CA, pp. 281–301.
10. Ainscough, J. F.-X., John, R. M., Barton, S. C., and Surami, M. A. (2000) A skeletal muscle-specific mouse IGF2 repressor lies 40Kb downstream of the gene. *Development* **127,** 3923–3930.
11 Schedl, A., Grimes, B., and Montoliu, L. (1996) YAC transfer by microinjection. *Meth. Mol. Biol.* pp. 293–306.

5

A Transgenic Approach to Studying Imprinted Genes

Modified BACs and PACs

Rosalind M. John, Justin F.-X. Ainscough, and Sheila C. Barton

1. Introduction

1.1. Escherichia coli*-Based Large Genomic Clones*

The advantages of using large genomic clones in the analysis of imprinted genes is described in Chapter 4 with particular reference to yeast artificial chromosomes (YACs). These contain on average 500–600 kb of DNA but can be much larger (>1 Mb). YACS are propagated in yeast and are therefore amenable to genetic modification by homologous recombination, and there are now many examples of their use to generate transgenic mice. This chapter describes a relatively new strategy for using large genomic clones that relies on *Escherichia coli*-based systems.

Bacterial artificial chromosomes (BACs) are single-copy plasmids based on the *E. coli* fertility plasmid (F factor). P-1 derived artificial chromosomes (PACs) are based on the bacteriophage P1. Similar to YACs, these vectors are capable of holding large inserts (up to 300 kb) and are stably maintained as 1 or 2 copies. A number of genomic libraries are now available, both within the academic community and commercially (e.g., Genome Systems, Research Genetics). The main advantages of using these bacteria-based systems is the stability of the clones in culture and the ease of preparation of high-quality, intact DNA.

1.2. Modification of BACs and PACs

Until recently there were no methods for modifying BACs or PACs, which substantially limited their use. However, homologous recombination-based

From: *Methods in Molecular Biology, vol. 181: Genomic Imprinting: Methods and Protocols*
Edited by: A. Ward © Humana Press Inc., Totowa, NJ

techniques for modifying *E. coli*-based clones have now been developed, two of which are summarized here. The protocols and plasmids can be obtained from the respective authors and are described there in detail.

The first technique *(1)*, was used to modify a BAC and presumably will work as well with other *E. coli*-based systems. It takes advantage of a temperature-sensitive shuttle vector carrying an *E. coli RecA* gene. The required modification (marker insertion, deletion, or substitution) is constructed in a high-copy-number plasmid such as pBluescript with at least 500 bp of genomic DNA, which is homologous to the target locus, on either side of the modification. This construct is in many ways similar to the type of construct used in homologous targeting of an endogenous locus in embryonic stem cells. This is cloned into the low-copy shuttle vector (pSV1.RecA), which is tetracycline resistant and can replicate only at 30°C. The shuttle construct is transformed into bacteria containing the target BAC (or PAC) and a series of selection procedures follows. In the first step, selection for tetracycline (shuttle vector) and chloramphenicol (BAC vector) at 30°C selects colonies in which both plasmids are present but replicating as two separate molecules. By raising the temperature to 43°C, only the replication origin of the BAC is active and colonies are selected in which the shuttle vector has recombined onto the BAC using the *RecA* pathway. Once the integration event is identified (1/2 to 1/10 clones in our hands), the colonies are allowed to grow in the absence of tetracycline selection (shuttle vector), and some of these will lose the shuttle vector sequence and either the modification locus or the unmodified target locus. When the shuttle vector is lost, the *RecA* gene is also lost and the bacteria are no longer able to recombine homologous sequences, so the modified BAC is stably maintained. These tetracycline-sensitive colonies can be selected for on chloramphenicol/fusaric acid plates at 37°C. The "resolved" BACs will either regenerate the original BAC or create the intended modification, which can be determined by analysis on an agarose gel.

The second technique, known as "ET cloning," also relies on homologous recombination in *E. coli* *(2)* and has been demonstrated to modify a number of targets including a P1 clone. This system allows recombination between a linear fragment with short (60-bp) homology arms and a circular target molecule. The advantage of this system is in the construction of the modification since, in theory, the homology arms can be added by PCR amplification without the requirement for convenient restriction sites and the fragment can be used directly in a transformation without a cloning step. In addition, only a small amount of sequence information at the target site is required; 120 bp would allow the design of two 60-bp arms of homology for an insertional modification. However, in contrast to the first method *(1)*, this protocol requires the final modification to carry a marker selectable in *E. coli*. This may be a major disadvantage if a subtle modification, such as a single-base-pair change, is planned. To remove the selectable marker necessitates one of two additional steps: (1) The marker may be flanked by either FRT or loxP sites and the modification exposed to the corresponding site-specific recombinase, which still leaves a small site-specific recombination target site at the point of DNA manipulation; (2) two rounds of ET cloning are required, combining positive and counterselection steps.

The first step in the ET protocol involves transformation of *E. coli* containing the target clone with the pBAD-ETγ vector. This plasmid carries a truncated *RecE* controlled by the arabinose-inducible promoter, a *recT* controlled by the EM7 promoter, and *red*γ controlled by the Tn5 promoter. The clones, which now carry both pBAD-ETγ and the target clone, are exposed to arabinose, which induces expression of *RecE* and renders the cells capable of using a recombination pathway. These cells are then made competent and transformed with the linear molecule that carries the modification plus a selectable marker flanked by the short homology arms. The third selection step is for the marker within the modification, which is now integrated onto the P1 clone. Presence of the intended modification can be determined by analysis on an agarose gel.

We have used the first protocol successfully ***(1)*** to target two loci carried on BACs. In both cases the modifications were quite large, an addition of 7 kb spread over 10 kb. The time required, after construction of the modification in pBluescript, was less than 3 wk, with a total of 10 h hands-on time. The main rate-limiting step is in the final "pop-out" reaction, where there seemed to be a bias toward recovery of the original BAC, which was most likely due to the size of our insertion. In one case only 1/24 clones examined retained the required modification.

With the availability of these techniques, it is now possible to use both YACs and BACs/P1 to generate transgenic animals. Two procedures can be used to get the transgene into mice: pro-nuclear injection (described in Chapter 4) and embryonic stem (ES) cell lines. Pro-nuclear injection is the more rapid way to generate a large number of unique transgenic lines and requires only small amounts of DNA (<1 μg). The ES cell method is more time-consuming and significantly more DNA is required, but full characterization of the integration events can be performed before chimeras are generated to identify single-copy integration events and confirm the intactness of the transgene.

The preparation of BAC DNA for pro-nuclear injection is described in the first section of the protocol, and the preparation of DNA for the generation of embryonic stem (ES) cell lines is described in the second section. Similar protocols can be applied to any *E. coli*-based system with appropriate modification of the procedure (e.g., antibiotic selection, linearization agent).

2. Materials

2.1. Pro-Nuclear Injection

2.1.1. Preparation of BAC DNA

1. LB (1 L): 10 g bactotryptone, 5 g yeast extract, 10 g NaCl (autoclave).
2. 12.5 mg/mL chloramphenicol. Store at –20°C.
3. 14-mL polypropylene-round bottomed tube (Falcon 2059).
4. Solution I: 50 m*M* Tris-Cl pH 8.0, 10 m*M* ethylenediaminetetraacetic acid (EDTA) pH 8.0, RNAse A to 100 µg/mL. Store at 4°C.
5. Solution II: 0.2 *M* NaOH, 1% sodium dodecyl sulfate (SDS). Prepare fresh.
6. Solution III: 3 *M* potassium acetate, pH 5.5. Store at 4°C.
7. 1 Phenol/1 (chloroform ***[24]***)/*iso*-amyl alcohol ***[1]***).
8. Chloroform/*iso*-amyl alcohol (v/v, 24/1).
9. Isopropyl alcohol.
10. 70% ethanol.
11. 10 m*M* Tris-HCl, pH 7.5.

2.1.2. Linearization of BAC DNA

1. *Srf*1 (Stratagene 501064) or lambda terminase (Epicentre Technologies LT44200).
2. Pulsed-field gel electrophoresis (PFGE) apparatus (Pharmacia Gene Navigator System).
3. Phenol.
4. Chloroform ***(24)***/*iso*-amyl alcohol ***(1)***.
5. 3 *M* sodium acetate, pH 5.2.
6. 100% ethanol.
7. 70% ethanol.
8. 10 m*M* Tris-HCl, pH 7.5.

2.1.3. Pro-Nuclear Injection

1. Injection buffer: 10 m*M* Tris-HCl pH 7.5, 0.1 m*M* EDTA pH 8.0, 100 m*M* NaCl, 30 µ*M* spermine, 70 µ*M* spermidine. Filter-sterilize and store at 4°C. Make fresh each week.

2.2. ES Cell Electroporation

2.2.1. Preparation and Linearization of BAC DNA

1. Approximately 30 µg of DNA is required for one electroporation. Materials are as under **Subheadings 2.1.1.** and **2.1.2.**
2. 50 mL polypropylene tubes (Falcon).

2.2.2. Electroporation of ES Cell Lines

1. 92-mm polylysine-coated plastic tissue culture plates (Nunc).
2. 0.1% gelatin.

3. ES cell media (1 × Dulbecco's modified Eagle's medium F-12/DMEM, 2 m*M* L-glutamine, 50 μg/mL penicillin/streptomycin, 1 × 'N' nucleosides, 1000 units/mL LIF (leukemia-inhibitory factor), 1 m*M* sodium pyruvate, 0.1 m*M* nonessential amino acids, 6% sodium bicarbonate, 15% fetal calf serum, 10^{-6} *M* β-mercaptoethanol. Add 150–260 μg/mL G418 or 130 μg/mL hygromycin for selection.
4. Bio-Rad gene pulser.
5. 4 mm electroporation cuvette (Bio-Rad).
6. Trypsin/EDTA (10 × solution supplied by Sigma T4174).
7. PBS without calcium and magnesium (1 L): 10 g NaCl, 0.25 g KCl, 1.44 g Na_2HPO_4, 0.25 g KH_2PO_4, pH to 7.2 with HCl. Autoclave and store at room temperature.
8. Humidified CO_2 incubator at 38°C (5% CO_2 in air).
9. Hemocytometer (Weber Scientific International Counting Chamber, depth 0.1 mm, 1/400 mm^2).
10. 20-mL tubes.

2.2.3. Picking and Analyzing Colonies

1. 24-well tissue culture plates.
2. 90-mm bacteriological Petri dish.
3. 0.1% gelatin.
4. ES cell media (prepared as under **Subheading 2.3.2.**).
5. Low-powered microscope.
6. 1 × trypsin/EDTA (10 × solution supplied by Sigma T4174).
7. Dimethyl sulfoxide (DMSO).
8. Lysis buffer: 100 m*M* Tris-HCl pH 7.8, 50 m*M* EDTA, 0.2% SDS, 200 m*M* NaCl.
9. 10 mg/mL proteinase K.
10. 1 phenol/1(Chloroform ***[24]****iso*-amyl alcohol ***[1]***).
11. Chloroform/*iso*-amyl alcohol (24/1).
12. 3 *M* NaOAc, pH 6.0.
13. Isopropyl alcohol.
14. 70% ethanol.
15. TE pH 8.0: 10 m*M* Tris-HCl pH 8.0, 1 m*M* EDTA pH 8.0.

2.2.4. Blocking Mouse Repeats for Southern Analysis of BAC Transgenic Clones

1. Church and Gilbert hybridization buffer (1 L): 62.3 g Na_2HPO_4, 23.4 g NaH_2PO_4, 70 g SDS, 2 mL EDTA pH 8.0.
2. G-50 sephadex spin column (Amersham).
3. TE, pH 8.0: 10 m*M* Tris-HCl pH 8.0, 1 m*M* EDTA pH 8.0.
4. 10% SDS.
5. Mouse genomic DNA, sheared by repeated passage through an 18-gage needle.

6. 2 × SSC/0.5% SDS.
7. 0.1 × SSC/0.5% SDS.

2.2.5. Generation of ES Cell Chimeras by Aggregation

1. 6- to 7-wk-old MF1 (Olac) female mice.
2. 6- to 7-wk-old F1 (C57BL6 × CBA) mice in estrus, and vasectomized F1 males.
3. PMS, HCG, T6 + BSA, PB1 + BSA, Petri dishes, forceps, Pasteur pipets, light mineral oil, and CO_2 incubator are as described in Chapter 4, **Subheading 2.3.**).
4. Acid tyrode's (AT) 800 mg NaCl, 20 mg KCl, 20 mg $CaCl_2{\cdot}2H_20$, 10 mg $MgCl_2{\cdot}6H2_0$, 5 mg $Na_2HPO_4{\cdot}H_20$, 100 mg glucose. Make up to 100 mL with sterile water and pH to 2.5 with concentrated HCl. Filter-sterilize through 0.22-µm filter. Store in 1.5-mL aliquots at 4°C.
5. 30-mm and 50-mm bacteriological Petri dishes (Sterilin).

3. Methods

3.1. Pro-Nuclear Injection

3.1.1. Preparation of BAC DNA

1. Pick a single colony into a 14-mL tube containing 5 ml LB and 12.5 µg/mL chloramphenicol. Incubate at 37°C with shaking for 8 h (*see* **Note 1**).
2. Centrifuge for 10 min at 1000*g* in a bench-top centrifuge.
3. Remove all supernatant and resuspend the pellet in 350 µL of solution I + 100 µg/mL RNAse A.
4. Add 350 µL of solution II and mix gently by inverting the tube several times. Place on ice for 5 min.
5. Add 350 µL of solution III, mix gently but thoroughly, and leave on ice for 20 min.
6. Transfer to a 2-mL microfuge tube and spin for 10 min in a microfuge at maximum speed.
7. Transfer supernatant (1 mL) to a fresh microfuge tube, avoiding the white precipitate.
8. Add an equal volume of phenol/chloroform and mix gently.
9. Spin for 10 min in the microfuge.
10. Transfer the aqueous layer to a fresh microfuge tube, avoiding the interface.
11. Add an equal volume of chloroform and mix gently.
12. Centrifuge for 10 min at maximum speed in the microfuge.
13. Remove the aqueous layer to a fresh 2-mL tube.
14. Add 1 vol of isopropyl alcohol and mix gently.
15. Centrifuge immediately for 1 min at maximum speed in the microfuge.
16. Decant alcohol and wash with 500 µL of 70% ethanol.
17. Remove all traces of alcohol with a pipet tip and resuspend the pellet in 10 µL of Tris-HCl pH 7.5.

3.1.2. Linearization of BAC DNA—For Example, pBeloBAC (see **Note 2***)*

1. Digest 1 μg of BAC DNA with either *Srf*I (Stratagene) or lambda terminase (Epicentre) as recommended by the manufacturers in a 50-μL final volume.
2. Check for complete linearization by running 10 μL on a PFG.
3. To the remaining 40 μL, add an equal volume of phenol and mix gently.
4. Spin for 10 min in the microfuge.
5. Remove the aqueous layer to a fresh microfuge tube, avoiding the interface.
6. Add an equal volume of chloroform and mix gently.
7. Centrifuge for 10 min in the microfuge.
8. Remove the aqueous layer to a fresh microfuge tube.
9. Add 1/10th vol of 3 *M* sodium acetate pH 5.2, 2 vol of 100% ethanol, and mix gently.
10. Store the precipitating DNA at –20°C until the PFG result is known.
11. If the DNA is fully linearized and intact on the PFG, centrifuge the sample for 10 min in the microfuge.
12. Decant alcohol and wash with 500 μL of 70% ethanol.
13. Remove the alcohol with a pipet tip and resuspend the pellet in 10 μL of 10 m*M* Tris-HCl pH 7.5.
14. Check the DNA concentration by running 1/10 of the sample on a small agarose gel against samples of known concentration.
15. Store the DNA at –20°C. Avoid repeated freeze/thawing.

3.1.3. Preparation of DNA for Injection

One day of pro-nuclear injection requires 100 μL of a 1- to 3-ng/μL solution. Take 100–300 ng of the linearized BAC DNA and make up the total volume to 100 μL with injection buffer. The DNA is ready to inject following the protocol described in Chapter 4.

3.2. ES Cell Electroporation

3.2.1. Preparation of BAC DNA for ES Cell Electroporation

1. Pick single colonies into ten 50-mL tubes containing 25 mL LB and 12.5 μg/mL chloramphenicol (total volume 250 mL). Incubate at 37°C with shaking for 8 h (*see* **Note 1**).
2. Centrifuge the culture for 10 min at 1600*g* in a bench-top centrifuge.
3. Remove all the supernatant and resuspend each bacterial pellets in 1 mL of solution I + 100 μg/mL RNAse A.
4. Add 1 mL of solution II and mix gently by inverting the tube several times. Place on ice for 5 min.
5. Add 1 mL of solution III, mix gently but thoroughly, and leave on ice for 20 min.

6. Combine the samples (30 mL total volume) in a 50-mL tube and spin for 10 min at 1600*g*.
7. Transfer the supernatant to a fresh 50-mL tube and spin again for 10 min at 1000*g*.
8. Transfer the supernatant to a fresh 50-mL tube, add an equal volume of phenol/chloroform, and mix gently.
9. Spin for 10 min at 1000*g*.
10. Transfer the aqueous layer to a fresh 50-mL tube, avoiding the interface.
11. Add an equal volume of chloroform and mix gently.
12. Centrifuge for 10 min at 12,000*g*.
13. Divide the aqueous layer equally among four 14-mL tubes.
14. Add 1 vol of isopropyl alcohol to each tube and mix gently.
15. Centrifuge immediately for 10 min at 12,000*g*.
16. Decant alcohol and wash with 1 mL of 70% ethanol.
17. Remove the alcohol with a pipet tip and resuspend the pellets in a total volume of 100 µL 10 m*M* Tris-HCl pH 7.5.

3.2.2. Linearization of BAC

The BAC DNA can be linearised as described under **Subheading 3.1.2.** by scaling up the reactions 10-fold. Digest the DNA in a 500-µL reaction volume and resuspend in 40 µL of sterile water on the day that it is to be used.

3.2.3. Electroporation of ES Cell Lines (*see* Note 3).

1. Grow ES cells under routine conditions *(3)* to obtain at least 1×10^7 cells, which is sufficient for a single electroporation. This is approximately equivalent to three semiconfluent 92-mm plates.
2. Prepare six gelatinized 92-mm polylysine-coated plastic tissue culture plates. Add 5 mL of 0.1% gelatin to each plate and swirl. Aspirate the gelatin and leave for 5 min. Add 9 mL of ES cell medium to each plate (*see* **Note 4**).
3. Chill PBS on ice.
4. Set Bio-Rad gene pulser to 250 V, 500 µF.
5. Chill a 4-mm electroporation cuvet on ice.
6. Wash ES cells with 10 mL of ice-cold PBS.
7. Add 3 mL 1 × trypsin/EDTA and swirl. Aspirate trypsin and incubate in CO^2 incubator for 3–5 min.
8. Add 2 mL of ES cell medium and pipet up and down to break up the clumps, using a 1-mL Gilson pipetman.
9. Transfer the ES cell-containing medium to a 20-mL tube and centrifuge for 500*g* for 5 min.
10. Aspirate medium and resuspend in 0.8 mL of cold PBS.
11. Count cells with hemocytometer. Adjust cell concentration to 1×10^7 in a final volume of 0.8 mL cold PBS. Keep on ice. Three semiconfluent 92 mm plates should be enough for one electroporation.

12. Mix 0.8 mL of cells with DNA in a maximum volume of 100 µL of H_2O on ice.
13. Transfer to the prechilled electroporation cuvet.
14. Deliver electric pulse (the time constant should be 8–14).
15. Quickly transfer the cells into 6 mL of ES cell medium + LIF in a 20-mL tube.
16. Divide the cell suspension among six 92-mm gelatinized plates and grow without selection for 1 d.
17. Add ES cell medium containing selection agent and grow cells under selection for 7–10 d in a CO_2 incubator.

3.2.4. Picking and Analyzing Colonies

1. Prepare one 24-well tissue culture dish. Add 0.5 mL of 0.1% gelatin to each well. Aspirate. After 5 min add 1 mL of ES cell medium (without selection) (*see* **Note 4**).
2. Set up tissue culture hood with a low-powered microscope.
3. Prepare a 92-mm lid of a bacterial Petri dish next to the microscope stand with 30-µL drops of trypsin in a row.
4. On the plate from which ES cell colonies are to be picked, replace the medium with PBS.
5. Adjust a 200-µL Gilson pipet to 30 µL and with a yellow tip detach the first ES cell colony from the plate by gently scraping the tip over the colony. Immediately suck up the colony into the yellow tip.
6. Expel the colony into the first drop of trypsin.
7. After 3–5 min, pipet up the drop containing the first colony and transfer it to the first well of the 24-well tissue culture plate. The trypsin and sucking force of the pipet should together cause the colony to disaggregate to a single-cell suspension.
8. Repeat until you have picked sufficient colonies.
9. Feed the cells with 1 mL of ES cell medium (+ LIF) for 2 d.
10. Prepare two fresh 24-well dishes with gelatin. Aspirate the gelatin and add 1 mL of ES cell medium to each well. Label plates A and B.
11. Trypsinize the cells and resuspend in 1 mL of ES cell medium.
12. Transfer 0.5 mL to the ES cell 24-well plate A. This will be the freezing plate (*see* **Note 4**).
13. Transfer 0.5 mL to the ES cell 24-well plate B. This will be the DNA plate (*see* **Note 4**).
14. Feed cells with 1 mL of ES cell medium for 2 d.
15. Trypsinize the cells in plate A and resuspend in 500 µL of ES cell medium. Add 50 µL of DMSO, wrap the plate in Parafilm, and freeze in a –80°C freezer.
16. When the cells in plate B are confluent, wash with PBS and add 0.75 mL of lysis buffer to each well.
17. Pipet up and down a few times to lyse the cells completely, then transfer each lysate to a 1.5-mL microfuge tube.
18. Add 7.5 µL of proteinase K.

19. Incubate at 55°C for 2 h or overnight at 37°C.
20. Add an equal volume of phenol/chloroform and mix gently for 5 min at room temperature.
21. Centrifuge for 10 min. Collect the upper phase and extract with chloroform.
22. Remove the aqueous phase and add 1/30th vol of 3 *M* sodium acetate pH 6.0 and 1 vol of isopropyl alcohol. Mix gently; the DNA will precipitate immediately. Hook out with a bent Pasteur pipet sealed at the end. Rinse in 70% ethanol. Air dry for 10 min and resuspend in 50 µL of TE (*see* **Note 5**).

3.2.5. Blocking Mouse Repeats for Southern Analysis of BAC Transgenic Clones (see **Note 6***)*

1. Prehybridize Southern membrane in Church hybridization buffer at 65°C for a minimum of 5 min *(7)*.
2. Radioactively label 0.5 µg of BAC DNA.
3. Purify labeled DNA by passage through a G-50 Sephadex spin column.
4. Add 250 µg of sheared mouse genomic DNA in a final volume of 250 µL T.E./0.5% SDS.
5. Boil for 10 min, then chill on ice for 10 s.
6. Add to 5 mL of hybridization buffer and incubate at 65°C for 1–2 h.
7. Add to membrane and hybridize for 14–16 h at 65°C.
8. Wash in 2 × SSC/0.5% SDS for 5 min, followed by 0.1 × SSC/0.5% SDS twice for 30 min.
9. Expose to autoradiography film overnight at –70°C.

3.2.6. Generation of ES Cell Chimeras by Aggregation (see **Note 7***).*

1. Day 1: Superovulate MF1 (Olac) females by injecting 150 µL of PMS intraperitoneally using a 27-gage needle (*see* **Note 8**).
2. Day 3: After 48 h, inject 150 µL of HCG intraperitoneally using a 27-gage needle. Set up matings with stud MF1 (Olac) males.
3. Day 4: Check females for vaginal plugs and transfer to a separate cage.
4. Day 5: Set up recipients by mating 12 F1 females (C57BL6 × CBA) in estrus with 12 vasectomized F1 males.
5. Day 6: Check recipients for vaginal plugs and transfer to separate cages with the plug date recorded.
6. Place a 30-µL drop of T6 + BSA in a 30-mm Petri dish. Cover with mineral oil. Incubate in CO_2 incubator until required.
7. Isolate morulae by standard procedures. Morulae should be at the 8-cell stage and beginning to compact. Using a mouth pipetting system, transfer the embryos to a drop of T6 + BSA in a 50-mm bacterial Petri dish to wash.
8. Transfer the embryos into the 30-µL drop of T6 + BSA in the 30-mm Petri dish. Incubate in the CO_2 incubator.

3.2.7. Preparation of ES Cells

1. Day 6: Trypsinize ES cells as described under **Subheading 3.2.3.** One well of a 50–75% confluent 24-well plate is more than sufficient. Resuspend thoroughly in ES cell medium to give a single-cell suspension.
2. Count cells on hemocytometer. Approximately 1×10^6 cells/mL are required. Adjust volume with ES cell medium.
3. Place nine 15-μL drops of ES cells in a 50-mm bacterial Petri dish in 3 × 3 rows. Flood the dish with 6 mL of mineral oil and transfer to CO_2 incubator to allow the cells to settle to form an even lawn.

3.2.8. Preparation of Embryos

1. Place a 100-μL drop of PB1 + BSA at the top of a 50-mm bacterial Petri dish. Place two 30-μL drops of AT to one side and nine 30-μL drops of ES cell medium to the other side.
2. Transfer well-formed embryos from the T6 + BSA drop to the drop of PB1 + BSA (*see* **Note 9**).
3. Under a dissecting scope, take one embryo into the pipet in a minimal volume of PB1 + BSA. Transfer to the top of the drop of AT. Quickly expel the remaining liquid from the pipet and then go back into the AT drop to catch the embryo before it hits the bottom of the dish (*see* **Note 10**). Transfer quickly to the first drop of ES cell medium.
4. Pick up the embryo and transfer it to the second drop of ES cell medium.
5. Repeat the process; 5–10 embryos can be taken through the AT drop at a time.
6. When 50–60 embryos have been processed through the AT, transfer all from the second ES cell medium drop through the fresh ES cell medium drops to the sixth drop to remove all traces of AT.
7. After 50–100 embryos have been treated, use a fresh drop of AT.
8. After all morulae have been processed, divide equally among *n* ES cell medium drops, where *n* = number of ES cell lines to be aggregated. About 20 embryos are sufficient for each line.
9. As soon as possible, transfer the embryos to the drops containing the ES cells (*see* **Note 11**).
10. Transfer to the CO_2 incubator to allow the ES cells to stick to morulae.
11. After 30 min, check the embryos (*see* **Note 12**). Ideally, about 10 or so ES cells should attach to each embryo. Transfer to a drop of T6 + BSA to rinse away the ES cell medium and any loose ES cells.
12. Transfer the embryos to a 50 mm dish containing nine 30-μL drops of T6 + BSA under mineral oil. Two to three embryos can be incubated in each drop, provided they are placed away from each other.
13. Culture for 2 d in the CO_2 incubator.

Table 1
Agents for Linearization of Large Genomic Clones[a]

Site	No. of sites in pBeloBAC	No. of sites in pBACe3.6	No. of sites in pPAC4	No. of sites in pCYPAC2
AscI	0	2	2	0
FseI	0	0	0	0
NotI	2	2	2	2
PacI	0	1	2	1
PmeI	0	1	2	1
SgrA1	2	2	0	1
SrfI	1	0	0	0
cosN	1	0	1	0
loxP	1	1	1	1

[a]All restriction enzymes recognize an 8-base sequence that is likely to occur infrequently within a BAC or PAC clone.

3.2.9. Transfer into Recipients

1. Day 8: Transfer the embryos into 30-μL drops of PB1 + BSA under mineral oil (prewarmed in the CO_2 incubator). Allow 1 drop per recipient (10 embryos). The embryos will have expanded into blastocysts by this stage.
2. Using standard surgical procedure, prepare recipients to receive the embryos.
3. Transfer 10–20 embryos per recipient.
4. Embryos are transferred into recipients at d 3 (d 1 is the day of vaginal plug), so the pups will be born 16–17 d later (*see* **Note 13**).

4. Notes

1. One 5-mL culture should yield approximately 1–2 μg DNA, which is sufficient for several days of pro-nuclear injection. The length of time in culture can affect the yield and quality of BAC DNA, and 8 h is optimal in our experience. A reliable alternative to the BAC DNA preparation method detailed here is provided by the Wizard PureFection (Promega Corporation) kit.
2. The DNA used in generating BAC or PAC transgenic mice must first be linearized. All the commonly used vectors include a choice of unique restriction sites for this purpose. For example, the pBELOBAC vector contains an *Srf*I site that is usually not present in the insert (this can be assessed by digesting the BAC clone with Srf1 and resolving the sample on pulse-field gel as described in the previous section). There are also two *Not*I sites that flank the cloning site, allowing for excision of the genomic sequence from the vector. Again, PFG electrophoresis is required to test for *Not*I sites within the genomic clone. If both *Srf*I and *Not*I sites occur within the genomic fragment, this vector also contains a *cos*N site, which can be opened using lambda terminase. **Table 1** summarizes potential

linearization sites in some of the vectors commonly used to clone large genomic fragments.

3. It is possible to incorporate the modification of the BAC or PAC clone to contain a neomycin or hygromycin gene, either under a constitutive promoter or under the promoter of a gene on the BAC/PAC. For large constructs without a selectable marker (unmodified cosmids, BACs, etc.), a neomycin or hygromycin gene under a constitutive promoter can be co-electroporated with the BAC. For co-electroporation we use a ratio of 10/1 construct to neomycin cassette. If expression of the neomycin or hygromycin gene is driven by an untested promoter, the concentration of the selective agent may vary from the standard conditions.
4. Feeders: ES cells can be cultured on a layer of feeder cells in the absence of LIF *(3)*. These are primary embryonic fibroblast cells obtained from 11.5–12.5 dpc embryos, where the nuclear material has been inactivated by mitomycin C. When under selection, feeder cells obtained from transgenic mice constituitively expressing neomycin or hygromycin-resistance can be used. We obtain our neomycin-resistant feeders from the ROSA-26 line *(4)*. We use 2×10^6 feeder cells per 92-mm plate and 1×10^5 cells per well of a 24-well tissue culture plate. Plate B, the DNA plate, should be grown without feeders from **Subheading 3.2.4., step 12**, onward.
5. Approximately 20 μg of DNA should be recovered from one confluent well of a 24-well dish. 10 μL of this DNA (4 μg) should be enough for digesting with appropriate restriction enzymes and analysis by Southern blotting.
6. It is sometimes possible to use the genomic BAC/PAC clone as a probe on a Southern blot to assess transgene integrity. This allows you to check for intactness of the integration event and approximate copy number in one hybridization step. We find that this protocol works well for constructs of up to 100 kb, but is less suitable for larger clones, where the complexity of a restriction digest is too great to allow all the individual fragments to be resolved on a standard agarose gel. In this case it is advisable to use smaller DNA fragments, such as vector sequences, probes from the ends of the genomic sequence or which detect the modification, which can be hybridized sequentially to ascertain integrity of the integration event.
7. There are two widely used techniques for generating ES cell chimeras, blastocyst injection and morula aggregation. These are described in **refs. *5*** and ***6***. We have used both techniques successfully. Overall, aggregation is the simpler technique and does not require any sophisticated equipment or specific training. It also has the advantage of allowing the aggregation of several lines simultaneously and therefore is the technique we recommend.
8. We use the albino MF1 (Olac) strain, as our ES cell lines are derived from mice with a pigmented coat color that allows for simple identification of chimeric animals. Eight to 12 females should produce enough embryos for aggregation of several ES cell lines. When chimeras are ready to mate, they can be test-bred with MF1 (Olac) females. Pups inheriting the ES cells will be pigmented at birth. Earlier analysis can be performed at d 11.5 of embryogenisis, as the pigmented eye color is easy to observe.

Table 2
Transgenic Lines Generated Using This Protocol

ES cell line	Chimeric animals born	ES cell coat color (%)	ES cell transmission (%)[a]
5D3	6	100 (n = 6)	100
5A2	4	100 (n = 3)	100
		30	n.t.
5A4	5	100 (n = 4)	100
		60	100
6C1	1	100	100
5B6	2	50	100
		50	n.t.
5C2	1	80	12.5

[a]n.t. = not tested.

9. Avoid embryos in which the cells are uneven in size or in which some cells are brownish in color, since these are not healthy. Make several glass drawn-out pipets to get the best size, which is not too much bigger than the morulae. Always draw some medium into the pipet before picking up the embryos. Never allow the embryos to go up as far as the widening part of the pipet, since either they will get stuck or they will make contact with air.
10. You should see the zona dissolving as the embryos sink in the AT, but it is important not to leave them here for more that 1–2 s, as they will fall apart. With practice, multiple embryos can be treated with AT simultaneously.
11. The ES cells will have formed an even, single-cell layer at the bottom of the drop. Try to keep the embryos fairly separate, since if they touch they will fuse. Therefore transfer only 3–4 embryos per drop.
12. The time of incubation with ES cells is crucial. Low-passage ES cells (6–10) need less time, as they seem to be more "sticky," so it may be necessary to experiment with different timing. If too many ES cells attach to the embryo, it will not survive; too few and there will be no or low-percentage chimeras.
13. Using this procedure we have generated chimeras for six ES cell lines over a 2-wk aggregation period. All of these lines resulted in germline transmission, with chimeras from four different lines transmitting the ES cells to 100% of their offspring (see **Table 2**).

References

1. Yang, X. W., Model, P., and Heintz, N. (1997) Homologous recombination based modification in *Escherichia coli* and germline transmission in transgenic mice of a bacterial artificial chromosome. *Nature Biotechnol.* **15,** 859–865.

2. Zhang, Y., Buchholz, F., Muyers, J. P. P., and Stewart, A. F. (1998) A new logic for DNA engineering using recombination in *Escherichia coli. Nature Genet.* **20,** 123–128.
3. Robertson, E. J. (1987) Embryo-derived stem cell lines. In: Teratocarcinomas and embryonic stem cells: a practical approach (Robertson, E. J., ed.), IRL, Oxford, U.K., pp. 71–112.
4. Soriano, P. (1999) Generalized *lacZ* expression with the ROSA26 Cre reporter strain. *Nature Genet.* **21,** 70–71.
5. Bradley, A. (1987) Production and analysis of chimaeric mice. In: Teratocarcinomas and embryonic stem cells: a practical approach (Robertson, E. J., ed.), IRL, Oxford, U.K., pp. 113–151.
6. Wood, S. A., Pascoe, W. S., Schmidt, C., Kemler, R., Evans, M. J., and Allen, N. D. (1993) Simple and efficient production of embryonic stem cell-embryo chimeras by coculture. *Proc. Natl. Acad. Sci. USA* **90,** 4582–4585.
7. Church, C. M. and Gilbert, W. (1984) Genomic sequencing. *Proc. Natl. Acad. Sci. USA* **81,** 1991–1995.

6

Methylation-Sensitive Genome Scanning

Izuho Hatada and Tsunehiro Mukai

1. Introduction

Imprinted genes in mammals are expressed exclusively from one of the parental alleles *(1–6)*. This is regulated by parental-allele-specific CpG methylation. For example, *H19* is methylated exclusively on the paternal allele, which is repressed, and is expressed exclusively from the maternal allele, which is unmethylated. Therefore, one way to find imprinted genes is searching for parental-allele-specific CpG methylation. Southern analysis using methylation-sensitive restriction enzymes could be used for such a purpose. However, usually only one gene can be analyzed by one Southern analysis. Moreover, Southern analysis requires one DNA probe for each analysis. These facts indicate at least 300 Southern analyses using 300 different probes are required to find only one imprinted gene, because the population of imprinted genes is estimated to be 0.3%. Therefore, this kind of analysis is not appropriate for searching for new imprinted genes, and the development of a new method that can simultaneously analyze thousands of genes was required.

1.1. Restriction Landmark Genome Scanning (RLGS) Method

We developed a powerful genome scanning method, termed the restriction landmark genome scanning (RLGS) method *(7)*, for analyzing thousands of genes simultaneously in higher organisms such as mammals. This method is based on end-labeling of genomic DNA at restriction sites and its separation by two-dimensional gel electrophoresis. Before the RLGS method was developed, the detection of genes from higher organisms by end-labeling had been thought to be impossible. There are two reasons for this: (1) The mammalian genome is so complex (for example, the human genome size is 3×10^9 bp and digestion

From: *Methods in Molecular Biology, vol. 181: Genomic Imprinting: Methods and Protocols*
Edited by: A. Ward © Humana Press Inc., Totowa, NJ

with *Eco*RI will give more than 10^6 DNA fragments) that DNA fragments digested by restriction enzymes cannot be separated even by two-dimensional electrophoresis. (2) Generally, genomic DNA is cleaved randomly in the preparation step and thus has nonspecific cleaved ends, nicks, and /or gaps. This leads to high background caused by the incorporation of radioisotopes into sites damaged in the labeling process. To overcome these problems we adopted two solutions: (1) To reduce the complexity, genomic DNA is cut and labeled at rare-cleaving restriction enzyme sites, for which there are thousands of recognition sites in the genome, before cutting with more frequently cutting restriction enzymes that will give more than 10^6 DNA fragments. Using this method, only DNA fragments that have rare restriction sites are detected and used as landmarks in the genome. (2) High background can be avoided by blocking damaged sites with enzymatically incorporated nucleotide analogs, because such analogs prevent exonucleolysis and/or the additional incorporation of nucleotides at blocked ends.

The procedure for the RLGS method is comprised of eight steps (**Fig. 1**).

1. Blocking: Damaged sites in genomic DNA are blocked by nucleotide analogs, 2′-deoxyribonucleoside 5′-[α-thio] triphosphate and 2′3′-dideoxyribonucleoside 5′-triphosphate, to reduce the background.
2. Landmark cleavage: Blocked genomic DNA is cleaved with a rare-cutting restriction enzyme (restriction enzyme A; average cleaved fragment size should be more than 100 kbp).
3. Labeling: Cleavage ends of the genomic DNA are labeled with radioisotope.
4. Fragmentation of labeled DNA with restriction enzyme B: Labeled genomic DNA is cleaved with restriction enzyme B (average cleaved fragment size should be between 1 kbp and 100 kbp) to reduce the fragment size to several kilobase pairs, kbp which is appropriate for agarose gel electrophoresis.
5. First fractionation: Genomic DNA fragments are fractionated in one dimension by thin-layer agarose gel electrophoresis.
6. Fragmentation of labeled DNA with restriction enzyme C: Fractionated genomic DNA fragments are cleaved in the gel with restriction enzyme C (average cleaved fragment size should be less than 10 kbp) to make fragments appropriate for polyacrylamide gel electrophoresis.
7. Second fractionation: Genomic DNA fragments in the agarose gel are fractionated in the second dimension by polyacrylamide gel electrophoresis.
8. Autoradiography: The labeled genomic DNA fragments are detected as spots.

Figure 2 shows a spot profile of mouse DNA obtained by RLGS. This profile contains about 2000 completely separated spots corresponding to sites for restriction enzyme A.

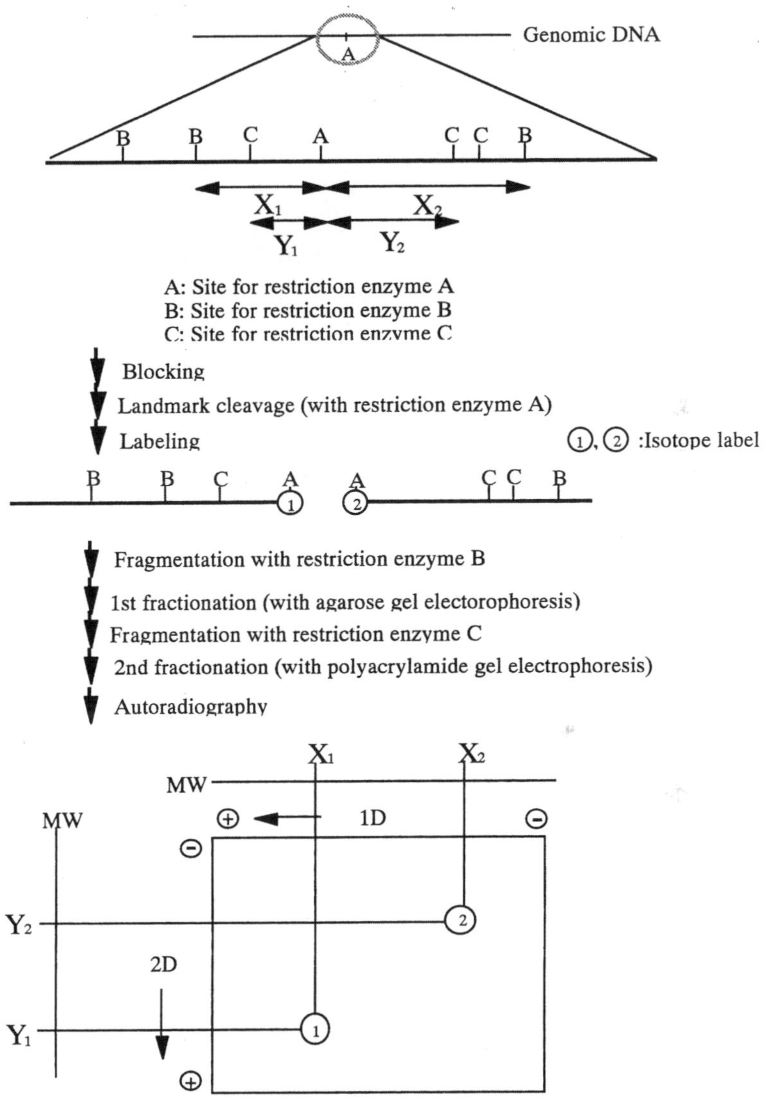

Fig. 1. Procedure for genome scanning by two-dimensional gel electrophoresis. RLGS consists of eight steps: (1) blocking, (2) landmark cleavage (with restriction enzyme A), (3) labeling (4) fragmentation with restriction enzyme B, (5) first fractionation (with agarose gel electrophoresis), (6) fragmentation with restriction enzyme C, (7) second fractionation (with polyacrylamide gel electrophoresis), and (8) autoradiography. X_1 or X_2 and Y_1 or Y_2 represent the distance from a restriction landmark to the neighboring site for restriction enzyme A or B, and that to the site for restriction enzyme C, respectively.

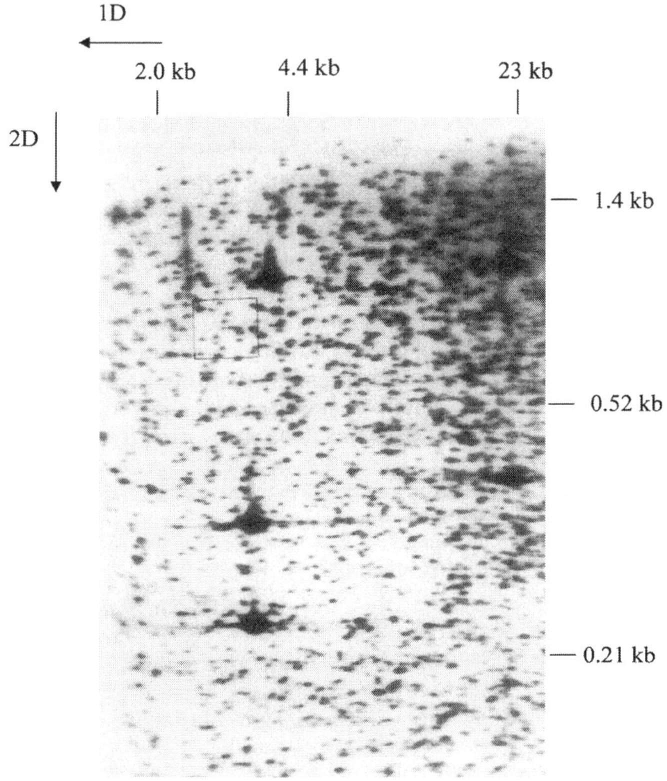

Fig. 2. The RLGS profile of genomic DNA from C57BL/6 mice. *Not*I, *Eco*RV, and *Mbo*I were used as restriction enzymes A, B, and C, respectively. The boxed region is the area shown in Fig. 6. Size markers are shown in kilobase pairs.

1.2. Searching for Parental-Origin-Specific CpG Methylation by the RLGS Method

Methylation status can be distinguished by the RLGS method when a CpG methylation-sensitive enzyme, such as *Not*I, is used for restriction enzyme A *(8)*. It is expected that the intensity of any spot varies in proportion to the population of methylated DNA molecules (**Fig. 3**). Lack of methylation of both alleles (**Fig. 3**, lower left panel), methylation of one allele (middle panel), and methylation of both alleles (right panel) will give full intensity, half-intensity, and no spot, respectively. To identify parental-origin-specific methylation, the spot profiles of F_1 progeny of reciprocal crosses between two inbred strains, for example (DBA2 × C57BL/6)F_1 and (C57BL/6 × DBA2)F_1, should be compared. If there is no parental-origin-specific methylation, both the maternal

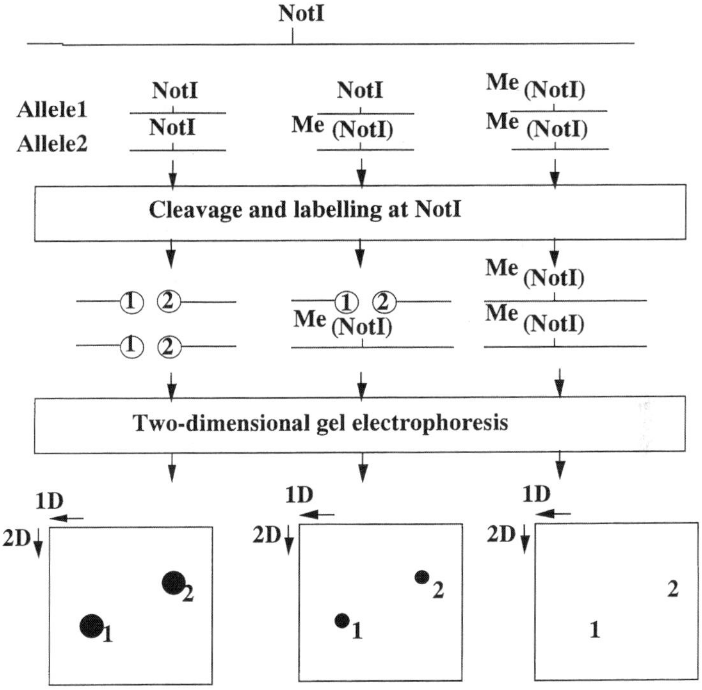

Fig. 3. Detection of methylation differences in genomes by RLGS. Maps show DNA-methylation state at the *Not*I site, which is used as restriction enzyme A. Three types of methylated state are expected, depending on which allele is methylated, as shown in the lower panels: no allele is methylated (left), one allele is methylated (middle), and both alleles are methylated (right). ^{Me}(*Not*I) represents the methylated site of this enzyme. As the CpG methylation status of the *Not*I site affects the cleavage of this site, the RLGS profile results in a change in spot intensity. Each sample would give a full intensity spot (left), half-intensity spot (middle), or no spot (right).

and the paternal allele can be cleaved with *Not*I and detected as spots, resulting in identical spot profiles of F_1 progeny (**Fig. 4**). However, if one parental allele is methylated, only one allele can be cleaved with *Not*I and detected as spots, resulting in different spot profiles of F_1 progeny (**Fig. 5**). We found four spots that appeared in one cross while not appearing in another when *Not*I, *Eco*RV, and *Mbo*I were used as restriction enzymes A, B, and C, respectively *(8)*. For example, spot 2 (specific to DBA2), appeared in one cross (C57BL/6 × DBA2)F_1, while not appearing in the opposite cross (DBA2 × C57BL/6)F_1 (**Fig. 6**).

Other parental-origin-specific methylation patterns can be sought using other restriction enzymes.

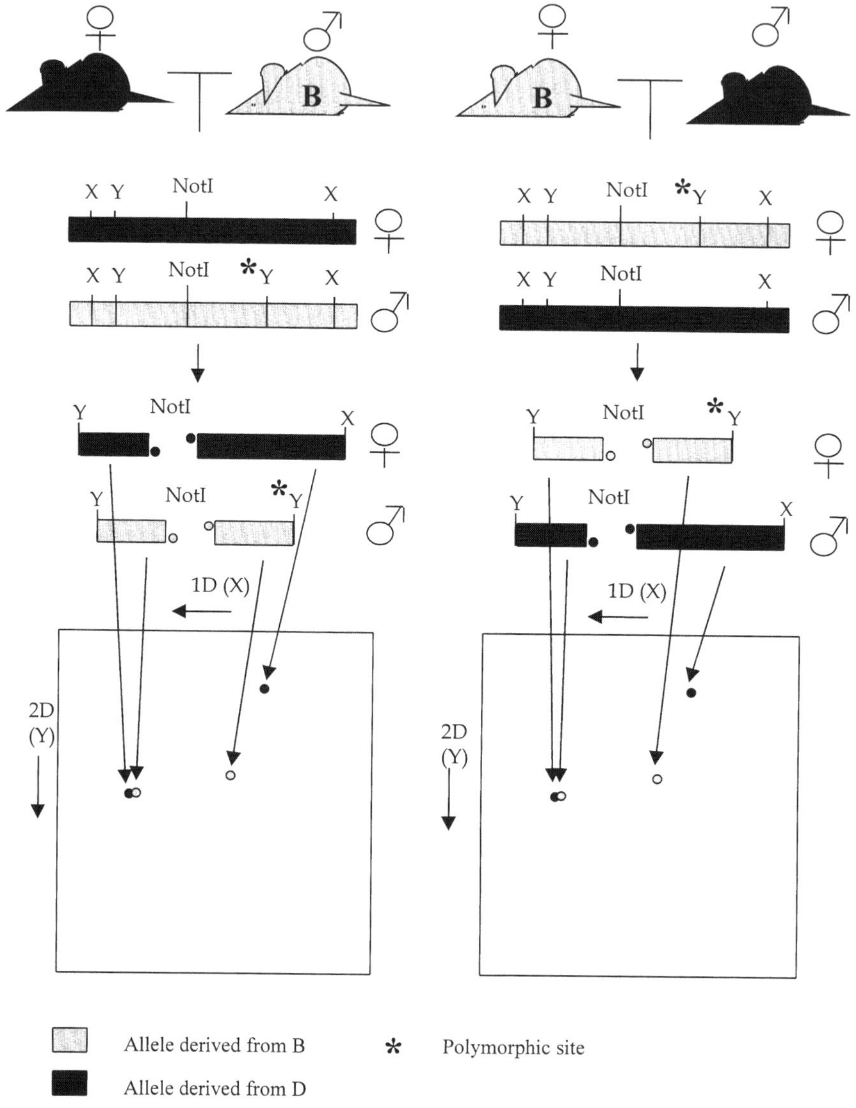

Fig. 4. Expected spot profiles derived from a gene without parental-origin-specific methylation. (DBA2 × C57BL/6)F_1 and (C57BL/6 × DBA2)F_1 are applied to the RLGS. Both the maternal and the paternal allele can be detected as spots, resulting in the identical spot profiles of F1 progeny of reciprocal crosses. B and D indicate C57BL/6 and DBA2, respectively.

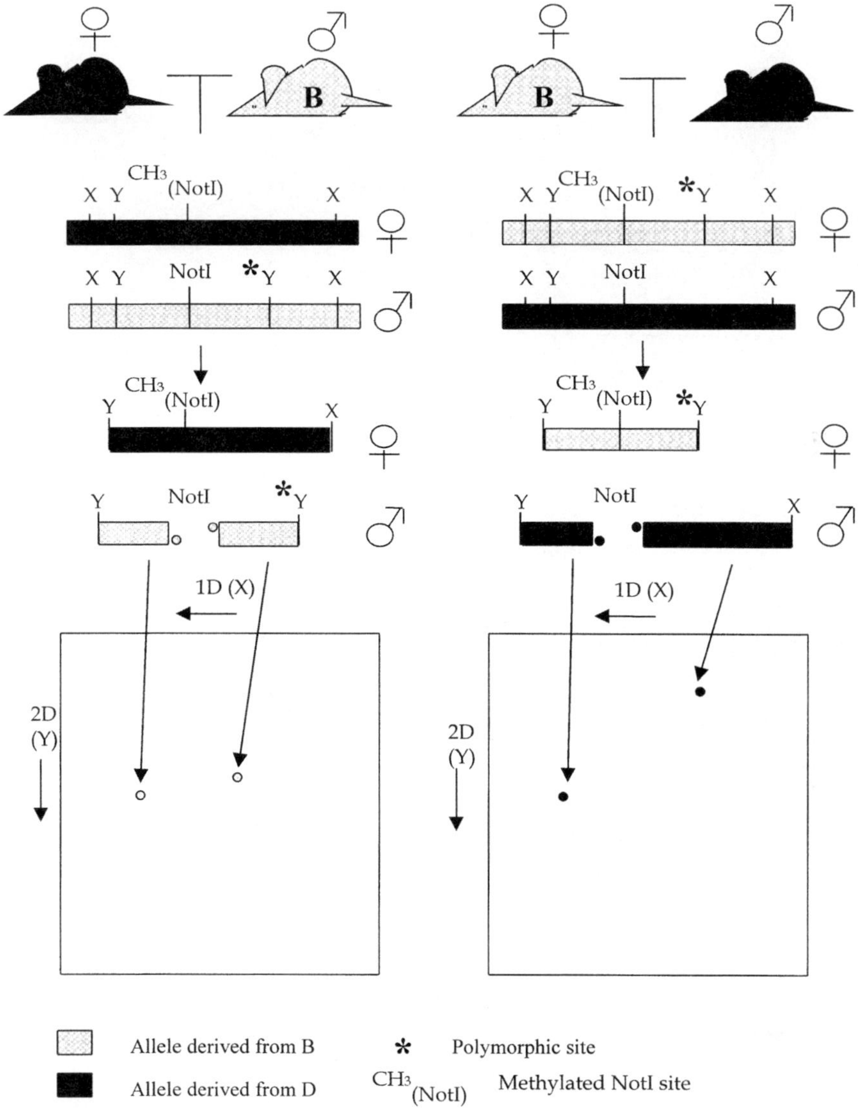

Fig. 5. Expected spot profiles derived from a gene with parental-origin-specific methylation. (DBA2 × C57BL/6)F_1 and (C57BL/6 × DBA2)F_1 are applied to the RLGS. The maternal allele is methylated and only the paternal allele can be detected, resulting in the different spot profiles of F_1 progeny of reciprocal crosses. B and D indicate C57BL/6 and DBA2, respectively.

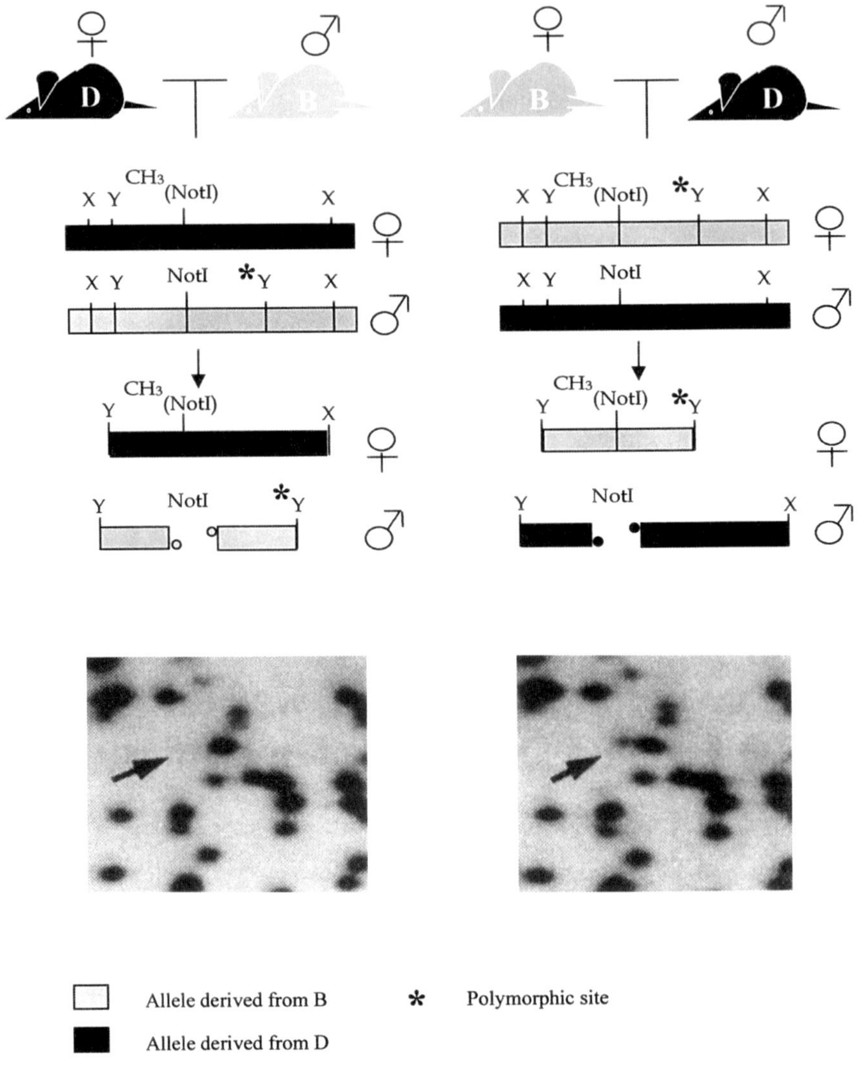

Fig. 6. Detection of the difference in spot intensity in F1 progeny of reciprocal crosses. (DBA2 × C57BL/6)F_1 and (C57BL/6 × DBA2)F_1 are applied to the RLGS and the boxed area from Fig. 2 is shown. Arrows indicate spot 2. Spot 2 did not appear in (DBA2 × C57BL/6)F_1 but appeared in (C57BL/6 × DBA2)F_1. B and D indicate C57BL/6 and DBA2, respectively.

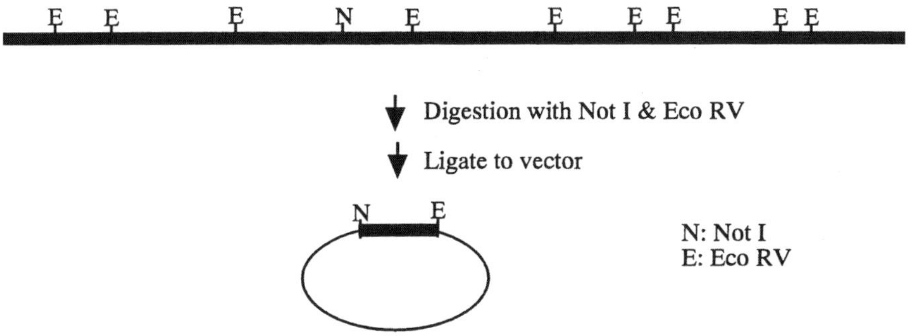

Fig. 7. Construction of a boundary library using *Not*I and *Eco*RV.

1.3. Cloning of Parental-Origin-Specific Methylated Spots

To clone the spot in which you are interested, a boundary library should be constructed by using restriction enzymes A and B (**Fig. 7**). For example, when *Not*I, *Eco*RV, and *Mbo*I are used as restriction enzymes A, B, and C, the library is constructed by using *Not*I and *Eco*RV. This library contains only *Not*I-*Eco*RV fragments and does not contain any *Eco*RV-*Eco*RV fragments. The fragment that must be cloned is concentrated several thousands times in this library. This library is applied to RLGS, except for the blocking and labeling steps, with labeled genomic DNA. The target spot is cut out from the gel and *Not*I-*Mbo*I fragments from this spot are electroeluted for cloning into *Not*I, *Bam*HI sites of the vector.

2. Materials

The materials described here are for a case using *Not*I, *Eco*RV, and *Mbo*I as restriction enzymes A, B, and C.

2.1. Extraction of Genomic DNA from the Sample

1. 10 m*M* Tris-HCl (pH 8.0), 0.5 *M* ethylenediaminetetraacetic acid (EDTA), 1% sarkosyl.
2. 10 mg/mL Proteinase K.
3. Phenol chloroform isoamyl alcohol (v/v/v, 25/24/1).
4. TE: 10 m*M* Tris-HCl (pH 7.4), 1 m*M* EDTA.
5. 1 mg/mL RNase A (boiled for 5 min).
6. Ethanol (70% and 100%).
7. Liquid nitrogen.
8. Aluminum foil.
9. Hammer.
10. Mortar and pestle.

2.2. Labeling Genomic DNA

1. 10 × buffer H: 500 m*M* Tris-HCl, pH 7.4, 100 m*M* $MgCl_2$, 1 *M* NaCl, 100 m*M* dithiothreitol.
2. 10 μ*M* dGTP[αS].
3. 10 μ*M* dCTP[αS].
4. 10 μ*M* ddATP.
5. 10 μ*M* ddTTP.
6. DNA polymerase I.
7. 0.1% BSA.
8. 0.1% Triton X-100.
9. *Not*I.
10. Control plasmid: linearized plasmid which has at least one recognition site for restriction enzymes A and B.
11. 0.25% bromophenol blue, 0.25% xylene cyanol FF. 30% glycerol.
12. 0.8% agarose gel: 0.8% Agarose, 1 × TBE.
13. 1M Dithiothreitol.
14. [α-32P] dGTP (3000 Ci/m*M*).
15. [α32P] dCTP (3000 Ci/m*M*).
16. Sequenase V2.0.
17. 1 m*M* ddGTP.
18. 1 mM ddCTP.
19. *Eco*RV.
20. 3 *M* NaOAc.
21. 0.25% bromophenol blue, 0.25% xylene cyanol FF.
22. 100% ethanol.
23. TE (*see* **Subheading 2.1.**).
24. Phenol/chloroform/isoamyl alcohol (25/24/1).

2.3. Two-Dimensional Electrophoresis

1. 1 × TAM: 50 m*M* Tris-HCl, pH 7.5, 0.7 m*M* MgAc.
2. 0.8% Seakem GTG (FMC), 1 × TAM.
3. 1 × buffer H: 50 m*M* Tris-HCl, pH 7.4, 10 m*M* $MgCl_2$, 100 m*M* NaCl, 10 m*M* dithiothreitol.
4. TBE: 90 m*M* Tris-borate, 2 m*M* EDTA, pH 8.0.
5. 0.8% SeaKem GTG, 1 × buffer H.
6. *Mbo*I.
7. 1-D small plate: 150 × 30 × 5 mm glass plate (**Fig. 8**).
8. 1-D upper plate: 470 × 150 × 5 mm glass plate (**Figs. 8** and **11**).
9. 1-D lower plate: 500 × 150 × 5 mm glass plate (**Figs. 8** and **11**).
10. Well plate: Two plastic sheet (16 × 1 × 0.7 mm) are attached to a glass plate (150 × 30 × 5 mm) for making wells.
11. Upper plate for enzyme digestion: A glass plate (150 × 30 × 5 mm) with two holes (ϕ2 mm) for pouring enzyme solution. Spacers (1.2 mm thick) are attached to four sides of the plate (**Fig. 9**).

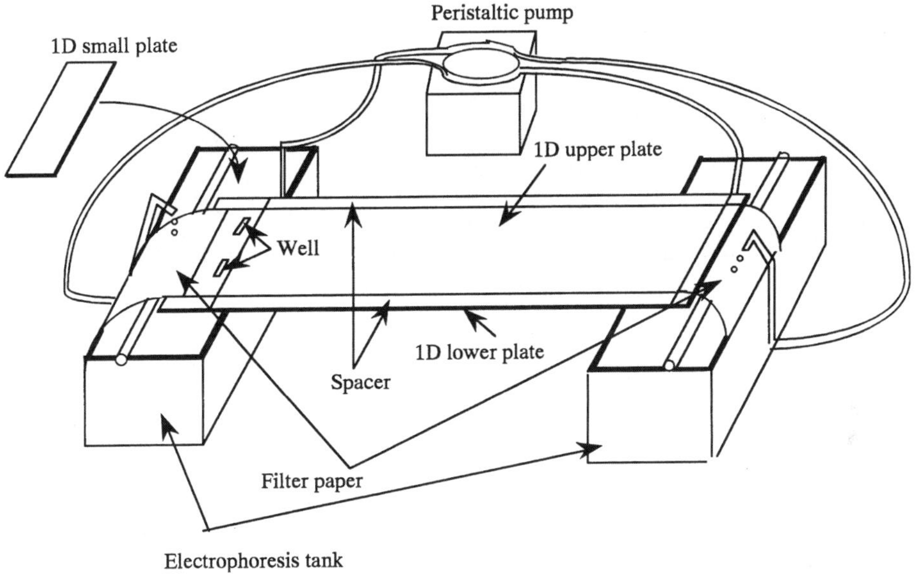

Fig. 8. Apparatus for one-dimensional agarose gel electrophoresis.

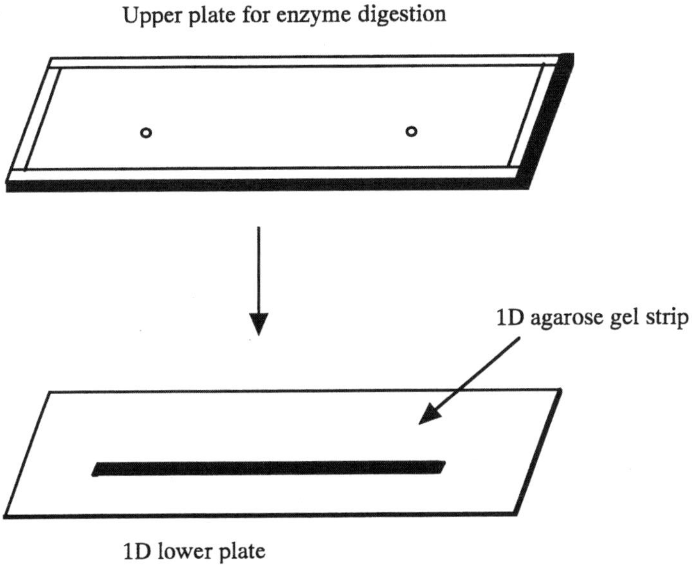

Fig. 9. Apparatus for enzyme digestion within the agaraose gel.

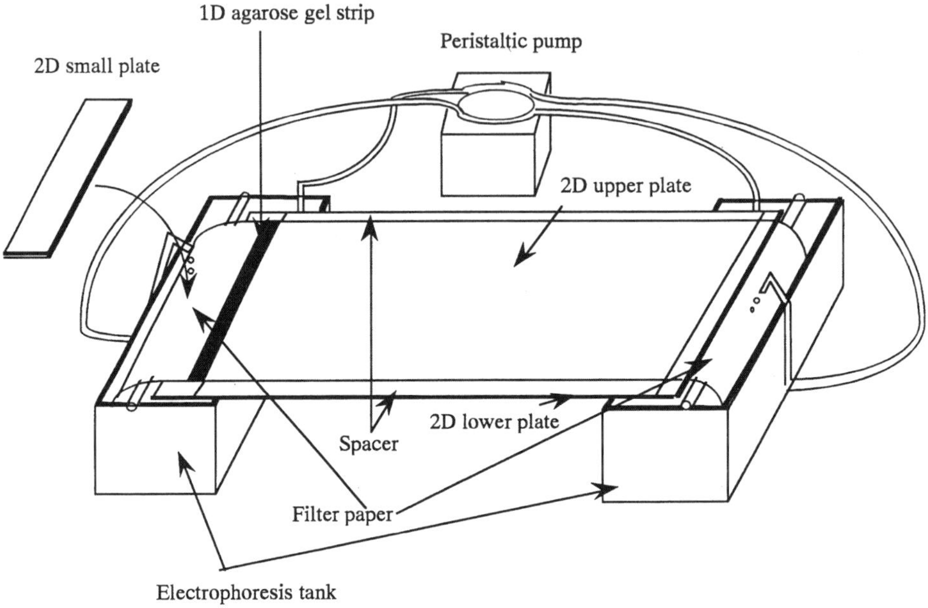

Fig. 10. Apparatus for two-dimensional polyacrylamide gel electrophoresis.

12. 2-D small plate: 415 × 30 × 5 mm glass plate (**Fig. 10**).
13. 2-D upper plate: 460 × 415 × 5 mm glass plate (**Fig. 10**).
14. 2-D lower plate: 490 × 415 × 5 mm glass plate (**Fig. 10**).
15. Electrophoresis tank (**Fig. 10**).
16. Spacer: Thickness is 1 mm.
17. Filter paper: Whatman 3MM.
18. X-ray film: XAR5, Kodak.

2.4. Cloning of Parental-Origin-Specific Methylated Spots

1. *Not*I.
2. *Eco*RV.
3. 10 × buffer H.
4. 0.1% BSA.
5. 0.1% Triton X-100.
6. 0.8% Seakem GTG(FMC), TBE.
7. Qiaquick gel extraction kit (Qiagen).
8. Vector plasmid: pBlueScript II.
9. Ligase and ligation buffer.
10. Competent cell for electrotransformation.
11. *Escherichia coli* pulser (Bio-Rad).
12. Agar plate with ampicillin.
13. Rubber policeman.

14. Qiaprep spin (Qiagen).
15. *Bam*HI.
16. L-Broth medium with ampicillin.

3. Methods

3.1. Extraction of Genomic DNA from the Tissue Sample

1. Precool 0.5 g of the sample, a folded aluminum foil, and a mortar and a pestle with liquid nitrogen.
2. Wrap the sample in folded aluminium foil and hammer it flat.
3. Quickly take out the crushed sample from the aluminum foil and transfer it to a mortar and grind it to a powder.
4. Transfer the powdered sample to a tube and add 12 mL of 10 m*M* Tris-HCl (pH 8.0), 0.5 *M* EDTA, 1% sarkosyl, and 1.2 mL of 10 mg/mL proteinase K.
5. Mix the solution and incubate it for 30 min at 65°C.
6. Cool the solution on ice. Add an equal volume of phenol/chloroform/isoamyl alcohol (25/24/1) and gently mix the two phases.
7. Separate the two phases by centrifugation at 5000*g* (3000 rpm) for 10 min at 4°C.
8. Transfer the viscous aqueous phase to a new tube using a wide-bore pipet.
9. Repeat **steps 6–8** once more.
10. Dialyze the aqueous phase at 4°C two times against 1 L of TE.
11. Transfer the dialysate to a new tube. Add 0.001 vol of 1 mg/mL RNase A and incubate for 2 h at 37°C.
12. Add 2 vol of ethanol and mix.
13. Centrifuge the solution for 10 min at 4°C.
14. Remove the supernatant.
15. Wash the pellet with 70% ethanol.
16. Centrifuge briefly and remove the supernatant.
17. Dissolve the pellet in 300 µL of TE.

3.2. Labeling Genomic DNA

The method described here is for a case using *Not*I, *Eco*RV, and *Mbo*I as restriction enzymes A, B, and C, respectively.

1. Make the following mixture for the blocking reaction (*see* **Note 1**):

Prepared genomic DNA	10 µg
10 × buffer H	5 µL
10 µM dGTP[αS]	1.7 µL
10 µM dCTP[αS]	1.7 µL
10 µM ddATP	1.7 µL
10 µM ddTTP	1.7 µL
DNA polymerase I	10 units
Distilled water to 50 µL	

2. Incubate the reaction for 30 min at 37°C.
3. Inactivate the enzyme by incubating the reaction at 65°C for 30 min.
4. Add the following reagents for digestion with *Not*I (*see* **Note 2**):

10× bufferH	5 μL
0.1% BSA	10 μL
0.1% Triton X-100	10 μL
*Not*I	100 units
Distilled water to 100 μL	

5. Transfer 5 μL to another tube and add 1 μL of control plasmid for monitoring complete digestion.
6. Incubate the sample reaction and the control reaction for 1 h at 37°C.
7. Remove 2 μL of the control and add 1 μL of 0.25% bromophenol blue, 0.25% xylene cyanol FF. 30% glycerol. Analyze by electrophoresis through 0.8% agarose gel for checking complete digestion.
8. Add the following reagents for labelling (*see* **Note 3**):

1 M Dithiothreitol	1 μL
10 μM ddATP	1.7 μL
10 μM ddTTP	1.7 μL
[α-32P] dGTP (3000Ci/mM)	5 μL
[α-32P] dCTP (3000Ci/mM)	5 μL
Sequenase V2.0	20 units

9. Incubate the reaction for 30 min at 37°C.
10. Add following reagents for stopping the reaction (*see* **Note 4**):

1 mM ddGTP	2 μL
1 mM ddCTP	2 μL

11. Add 100 units of *Eco*RV (*see* **Note 5**).
12. Transfer 5 μL to another tube and add 1 μL of control plasmid for monitoring complete digestion.
13. Incubate the sample reaction and the control reaction for 1 h at 37°C.
14. Remove 2 μL of the control and add 1 μL of 0.25% bromophenol blue, 0.25% xylene cyanol FF. 30% glycerol. Analyze it by electrophoresis through 0.8% agarose gel, checking for complete digestion.
15. Add an equal volume of phenol/chloroform/isoamyl alcohol (25/24/1) and gently mix the two phases.
16. Separate the two phases by centrifugation 12,000*g* (15,000 rpm) for 5 min at room temperature.
17. Transfer the aqueous phase to a new tube.
18. Add 0.1 vol of 3 *M* NaOAc and 2 vol of ethanol and mix.
19. Chill the solution at –70°C for 10 min.
20. Centrifuge the solution for 10 min at 4°C.
21. Remove the supernatant.
22. Wash the pellet with 70% ethanol.
23. Centrifuge briefly and remove the supernatant.
24. Dissolve the pellet in 10 μL of TE.

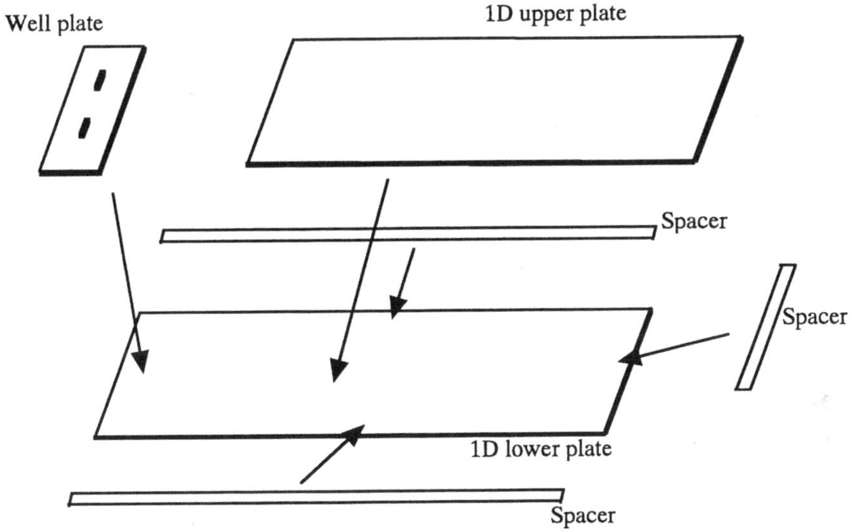

Fig. 11. Preparation of the 1-D plate.

25. Use 1 μL for the estimation of concentration.
26. Transfer 2 μg to a new tube and add 2 μL of 0.25% bromophenol blue, 0.25% xylene cyanol FF and TE to 10 μL for loading.

3.3. Two-Dimensional Electrophoresis

The method described here is for a case using *Not*I, *Eco*RV, and *Mbo*I as restriction enzymes A, B, and C.

1. Prepare the 1-D glass plates and spacer for pouring the gel (**Fig. 11**) and prewarm it at 37°C in a oven.
2. Pour 80 mL of melted 0.8% SeaKem GTG, 1 × TAM, by using a syringe with 20G needle.
3. Lay the glass plates.
4. After the gel is completely set, carefully remove the well plate and mount the gel in an electrophoresis tank filled with 1 × TAM.
5. Cut out the gel 1 cm from the top and mount the Whatman 3MM filter papers as shown (**Fig. 8**).
6. Moisten the filter papers and set up the peristaltic pump as shown (**Fig. 8**).
7. Load the sample in one well and load size markers in the other well.
8. Apply a voltage of 500 V for 10 min. Keep the filter papers wet during this period.
9. Pour 1 × TAM around the wells and attach the 1-D small plate.
10. Apply a voltage of 370 V for 18 h. Keep the filter papers wet during this period by peristaltic pump.

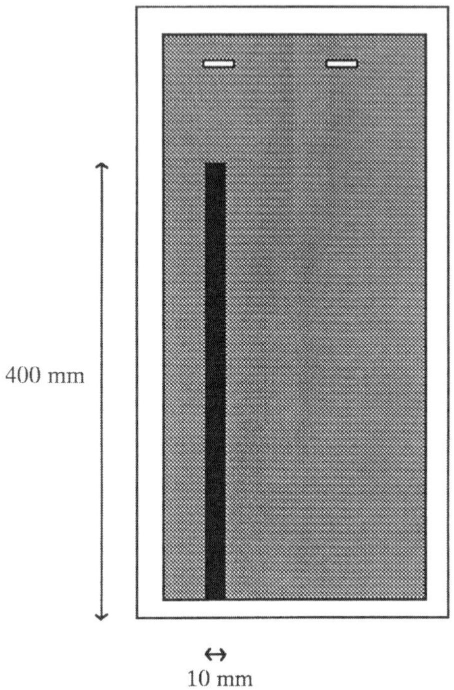

Fig. 12. The agarose gel portion used for digestion with restriction enzyme C is indicated by the black area.

11. Put the glass plates on a graph paper and remove the 1-D small plate and 1-D upper plate.
12. Cut and leave a 10 × 400 mm DNA-containing portion (**Fig. 12**) and remove the remaining gel.
13. Soak the gel on the plate for 20 min in 1 × buffer H (*see* **Note 6**).
14. Remove 1 × buffer H completely.
15. Mount the prewarmed upper plate for enzyme digestion and fix using a clip (**Fig. 9**).
16. Pour the following solution from two pores to spread over the gel (*see* **Note 7**):

1 × buffer H	1.3 μL
*Mbo*I	1500 units

17. Seal the pores with tape.
18. Incubate for 2 h at 37°C in the oven.
19. Prepare polyacrylamide gel (5–6% polyacrylamide to acrylamide/bisacrylamide, 29/1) in 2-D glass plate.
20. Remove the upper plate and soak in 1 × TBE for 10 min.
21. Remove 1 × TBE completely.
22. Transfer the gel strip to the 2-D gel plates as shown (**Fig. 10**).

23. Pour melted 0.8% Seakem GTG, 1 × buffer H, using a syringe with 24G needle, to fuse the agarose gel strip and polyacrylamide gel.
24. Set up the Whatman 3MM filter paper as shown (**Fig. 10**).
25. Apply a voltage of 900 V for 30 min. Keep the filter papers wet during this period.
26. Pour 1 × TBE around the agarose gel strip and attach the 2-D small plate (**Fig. 10**).
27. Apply a voltage of 900 V for 6 h. Keep the filter papers wet during this period using a peristaltic pump.
28. Transfer the gel to Whatman 3MM filter paper and cover with Saran Wrap.
29. Dry the gel under vacuum on a gel dryer set at 80°C.
30. Autoradiograph by exposing the gel to X-ray film (Kodak XAR5) with intensifying screen (QuantaIII) at –70°C for 1 wk.

3.4. Cloning of Parental-Origin-Specific Methylated Spots

1. Estimate the one-dimensional size of the target spot.
2. Digest 100 µg of genomic DNA with *Not*I and *Eco*RV.
3. Size-fractionate the digested DNA by 0.8% agarose gel electrophoresis.
4. Cut out the portion corresponding to the target spot and recover the DNA with Qiaquick gel extraction kit.
5. The recovered DNA is ligated to the *Not*I-*Eco*RV cleaved vector purified by agarose gel electrophoresis.
6. The ligated mix is applied to electrotransformation.
7. Spread the transformed *E. coli* on to agar plates with ampicillin.
8. Incubate at 37°C for 12 h.
9. Recover and mix all colonies from the plate with a rubber policeman.
10. Purify plasmid DNA from the colony mixture with Qiaprep spin.
11. Apply 1 µg of boundary library to the RLGS, except for the blocking and labeling steps, using 1 µg of labeled genomic DNA.
12. Cut out the 2-D gel portion corresponding to the target spot from the gel.
13. Electroelute the *Not*I-*Mbo*I fragment of the target spot.
14. Ligate the eluted DNA to the *Not*I-*Bam*HI cleaved vector purified by agarose gel electrophoresis.
15. Apply the ligated mix to electrotransformation.
16. Spread the transformed *E. coli* onto an agar plate with ampicillin.
17. Incubate at 37°C for 12 h.
18. Pick several colonies and culture each colony in L-broth medium with ampicillin.
19. Purify plasmid DNA from each culture and check each DNA.

4. Notes

1. The method described here is for a case using restriction enzyme A, which can make protruding cohesive 5′ termini. Use 2′-deoxyribonucleoside 5′-[α-thio] triphosphate, which can be incorporated into the cleavage site of restriction enzyme A by fill-in reaction. Use 2′,3′-dideoxyribonucleoside 5′-triphosphate,

which cannot be incorporated into the cleavage site of restriction enzyme A by a fill-in reaction.
2. Use restriction enzyme A and an appropriate buffer for it.
3. Use 2′,3′-dideoxyribonucleoside 5′-triphosphate, which cannot be incorporated into the cleavage site of restriction enzyme A by fill in reaction. Use labeled 2′-deoxyribonucleoside 5′- triphosphate, which can be incorporated into the cleavage site of restriction enzyme A by a fill-in reaction.
4. Use 2′,3′-dideoxyribonucleoside 5′-triphosphate, which can be incorporated into the cleavage site of restriction enzyme A by a fill-in reaction.
5. Use restriction enzyme B and the appropriate buffer for it.
6. Use the buffer appropriate for restriction enzyme C.
7. Use restriction enzyme C and the appropriate buffer for it.

References

1. Solter, D. (1988) Differential imprinting and expression of maternal and paternal genomes. *Annu. Rev. Genet.* **22,** 127–146.
2. Barlow, D. P. (1994) Imprinting: a gamete's point of view. *Trends Genet.* **10,** 194–199.
3. Ohlsson, R., Barlow, D. P., and Surani, A. (1994) Impression of imprints. *Trends Genet.* **10,** 415–417.
4. Razin, A. and Cedar, H. (1994) DNA methylation and genomic imprinting. *Cell* **77,** 473–476.
5. Efstratiadis, A. (1994) Parental imprinting of autosomal mammalian genes. *Curr. Opin. Genet. Dev.* **4,** 265–280.
6. Nicholls, R. D. (1994) New insights reveal complex mechanisms involved in genomic imprinting. *Am. J. Hum. Genet.* **54,** 733–740.
7. Hatada, I., Hayashizaki, Y., Hirotsune, S., Komatsubara, H., and Mukai, T. (1991) A genomic scanning method for higher organisms using restriction sites as landmarks. *Proc. Natl. Acad. Sci. USA* **88,** 9523–9527.
8. Hatada, I., Sugama, T., and Mukai, T. (1993) A new imprinted gene cloned by a methylation-sensitive genome scanning method. *Nucleic Acids Res.* **21,** 5577–5582.

7

Subtraction-Hybridization Method for the Identification of Imprinted Genes

Fumitoshi Ishino, Yoshimi Kuroiwa, Naoki Miyoshi, Shin Kobayashi, Takashi Kohda, and Tomoko Kaneko-Ishino

1. Introduction

Imprinted genes show monoallelic expression from either the paternal or maternal genome *(1,2)*, and their regulated expression is usually associated with the existence of parentally differentially methylated regions on genomic DNAs *(3,4)*. Because of this, essentially two different approaches, using either cDNA or genomic DNA as starting material *(5)* have been developed for systematic isolation of imprinted genes. In this chapter, we describe a subtraction-hybridization method *(6–8)* as an example of the former approach. Both parthenogenetic embryos and androgenetic embryos *(9,10)* are the most suitable biological materials for the subtraction or detection of imprinted genes. However, it is difficult to obtain a large amount of such special materials because only a small number of these embryos develop to the d 10 stage *(9,10)*. Thus, polymerase chain reaction (PCR)-based techniques, such as the differential display *(11–13)* and subtraction-hybridization methods, are necessary to accomplish this experiment. The subtraction-hybridization method has been successfully applied for isolation of both paternally expressed genes (*Pegs*) *(6,14,15)* and maternally expressed genes (*Megs*) *(7)*, and it allows cDNA libraries to be made from a very small amount of biological material. We are convinced that this method can be applied in many fields of biological science.

From: *Methods in Molecular Biology, vol. 181: Genomic Imprinting: Methods and Protocols*
Edited by: A. Ward © Humana Press Inc., Totowa, NJ

1.1. PCR Methods for Making Representative cDNA Libraries

Invention and improvement of the PCR method has enabled us to detect single copies of genes using genomic DNA and mRNA from a single cell. However, to make representative cDNA libraries using the PCR method, special PCR conditions needed to be developed that simultaneously satisfied conditions of both long and accurate PCR (LA PCR) *(16)* and unbiased amplification of different sizes of cDNAs from <1 kb to 10 kb. We chose *Pfu* DNA polymerase in our experiments because it has the lowest mutation rate among known thermostable DNA polymerases *(17)*. Although it has been reported that the *Pfu* enzyme is not suitable for amplification of DNA fragments longer than 2 kb, we have improved the PCR conditions to resolve this problem. Using a model amplification system in which λ *Bst*I fragments (117 bp–8453 bp) were attached to PCR linkers (described below), we confirmed that DNA fragments from <1 kb to 7–8 kb were amplified quantitatively for at least the first 10–15 PCR cycles. Although 1–5 μg of cDNA have been used to make conventional cDNA libraries, it is now possible to make PCR cDNA libraries and carry out subtraction experiments from as little as 5 ng of cDNA.

Because a very small amount of mRNA was used for cDNA synthesis, it was sometimes impossible to quantify the precise amount of cDNA with UV spectrometry, thus, an excess amount (about 1000-fold) of lone-linker DNA *(18,19)* was used for linker ligation in order to accomplish the ligation reaction of cDNAs with lone-linker DNAs (**Fig. 1**). The lone-linker DNAs are designed to form only linker dimers by themselves, and self-ligation reactions could not then proceed because of the resulting 3′ protruding structures. Therefore, the resultant dimers (40-mer) and the unreacted monomer linkers (20-mer) could be removed by gel filtration columns. It is important to remove the linker dimers completely because they are amplified very easily and disturb the efficient amplification of cDNAs in the next step. 5′-Biotinylated 20-mer DNA was used for both linker DNAs and PCR primers for complete modification of one of the cDNAs. We previously checked that the amplification efficiencies between primer sets attached to two cDNA samples used for subtraction experiments were equivalent under the PCR conditions used and described below. Therefore, it is highly recommended that the amplification efficiency be checked again when oligonucleotide linkers and PCR primers other than those in **Fig. 1** are used.

1.2. PCR Conditions for Making cDNA Libraries and Subtraction Experiments

Both selection of linkers (primers) and the PCR conditions are important for nonbiased amplification of total cDNAs because they are composed of cDNAs of different lengths. We chose the *Pfu* enzyme for its high thermostability and

Linker-1 (for cDNA from fertilized embryos)

```
 5'                      3'
 TCGACTCGAGTATACTTACA   20mer
 TGAGCTCATATCAATGT-P    17mer
 3'                 5'
```

Linker-2 (for cDNA from androgenetic embryos)

```
     5'                      3'
 (B)-GATTACTCGAGACTAATATC   20mer
     ATGAGCTCTGATTATAG-P    17mer
     3'                5'
```

(B) : biotin residue

Fig. 1. Structure of linker DNAs. Two sets of linker DNAs and PCR primers are prepared for amplification of two different cDNAs, 1 and 2. 5′-Biotinylated 20-mer oligonucleotide is used for making the cDNA-2.

low mutation rate. Because it takes longer for the extension of long cDNAs, an extention time of 10 min was decided on for each amplification cycle, resulting in a longer PCR time than for usual PCR protocols. Therefore, thermostability is one very important factor in enzyme selection. Because one of our purposes was to make cDNA libraries from very small amounts of biological materials, high fidelity of DNA synthesis was also required for amplification. Recently, several new high-performance thermostable DNA polymerases have become available; thus, some other enzymes may be suitable for this purpose. Usually, a combination of two different types of thermostable DNA polymerases (a high-performance type, such as *Taq* DNA polymerase, and a high-fidelity type, such as *Pfu*) is used for long and accurate PCR (LA PCR) ***(16)***. These are also a good choice for amplification in subtraction experiments. However, it should be noted that the mutation rate is reported to be similar to that of *Taq* DNA polymerase rather than that of *Pfu* DNA polymerase.

The amplification efficiency varies with the PCR machines used. Total performance of the machines, such as heating and cooling ability, may be impotant. In our experience, the PCR GeneAmp PCR System 9600 (PE Applied Biosystems) demonstrated the best performance in these experiments. The number of PCR cycles is another important factor for amplification. Because smaller fragments tend to be predominantly amplified, excess PCR cycles should be avoided to achieve unbiased amplification. Usually, 10–15 cycles are recommended for one amplification experiment, depending on the amount of cDNA used. Amplified cDNA can be checked by Southern hybridization using several gene probes. If smaller bands appear in the first amplification experiment, the number of PCR cycles should be reduced. We have carried out the second amplification using 1/30 aliquots of the first amplification sample, in 30 reaction tubes, to obtain the total cDNA used in the subtraction study (1–3 μg).

1.3. Setting the Subtraction Conditions

It should be noted that the mixing ratio of the two cDNAs is very important. We have mixed cDNAs from androgenetic (or parthenogenetic) embryos (cDNA-2 in **Fig. 2**) with those from normal fertilized embryos (cDNA-1 in **Fig. 2**) in a ratio of 100 to 1. When imprinted genes are completely expressed monoallelially, this ratio works very well. However, in some imprinted genes that show low levels (about one-hundredth) of expression from the "silent" allele, such as *H19*, efficient concentration was not achieved (*see* **Fig. 3**). Therefore, the expression levels of the genes to be examined in the two biological materials used in the subtraction experiment must be carefully considered when setting up the subtraction conditions.We also recommend the use of about 1 μg of total cDNA in each subtraction experiment in each 10-μL volume of reaction mixture, because the DNA concentration and renaturation time (Cot value) are very important factors for efficient hybridization.

2. Materials

2.1. Linker Ligation

1. Chemically synthesized 5′-phosphorylated 17-mer oligonucleotides-1 and -2 (*see* **Fig. 1** and **Note 1**).
2. Chemically synthesized 20-mer oligonucleotide complementary to the 17-mer oligonucleotide-1 (the same as primer-1 in **Subheading 2.2.**).
3. Chemically synthesized 5′-biotinylated 20-mer oligonucleotide complementary to the 17-mer oligonucleotide-2 (the same as primer-2 in **Subheading 2.2.**).
4. TE buffer: 10 m*M* Tris-HCl, pH 8.0, 1 m*M* ethylenediaminetetraacetic acid (EDTA).
5. Water baths (set at 75 and 4°C).
6. T4 DNA ligase.
7. CHROMA SPIN-400 (Clontech).
8. 20x SSC: 3.0 *M* NaCl, 3.0 *M* sodium citrate, ph 2.0.

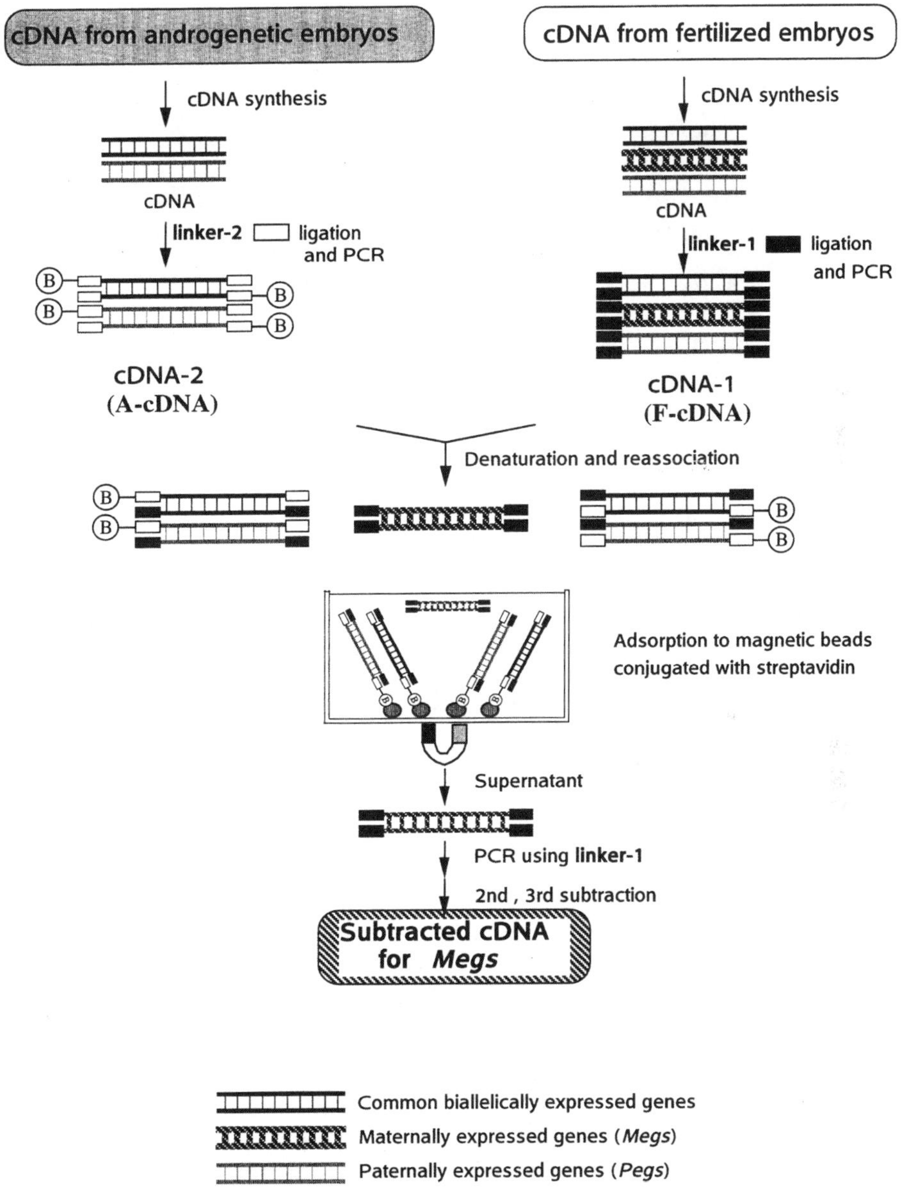

Fig. 2. Subtraction between cDNAs from androgenetic and normal fertilized embryos. Because *Meg* genes are not expressed in androgenetic embryos, they are concentrated by a series of gene subtractions of the androgenetic embryos from normal fertilized embryos. The subtracted cDNA can be amplified with PCR using primer-1 and successive subtractions are possible until an efficient concentration is achieved.

2.2. Quantitative Amplification of cDNAs by PCR

1. Native *Pfu* DNA polymerase (Stratagene).
2. 10 × PCR buffer (supplied with *Pfu* emzyme): 200 m*M* Tris-HCl, pH 8.3, 100 m*M* KCl, 60 m*M* $(NH_4)_2SO_4$, 20 m*M* $MgCl_2$, 1% Triton X-100.
3. 50 m*M* $MgCl_2$.
4. dNTP mixture (1 m*M* each) (Pharmacia).
5. PCR Primers-1 and -2 (the same as 20 mer oligonucleotides-1 and -2 in **Subheading 2.3.**).
6. Double-distilled water (DDW).

2.3. Subtraction Between Two Different cDNAs

1. HE buffer: 10 m*M* HEPES, pH7.5, 1 m*M* EDTA.
2. 2 × hybridization buffer: 1.5 *M* NaCl, 50 m*M* HEPES (pH7.6), 10 m*M* EDTA, 0.2% sodium dodecyl sulfate (SDS).
3. Mineral oil.
4. Heating buffer: 75 m*M* NaCl, 2.5 m*M* HEPES (pH 7.5), 0.5 m*M* EDTA, 0.01% SDS.
5. DynaBeads M-280 streptavidin (Dynal).
6. Magnetic stand MPC-E (Dynal).
7. CHROMA SPIN TE-400 (Clontech).

2.4. Isolation of Imprinted Genes by Differential Hybridization

1. Hybond-N+ membrane (Amersham).
2. Denaturation solution: 1.5 *M* NaCl, 0.5 *M* NaOH.
3. Neutralization buffer: 1.5 *M* NaCl, 0.5 *M* Tris-HCl, pH 7.5, 1 m*M* EDTA.
4. UV crosslinker.
5. Random priming DNA labeling kit (TaKaRa).
6. [α-32P]dCTP (1.85 MBq).
7. Church's solution: 0.5 *M* phosphate buffer, pH 7.0, 7 % SDS, 1 m*M* EDTA.
8. Church's wash solution: 45 m*M* phosphate buffer, pH 7.0, 1% SDS.
9. 0.1 × SSC, 0.1% SDS.
10. BioImaging analyzer BAS 2000 (Fuji film).

3. Methods

There are several mRNA isolation kits and cDNA synthesis kits available for small-scale experiments, such as the FastTrack mRNA isolation kit (Invitrogen) and the λZapII cDNA synthesis kit (Stratagene). We have carried out these steps according to the manufacturer's protocol except for using a reduced amount of oligo dT cellulose resin sufficient for the absorption of mRNAs from initial materials (one-half to one-third of the recommended amount), the addtion of 2 µg of glycogen (Boeringer-Manheim) in the ethanol precipitation steps, and the usual oligo dT primer (Boeringer-Manheim) for the first RT reaction, instead of the manufacturer's primers, which added a different sequence incorporating a restriction site on its 5′-end. Sometimes we cannot quantify the amount of the isolated mRNA or synthesized cDNA by UV spectrophotometry because of the limited amount of material. Therefore, it is necessary to check that the final cDNAs can be recovered after amplification of cDNAs by the PCR method described below.

3.1 Making Lone Linkers and Linker Ligation

1. Dissolve each oligonuleotide in TE buffer (1.2 nmol each of 5′-phosphorylated 17-mers, oligonucleotide-1 and -2, and complementary 20-mers, oligonuculetide-1 and -2).
2. Mix equal molar amounts of 5′-phosphorylated 17-mer oligonucleotide-1 (or -2) and complementary 20-mer oligonuculetide-1 (or -2) and make up to 60 µL with TE buffer.
3. Incubate at 75°C for 5 min and cool to room temperature (RT) for 2–3 h to allow formation of lone linker-1 (or -2).
4. Add 100 pmol of lone linker-1 (or -2) (4.5 µL) and 1 µL of T4 DNA ligase to each synthesized cDNA-1 (or -2) (*see* **Note 2**).
5. Incubate at 4°C overnight and stop the reaction by adding 1 m*M* (final concentration) EDTA.
6. Remove the unreacted lone linkers and formed dimers by gel filtration using CHROMA SPIN-400 (*see* **Note 3**).

3.2. Quantitative Amplification of cDNAs by PCR

One-fifth to one-tenth of the linker-bound cDNAs are used for PCR amplification for the first time to check the number of PCR cycles. The length of the amplified cDNAs should be checked by Southern hybridization of the PCR product using probes of known genes (such as β-*actin*).

1. Prepare the reaction mixture:

10 × reaction buffer	10 µL
50 mM $MgCl_2$	3 µL (final concentration 3.5 mM; *see* **Note 4**).
dNTP mixture (1 mM each)	12 µL (final concentration 120 mM each)
primer-1 (or -2)	80 pmol
linker bound cDNA -1 (or -2)	1/5–1/10 vol from **Subheading 3.1.**
DDW to make up the final volume of 100 µL	

2. The reaction mixtures are incubated at 75°C for 1 min first, and then 2.5 U of native *Pfu* DNA polymerase is quickly added and kept at 75°C for another 1 min (*see* **Note 5**).
3. After the denaturation step at 96°C for 1 min, perform 10 cycles of PCR: 65°C for 1 min, 72°C for 10 min, and 96°C for 5 s for the first nine cycles, and 15 min is recommended for the last elongation step in the tenth cycle (*see* **Note 6**).

4. Each one-thirtieth aliquot of the mixture is used for another 10 amplification cycles as described **steps 1–3** (total 30 tubes) for cDNAs.
5. Collect the 30 reaction mixtures in one Eppendorf tube.
6. Add proteinase K, SDS, and EDTA (final concentration: 50 μg/mL, 0.5%, and 5 m*M*, respectively) and incubate at 45°C for 3 h.
7. Add the same volume of phenol-chloroform and mix vigorously.
8. Recover the water phase and add twice the volume of EtOH.
9. After incubation at –20°C for 10 min, collect the cDNA pellets by centrifugation (15,000*g* or 15,000 rpm).
10. Resolve the pellets with 20 μL of TE.
11. Apply to a gel filtration column (CHROMA SPIN-TE400) and remove the primers and dNTPs.
12. Quantify the cDNAs by UV spectrophotometry (*see* **Note 7**).

3.3. Subtraction Between Two Different cDNAs (Fig. *2*)

1. Mix 1 μg of biotinylated cDNA-2 (cDNA from androgenetic embryos) and 10 ng of cDNA-1 (cDNA from normal fertilized embryos) and add HE buffer to make up to 5 μL. Then add 5 μL of 2 × hybridization buffer (final concentration: 0.75 *M* NaCl, 25 m*M* HEPES, 5 m*M* EDTA, 0.1% SDS).
2. Add 10 μL of mineral oil and heat denature at 100°C for 10 min. Then renature at 68°C for 24 h (*see* **Note 8**).
3. Cool immediately on ice and recover the water phase into a new tube.
4. Add 27.5 μL of HE buffer, 2 μg of glycogen, and 80 μL of EtOH and keep at –80°C for 10 min.
5. Centrifuge at 15,000*g* (15,000 rpm) at 4°C for 1h.
6. Resuspend the pellet with 20 μL of heating buffer and incubate at 68°C for 20 min.
7. Replace the solution into a 1.5-mL Eppendorf tube and add 4 mg of DynaBeads M-280 streptavidin.
8. Mix gently while tapping for 1 h at room temperature.
9. Stand the tube on a magnet stand (MPC-E) for 10 min and recover the supernatant.
10. Amplify half of the supernatant by PCR using primer-1 under the conditions as described under **Subheading 3.2.** (first subtracted cDNA).
11. Mix 10–20 ng of the first subtracted cDNA with 1 μg of biotinylated cDNA-2 and repeat **steps 1–9** three times, until the control genes are completely removed.
12. Purify the third subtracted cDNA by gel filtration using CHROMA SPIN TE-400.

3.4. Screening of Imprinted Genes Using Subtracted DNAs as Selection Probes (Fig. 3)

It is recommended that you prepare two kinds of control genes, one expressed equally and removed by subtraction, such as β-actin, and the other specific

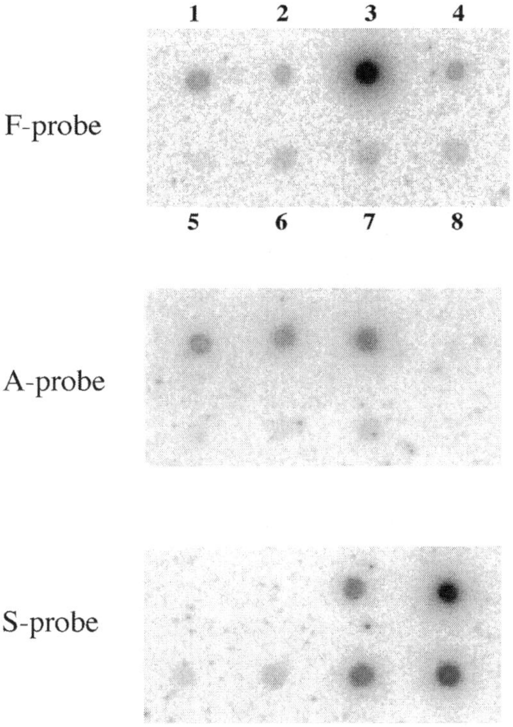

Fig. 3. Dot hybridization of *Meg* genes. Plasmid DNAs (about 50 ng) isolated from candidate λ plaques were spotted on three filter membranes and hybridized with cDNA probes from the androgenetic embryos (A-probe), normal fertilized embryos (F-probe), and the third subtracted cDNA (S-probe). Two control genes, β-actin (no. 1) and *Peg1/Mest* genes (no. 2), were completely removed in the subtracted probe. Compared with *Meg1/Grb10* (no. 4) and *Meg3* (no. 8), concentration of *H19* (no. 3) and $p57^{Kip2}$ (no. 6) was not so efficient because of leaky expressions from the paternal alleles. Although no. 7 was greatly concentrated in the subtracted probe, it was expressed at the same levels in the two starting cDNAs and proved not to be a real imprinted gene. 1, *β-actin*; 2, *Peg1/Mest*; 3, *H19 (Meg2)*; 4, *Meg1/Grb10*; 5, unknown gene (not imprinted), 6, $p57^{Kip2}$ (*Meg5*); 7, unknown gene (not imprinted); 8, *Meg3*.

to one of the cDNAs. Differential hybridization among two starting cDNAs and the subtracted cDNA is the most efficient way to isolated imprinted genes (*Pegs* or *Megs*), according to our exprerience. Otherwise about 80% of clones that gave positive signals proved not to be genuine imprinted genes, but those expressed at the same level in the two starting cDNAs. Almost all of the known imprinted genes have been isolated in the remaining 20% of clones. When expression levels of the genes in question are very low, we recommend the use of membrane filters on which plasmid DNAs isolated from candidate plaques are dotted (**Fig. 3**).

1. Prepare each of three membrane filters from one plate of cDNA library screened according to the molecular cloning.
2. Soak the membrane filters in denaturation solution for 5 min and then in neutralization buffer for 7 min.
3. Air-dry and UV crosslink for fixation of DNAs on the membranes.
4. Label each of the three cDNA with ^{32}P CTP using a DNA labeling kit.
5. Prehybridize the membranes with 20 mL of Church's solution at 69°C for 5 min.
6. Discard 10 mL of the solution, add each labeled DNA probe, and hybridize at 69°C overnight.
7. Wash the membranes with 30 mL of Church's wash solution at 69°C for 5 min, with 50 mL of Church's wash solution at 69°C for 30 min, twice, and then with 0.1 × SSC containing 0.1% SDS at 69°C for 30 min, twice.
8. Detect radioactivity with BioImaging analyzer.
9. Compare signals among the three filters and select the clones that give positive signals on both cDNA-1 and subtracted cDNA and negative on cDNA-2.

4. Notes

1. 5′-Phosphorylated 17-mer oligonucleotieds can be purchased from several companies. However, the 5′-phosphorylation reaction of 17-mer oligonucleotides can be carried out in the laboratory by T4 DNA polynucleotide kinase reaction.
2. Total volume should not exceed 20 μL.
3. It is necessary to remove monomers and dimers completely for the next PCR steps. When gel filtration columns (such as CHROMA SPIN-TE100) are used in this step, for removal of smaller-sized DNAs, a small amount of the monomers and dimers will remain in the elutions and will disturb the next PCR reaction.
4. 20 m*M* of $MgCl_2$ is contained in the 10 × reaction buffer.
5. The hot-start method is recommeded in this step.
6. Because a *XhoI* site is integrated in the linker-primer, the amplified cDNAs are used for making cDNA libraries after *XhoI* digestion and ligation with adequate λ phage vectors, such as λZAPII (Stratagene).
7. 2–3 μg of amplified cDNAs are usually obtained. This is enough for the next subtraction experiment.
8. The denaturation time should not exceed for 10 min because of degradation of cDNAs.

References

1. Bartolomei, M. S., Zemel, S., and Tilghman, S. M. (1991) Parental imprinting of the mouse H19 gene. *Nature* **351,** 153–155.

2. DeChiara, T. M., Robertson, E. J., and Efstratiadis, A. (1991) Parental imprinting of the mouse insulin-like growth factor II gene. *Cell* **64,** 849–859.
3. Ferguson-Smith, A. C., Sasaki, H., Cattanach, B. M., and Surani, M. A. (1993) Parental-origin-specific epigenetic modification of the mouse H19 gene. *Nature* **362,** 751–755.
4. Brandeis, M., Kafri, T., Ariel, M., Chaillet, J. R., McCarrey, J., Razin, A., and Cedar, H. (1993) The ontogeny of allele-specific methylation associated with imprinted genes in the mouse. *EMBO J.* **12,** 3669–3677.
5. Hayashizaki, Y., Shibata, H., Hirotsune, S., Sugino, H., Okazaki, Y., Sasaki, N., et al. (1994) Identification of an imprinted U2af binding protein related sequence on mouse chromosome 11 using the RLGS method. *Nature Genet.* **6,** 33–40.
6. Kaneko-Ishino, T., Kuroiwa, Y., Miyoshi, N., Kohda, T., Suzuki, R., Yokoyama, M., Viville, S., Barton, S. C., Ishino, F., and Surani, M. A. (1995) *Peg1/Mest* imprinted gene on chromosome 6 identified by cDNA subtraction hybridization. *Nature Genet.* **11,** 52–59.
7. Miyoshi, N., Kuroiwa, Y., Kohda, T., Shitara, H., Yonekawa, H., Kawabe, T., Hasegawa, H., Barton, S. C., Surani, M. A., Kaneko-Ishino, T., and Ishino, F. (1998) Identification of the *Meg1/Grb10* imprinted gene on mouse proximal chromosome 11, a candidate for the Silver-Russell syndrome gene. *Proc. Natl. Acad .Sci. USA* **95,** 1102–1107.
8. Kaneko-Ishino, T., Kuroiwa, Y., Kohda, T., Surani, M. A., and Ishino, F. (1997). Systematic approaches for the identification of imprinted genes, in *Genomic Imprinting: Frontiers in Molecular Biology* (Reik, W. and Surani, M. A., eds.), IRL, Oxford, New York, Tokyo, pp. 146–164.
9. Surani, M. A., Barton, S. C., and Norris, M. L. (1984) Development of reconstituted mouse eggs suggests imprinting of the genome during gametogenesis. *Nature* **308,** 548–550.
10. McGrath, J. and Solter, D. (1984) Completion of mouse embryogenesis requires both the maternal and paternal genomes. *Cell* **37,** 179–183.
11. Liang, P. and Pardee, A. B. (1992) Differential display of eukaryotic messenger RNA by means of the polymerase chain reaction. *Science* **257,** 967–971.
12. Liang, P., Zhu, W., Zhang, X., Guo, Z., O'Connell, R. P., Averboukh, L., Wang, F., and Pardee, A. B. (1994) Differential display using one-base anchored oligo-dT primers. *Nucleic Acids Res.* **22,** 5763–5764.
13. Ito, T., Kito, K., Adati, N., Mitsui, Y., Hagiwara, H., and Sakaki, Y. (1994) Fluorescent differential display: arbitrarily primed RT-PCR fingerprinting on an automated DNA sequencer. *FEBS Lett.* **351,** 231–236.
14. Kuroiwa, Y., Kaneko-Ishino, T., Kagitani, F., Kohda, T., Li, L. L., Tada, M., Suzuki, R., Yokoyama, M., Shiroishi, T., Wakana, S., Barton, S. C., Ishino, F., and Surani, M. A. (1996) *Peg3* imprinted gene on proximal chromosome 7 encodes for a zinc finger protein. *Nature Genet.* **12,** 186–190.
15. Kagitani, F., Kuroiwa, Y., Wakana, S., Shiroishi, T., Miyoshi, N., Kobayashi, S., Nishida, M., Kohda, T., Kaneko-Ishino, T., and Ishino, F. (1997) *Peg5/Neuronatin*

is an imprinted gene located on sub-distal chromosome 2 in the mouse. *Nucleic Acids Res.* **25,** 3428–3432.
16. Barnes, W. M. (1994) PCR amplification of up to 35-kb DNA with high fidelity and high yield from lambda bacteriophage templates. *Proc. Natl. Acad. Sci. USA* **91,** 2216–2220.
17. Mathur, E., Sheomaker, D., Scott, B., Rombouts, J., Bergseid, M., and Nielson, K. (1992) *Pfu* DNA polymerase update. *Strategies* **5,** 11.
18. Ko, M. S., Ko, S. B., Takahashi, N., Nishiguchi, K., and Abe, K. (1990) Unbiased amplification of a highly complex mixture of DNA fragments by "lone linker"-tagged PCR. *Nucleic Acids Res.* **18,** 4293–4294.
19. Akowithz, A. and Manuelidis, L. (1989) A novel cDNA/PCR strategy for efficient cloning of small amounts of undefined RNA. *Gene* **81,** 295–306.

8

Identification of Imprinted Loci by Methylation

Use of Methylation-Sensitive Representational Difference Analysis (Me-RDA)

Rachel J. Smith and Gavin Kelsey

1. Introduction

The technique of representational difference analysis (RDA) was originally described by Lisitsyn et al. (1993) as a means of identifying differences between complex genomes by the use of subtractive hybridisation *(1)*. This protocol for methylation-sensitive RDA (Me-RDA) describes adaptations to the original technique that, by the use of oligonucleotides for *Hpa*II or *Hin*6I sites, allow the identification of sequences whose methylation differs between two sources of DNA. Differences in the methylation of the maternal and paternal alleles of imprinted genes have been described for most imprinted genes studied to date *(2)*. The importance of methylation in the control of imprinted gene expression has been clearly demonstrated by the perturbation of allelic expression in mice deficient in *Dnmt1*, the major mammalian DNA methyltransferase *(3)*, and methylation has been suggested to fulfill the necessary requirements of an "imprint" *(4)*.

The use of allelic methylation differences to identify novel imprinted genes represents an important complement to expression-based screens, as methylation differences are not limited by tissue or stage specificity of expression *(5)*. Practical use of such methylation differences has previously led to the identification of the imprinted *U2af1-rs1* and *Ras-grf1* genes in restriction landmark genome scanning (RLGS) screens *(6,7; see* Chapter 6*)*. RLGS may be limited by its use of methylation at sites for rare cutter restriction enzymes, and its dependence on restriction fragment length polymorphisms (RFLPs) to assign parental origin or large quantities of uniparental DNA. In contrast,

From: *Methods in Molecular Biology, vol. 181: Genomic Imprinting: Methods and Protocols*
Edited by: A. Ward © Humana Press Inc., Totowa, NJ

Me-RDA detects parent-of-origin methylation differences in a fundamentally different way that may permit a more comprehensive screen for imprinted methylation. It is based on methylation differences at sites for frequent cutting restriction enzymes, such as *Hpa*II or *Hin*6I, with the expectation that each imprinted locus will be represented as multiple *Hpa*II and/or *Hin*6I fragments. Parental origin is assigned by the use of DNA sources in which the entire parental genomes (uniparental embryos) or individual chromosomes (uniparental disomies) are separated. Me-RDA has been used to identify novel imprinted loci on mouse distal chromosome 2, and led to the isolation of the imprinted transcripts *Nesp* and *Gnasxl* *(8)*.

1.1. Theory of RDA

RDA is a technique that has been developed to permit the identification of differences between two populations of DNA by the use of subtractive hybridization. In common with all subtractive hybridization methodologies, RDA uses a large excess of one DNA population (the driver), which is hybridized against a second DNA population (the tester), to enrich sequences that are unique to the tester (the target).

Subtractive hybridization is not very effective when applied to complex genomes, such as those of human and mouse, due to the low likelihood of reassociation of single-copy sequences. RDA overcomes this by producing reduced-complexity representations of the genome in which reassociation kinetics are more favourable. The tester and driver DNAs are digested with the same restriction endonuclease, adaptor oligonucleotides are added, and whole-genome amplification by the polymerase chain reaction (PCR) is performed. Since PCR under standard conditions favors the amplification of small fragments, the amplification products, referred to as "amplicons," will comprise only a small proportion of the restriction fragments, generally those smaller than 1.5 kb. It is these reduced complexity representations that are used in the subtractive hybridization. Before subtraction, the adaptors on the tester amplicons are removed and exchanged for a new set. Subtraction is then performed by denaturation and reassociation of the amplicons. A large excess of driver amplicon is used to force sequences common to tester and driver into tester–driver heteroduplexes. Kinetic enrichment of targets is achieved by PCR amplification of the subtraction using a primer corresponding to the tester-specific adaptors. Only tester–tester annealed fragments will have primer sites at both ends and are therefore the only molecules that will be amplified exponentially. Repeated rounds of subtraction may be carried out until the major components of the difference products (DPs) are tester-specific fragments, i.e., a high abundance of targets. The technique of RDA is illustrated in **Fig. 1**, and an example of a Me-RDA experiment is given in **Fig. 2**.

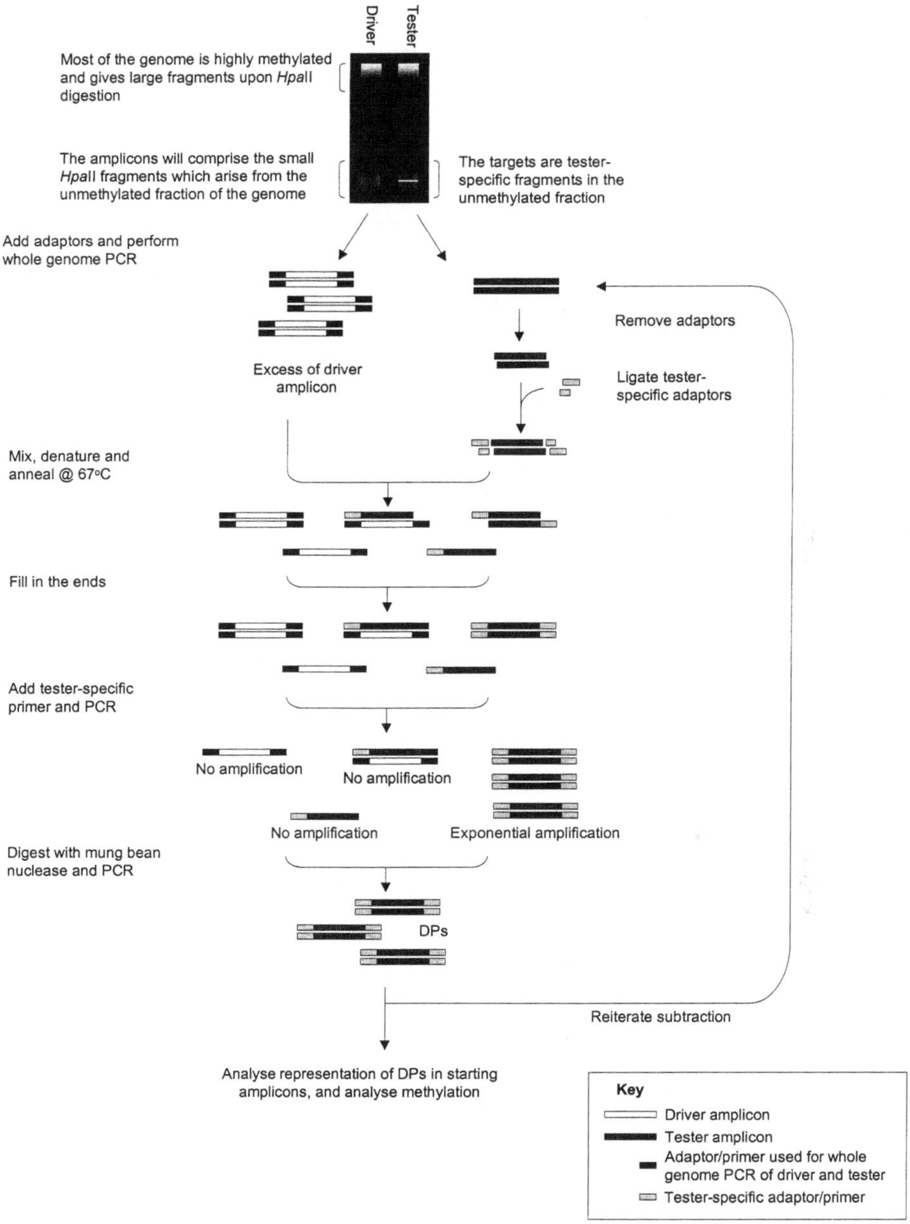

Fig. 1. Schematic representation of RDA. Amplicons are produced from tester and driver DNAs by restriction digestion, ligation of adaptor molecules, and whole-genome PCR. The adaptors on the tester amplicon are removed and new adaptors ligated. Subtractive hybridization is carried out using a large excess of driver amplicon and the tester–tester homoduplexes are recovered by a PCR approach, using PCR primers corresponding to the tester-specific adaptors. This technique is applied to the identification of methylation differences by the use of a methylation-sensitive restriction enzyme in the initial restriction digest, which is represented schematically at the top of the figure. (Adapted from **ref. *1***.)

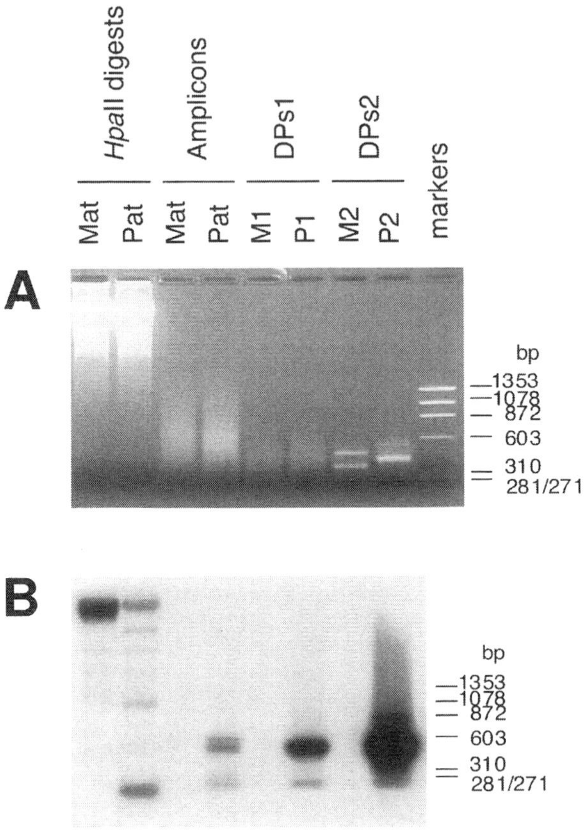

Fig. 2. Evaluation of Me-RDA involving DNAs from mouse embryos with paternal or maternal UPD for distal chromosome 2. (**A**) The agarose gel shows, from the left: *Hpa*II digests of DNAs from embryos with maternal disomy (Mat) and paternal disomy (Pat); the amplicons produced by whole-genome PCR from the *Hpa*II digests; and the difference products after a first round (DPs1) and second round (DPs2) of subtraction. DPs1 identified M1 refer to a subtraction in which the Mat amplicon was used as tester and the Pat amplicon used as driver, in which case the DPs are enriched in *Hpa*II fragments specific to the Mat amplicon and therefore unmethylated in the Mat UPD DNA. DPs1 identified P1 are the products of the reciprocal subtraction. DPs1 resubtracted against the respective driver amplicons result in the second round DPs: M2 and P2. Because of the high level of methylation of the mouse genome, much of the DNA in the *Hpa*II digests remains high molecular weight. The whole-genome PCR generates *Hpa*II amplicons that represent small (<1500 bp) unmethylated *Hpa*II fragments. DPs1 remain a complex population of fragments, but DPs2 appear as a simple pattern of bands corresponding to enrichment of the small number of *Hpa*II fragment differences that exist between the Mat and Pat DNAs (i.e., the targets). (**B**) The gel in **A** hybridized to a probe for the imprinted gene *neuronatin*, which is located on distal chromosome 2 and is methylated on the Mat chromosome. The hybridization illustrates the difference in methylation in the Mat and Pat DNAs, and how this translates to representation of *neuronatin* gene fragments specifically in the Pat amplicon. *Neuronatin* fragments become highly enriched in the P2 DPs.

In Me-RDA a methylation-sensitive enzyme, such as *Hpa*II, is used in the initial digest of the starting DNAs. Methylated sites will not be digested and the resulting fragments will be too large to amplify in the whole-genome PCR, but where the DNA is unmethylated the resulting small restriction fragments will be included. The targets of the Me-RDA comprise fragments that are specific to the tester, and are therefore sequences that are specifically unmethylated in the tester DNA. The amplicons produced will contain only a small fraction of the genome, but because of the high level of methylation of genomic DNA, are likely to contain a high proportion of the unmethylated fraction of the genome, in particular being enriched in CpG island sequences.

1.2. Choice of Starting Material

In order to identify differences in allelic methylation by Me-RDA, it is necessary to have designated tester and driver DNAs in which the parental genomes are differentially represented. Because Me-RDA is PCR based, it is feasible to apply it to limiting amounts of material, and the technique could be completed with less than 1 μg of DNA. For studying subchromosomal regions of the genome at which imprinting effects map, uniparental disomies (UPDs) provide an ideal source of material for Me-RDA (for further information on UPDs, refer to Chapter 3). The driver DNA used will be uniparental and, upon restriction digestion, will lack the small, unmethylated fragments that comprise the target in the tester DNA. If both maternal and paternal UPDs are available for the same subchromosomal region, these may each be used as the driver in reciprocal subtractions, with the other UPD used as the tester. This will permit the identification of regions both of paternal- and maternal-specific methylation. To examine for allele-specific methylation over the whole genome, material that is entirely uniparental in origin is required. In mice, this may be achieved using parthenogenetic or androgenetic uniparental embryos as the driver DNA and either the reciprocal uniparental type, or fertilized embryos, to provide the tester DNA. For further information on parthenogenetic and androgenetic mouse embryos, refer to Chapter 1.

It is important to ensure that, other than the targets of the RDA, the sources of DNA are as identical as possible. RFLPs between the DNAs could be targets for enrichment in the Me-RDA, and these should be eliminated as far as possible by choice of strains of mice used and pooling of material. It is important to ensure that methylation differences between the DNA samples used are limited to differences in allelic methylation. This should be achieved by matching the tissue types and developmental stages used in order to avoid tissue-specific methylation differences.

1.3. Choice of Restriction Endonuclease

The recommended restriction enzymes for Me-RDA have 4-bp recognition sites and are sensitive to methylation of the CpG dinucleotide; for example, we use *Hpa*II (CCGG) and *Hin*6I (GCGC). We use *Hin*6I in place of the more commonly encountered isoschizomers *Hha*I or *Cfo*I because *Hin*6I digestion produces the same 5′ overhang as *Hpa*II, allowing *Hpa*II and *Hin*6I DPs to be cloned into the same vector. The presence of only C and G nucleotides in the recognition sequence means that the sites for these enzymes tend to cluster in CG rich CpG island-like regions. These regions are of particular interest in the isolation of novel imprinted genes, as the differentially methylated regions (DMRs) of imprinted genes examined to date comprise CpG island, or island-like, regions. Other 4-base methylation-sensitive restriction enzymes (e.g., *Mae*III [ACGT]) could also be used, although these sites are less likely to cluster in CpG island regions. It is preferable to avoid enzymes that give blunt-ended digestion products, as the restriction fragments may not ligate efficiently to the adaptors.

2. Materials

Many materials and pieces of equipment are required for various procedures throughout the protocol. These are standard for a well-equipped molecular biology laboratory and are:

1. Incubator at 37°C in which to carry out restriction digests.
2. Two variable-temperature heating blocks.
3. Ice bath.
4. Thermocycler, which is used for PCR and other incubations.
5. Bench-top centrifuge to accommodate 1.5-mL and 0.5-mL microcentrifuge tubes. All centrifugation steps are carried out at 15,000*g*.
6. Equipment for gel electrophoresis: electrophoresis apparatus, UV transilluminator, and video camera (preferably with attached gel documentation software).
7. Reagents for gel eletrophoresis: agarose, 0.5× TBE (10× TBE: 0.9 *M* Tris-borate, 0.02 *M* ethylenediaminetetraacetic acid (EDTA)), ethidium bromide (EtBr, stock solution of 2.5 mg/mL, stored in a dark bottle as it is light sensitive). EtBr is a suspected mutagen, so handle all EtBr-containing reagents wearing gloves.
8. PCR reagents: appropriate oligonucleotide primers (stock solution of 40 pmol/μL); dNTPs (stock solution of 2 m*M*); STD (10× STD: 100 m*M* Tris-HCl, pH 8.8 (at 25°C), 500 m*M* KCl, 15 m*M* $MgCl_2$) and TRI (10× TRI: 300 m*M* Tricine, pH 8.4 (at 25°C), 20 m*M* $MgCl_2$) PCR buffers; and *Taq* DNA polymerase.
9. 3 *M* sodium acetate, pH 5.2, and 100% ethanol for precipitating DNA, 70% ethanol for washing DNA pellets.
10. TE: 10 m*M* Tris-HCl, 1 m*M* EDTA, pH 8.0.

Other materials required are indicated for the appropriate step in the protocol. All reagents should be of molecular-biology grade, and made up to the correct concentration using double-distilled or ultrapure H_2O. All appropriate solutions, such as H_2O, should be autoclaved before use. PCR primers, dNTPs, PCR buffers, all enzymes, and the buffers supplied with them should be stored at –20°C. All intermediates in the procedure should be stored at –20°C, at which temperature they may be stored indefinitely.

This protocol describes Me-RDA using *Hpa*II digested DNAs and the *Hpa*II set of adaptors/primers (*see* **Table 1**). Me-RDA with *Hin*6I-digested DNA and *Hin*6I adaptors/primers would be performed in an identical fashion.

2.1. Preparation of Driver and Tester Amplicons

2.1.1. Restriction Digest of Genomic DNA

1. 5 µg tester and driver DNAs of choice.
2. *Hpa*II restriction enzyme and appropriate buffer as supplied by the manufacturer.

2.1.2. Ligation of Adaptors

1. R*Hpa*24 and R*Hpa*12 adaptor oligonucleotides at a concentration of 200 pmol/µL (*see* **Table 1** and **Note 1**).
2. T4 DNA ligase and appropriate buffer as supplied by the manufacturer.
3. Water bath at 50°C that may be cooled to 10°C over >60 min (*see* **Note 2**).

2.1.3. Production of Amplicons

1. PCR reagents and additions of choice (e.g., glycerol has been used with R*Hpa*24 primer and TRI PCR buffer).
2. DNA concentration standard comprising *Sau*3A-digested genomic DNA of known concentration (e.g., 100 ng/µL; *see* **Note 3**).

2.2. Processing Tester and Driver Amplicons

1. *Hpa*II restriction enzyme and appropriate buffer as supplied by the manufacturer.
2. Low-melting-point (LMP) agarose.
3. 1× TAE (50× TAE: 2 *M* Tris-acetate, 0.05 *M* EDTA).
4. Molecular weight marker such as (ϕX174 DNA digested with *Hae*III.
5. β-Agarase enzyme and buffer supplied by the manufacturer.
6. M*Hpa*24 and M*Hpa*12 adaptor oligonucleotides at 200 pmol/µL (*see* **Table 1**).
7. Materials as indicated under **Subheading 2.1.2.**

2.3. Subtraction

2.3.1. Subtractive Hybridization

1. 25/24/1 aqueous phenol/chloroform/isoamyl alcohol (IAA). This solution should be kept in the dark and handled with caution in a fume hood, as it causes burns to eyes and skin.

Table 1
Oligonucleotides Used in Me-RDA with the Restriction Enzymes *Hpa*II and *Hin*6I

Enzyme	Site		oligonucleotides	12/24 adaptors
*Hpa*II	C'CGG GGC'C	R*Hpa*24	AGCACTCTCCAGCCTCTCAGCGAC	AGCACTCTCCAGCCTCTCAGCGAC AGAGTCGCTGGC
		R*Hpa*12	CGGTCGCTGAGA	
		M*Hpa*24	AGCCAACTGTGCTATGCGAGGAAC	AGCCAACTGTGCTATGCGAGGAAC ACGCTCCTTGGC
		M*Hpa*12	CGGTTCCTCGCA	
		J*Hpa*24	ACCGACGTCGACTATCCATGAAGC	ACCGACGTCGACTATCCATGAAGC AGGTACTTCGGC
		J*Hpa*12	CGGCTTCATGGA	
*Hin*6I	G'CGC CGC'G	R*Hin*24	AGCACTCTCCAGCCTCTCAGCGAG	AGCACTCTCCAGCCTCTCAGCGAG AGAGTCGCTCGC
		R*Hin*12	CGCTCGCTGAGA	
		M*Hin*24	AGCCAACTGTGCTATGCGAGGAAG	AGCCAACTGTGCTATGCGAGGAAG ACGCTCCTTCGC
		M*Hin*12	CGCTTCCTCGCA	
		J*Hin*24	ACCGACGTCGACTATCCATGCAAG	ACCGACGTCGACTATCCATGCAAG AGGTACGTTCGC
		J*Hin*12	CGCTTGCATGGA	

The 12-mer and 24-mer are used together to form an adaptor, which may be ligated to DNA digested with the enzyme indicated. The 24-mer is used as a primer in PCR reactions to amplify fragments that are ligated to the appropriate adaptor. The R*Hpa*/R*Hin* adaptors/primers are used on both driver and tester DNAs to allow whole-genome PCR to produce amplicons. The M*Hpa*/M*Hin* and J*Hpa*/J*Hin* adaptors are ligated to tester amplicons or DPs. They provide tester-specific primer annealing sites to allow recovery of tester-specific sequences following subtractive hybridization. To carry out three rounds of Me-RDA, 10 nmol of R*Hpa*24 and 1 nmol of all other primers are required (*see* **Note 1**).

2. 24/1 chloroform/IAA. This should be handled in a fume hood to avoid inhalation.
3. 2.5× EE (EE: 10 m*M* N-[2-hydroxyethyl]piperazine-N′-[3-propanesulfonic acid] (EPPS) pH 8.25 (at 25°C), 1 m*M* EDTA).
4. Minearl oil.
5. 5 *M* sodium chloride.

2.3.2. Recovery of DPs by PCR

1. M*Hpa*24 primer (stock solution of 40 pmol/µL).
2. 25/24/1 aqueous phenol/chloroform/IAA.
3. 24/1 chloroform/IAA.
4. Mung bean nuclease (MBN) and buffer supplied by the manufacturer.
5. 50 m*M* Tris-HCl, pH 8.8.

2.3.3. Subsequent Rounds of Subtraction

1. J*Hpa*24/12 adaptors/primers instead of R*Hpa* or M*Hpa* adaptors/primers (stock solutions of 200 pmol/µL and 40pmol/µL). Otherwise all other materials are as used in the first round of subtraction.

2.4. Analysis of DPs

2.4.1. Cloning DPs

1. *Hpa*II restriction enzyme and appropriate buffer as supplied by the manufacturer.
2. Vector of choice (e.g., pBS+ [Stratagene]) restriction digested and treated with alkaline phosphatase, in order to accept the cohesive ends generated by the restriction enzyme used in the subtraction (e.g., *Hpa*II and *Hin*6I fragments may be cloned into an *Acc*I site).
3. T4 DNA ligase and appropriate buffer as supplied by the manufacturer.
4. Competent cells of a suitable strain of *Escherichia coli* allowing blue/white selection, and standard materials for culturing bacteria: LB, LB agar plates with ampicillin, X-gal, IPTG ***(9)***.

2.4.2. Determining Efficiency of Subtraction

1. Colony lysis buffer: 10 m*M* Tris-HCl pH 7.5, 1 m*M* EDTA, 50 µg/mL proteinase K.
2. QIAquick PCR purification kit (QIAgen), or equivalent.
3. Concentration standards (e.g., differing dilutions of molecular weight marker).
4. Southern blots of starting tester and driver amplicons (e.g., 1 µg each amplicon, electrophoresed and blotted onto nylon membrane using standard protocols).
5. [α^{32}P]dCTP and preferred materials for random priming probe labeling and hybridization of probe to Southern blots.
6. 7.5 *M* ammonium acetate.

2.4.3. Determining Differential Methylation

1. Southern blots of suitable DNAs for analysis of allele-specific methylation (e.g., from uniparental material or hybrids between different species/subspecies).
2. Preferred conditions for hybridization of probes to Southern blots.

3. Methods

This protocol describes Me-RDA using *Hpa*II-digested DNAs and the *Hpa*II set of adaptors/primers. Me-RDA with *Hin*6I-digested DNA and *Hin*6I adaptors/primers would be performed in an identical fashion (*see* **Table 1**).

It is essential that throughout the protocol, care should be taken to ensure that there are no means by which contamination may be introduced. The reliance of this technique on PCR may lead to the amplification of contaminants, and spurious results. To this purpose, it is recommended that precautionary measures are taken, such as the use of barrier tips or positive-displacement pipets. Reagents should be kept in small aliquots, and those at working dilution should be renewed frequently.

3.1. Preparation of Tester and Driver Amplicons

Tester and driver amplicons are produced from restriction-digested genomic DNAs by the addition of adaptor oligonucleotides, and the use of primers corresponding to these adaptors in whole-genome PCR under optimized conditions. Only fragments within a particular size range, which is in part determined by the PCR conditions used, are amplified in the reaction. The resulting amplicons constitute a reduced-complexity representation of the genome. Since it is these amplicons that are used in the subtractive hybridization, it is important that the tester and driver amplicons are equivalent, for example, in size range. This is required to reduce the possibility of differences between the starting amplicons that are not the result of methylation differences, which might give rise to spurious DPs.

3.1.1. Restriction Digest of Genomic DNA

1. Separately digest 5 µg tester and driver DNAs with ≥4 U/µg of *Hpa*II to completion (overnight) under the recommended conditions.
2. Check 300 ng aliquot of digested against undigested DNA on a 1% agarose/0.5× TBE gel containing EtBr at 25 ng/mL (*see* **Note 4**).
3. Inactivate enzyme by incubating at 72°C for 15 min.

3.1.2. Ligation of Adaptors

1. Mix together in a 0.5-ml microcentrifuge tube 1 µg *Hpa*II-digested DNA, 500 pmol each of adaptors R*Hpa*12 and R*Hpa*24, ligase buffer to 1×, and H_2O to a final volume of 30 µL.

2. Heat to 50°C and allow to cool to 10°C over >60 min to anneal the adaptors (*see* **Note 2**).
3. Add 5 U (Weiss units) T4 DNA ligase and ligate overnight at 16°C.

3.1.3. Production of Amplicons

1. Set up analytical PCRs to check ligation and PCR conditions using 4 ng DNA from ligation reaction, 200 μ*M* each dNTP, 1 μ*M* R*Hpa*24 primer, 1× PCR buffer in final volume of 20 μL (*see* **Note 5**).
2. Bring to 72°C (to melt off the 12-mer) and add 0.4 U *Taq* DNA polymerase.
3. Incubate at 72°C for 5 min to fill in overhangs.
4. Then incubate at 95°C for 1 min 30 s and amplify through 20 cycles of 95°C for 30 s, 68°C for 30 s, and 72°C for 1 min 30s (final 72°C for 5 min).
5. Examine an aliquot of the reaction on a 1.2% agarose/0.5× TBE gel containing EtBr at 25 ng/mL.
6. Set up a series of analytical PCRs to examine the effects of DNA input (2–8 ng), primer concentration (1–4 μ*M*), and *Taq* DNA polymerase input (0.4–1 U) on the size range and yield of the amplicons (*see* **Note 6**). Increasing any of these results in amplicons of smaller fragment size. The effects of other additions (e.g., up to 10% glycerol is often used with R*Hpa*24 in conjunction with the TRI PCR buffer) could also be examined.
7. Choose conditions that produce amplicons of a size range that is deemed to contain the fragments of interest, and ensure that the tester and driver amplicons are as similar as possible.
8. Set up preparative (e.g., 200-μL) PCRs, scaling up exactly from the analytical PCR conditions that produce the desired amplicon.
9. Precipitate each 200-μL preparative PCR with 0.3 *M* sodium acetate pH 5.2 and 2.5 vol ethanol on ice >1 h. Centrifuge for 7 min to pellet the DNA. Remove supernatant and wash with 70% ethanol. Resuspend each pellet in 25 μL TE, pH 8, then pool.
10. Examine 1 μL on a 1.2% agarose/0.5× TBE gel to check size range and estimate concentration, using 100–500 ng genomic DNA cut with *Sau*3A as concentration standards (*see* **Note 3**). Expect yields of ~10 μg per 200 μL PCR reaction in TRI buffer and ~4 μg per 200 μL PCR reaction in STD buffer.
11. Set up sufficient preparative PCRs to produce 40 μg of driver amplicon for each subtraction step and 5–10 μg tester amplicon.

3.2. Processing Tester and Driver Amplicons

It is necessary to remove the R*Hpa* adaptors from the tester amplicon and replace with new (M*Hpa*) adaptors, to provide tester-specific primer sites to allow the recovery specifically of tester sequences following subtraction. This involves gel purification and recovery of the DNA, followed by the annealing and ligation of the new adaptors.

The driver amplicon has adaptor sequences at both ends, as well as containing significant quantities of the corresponding free primer. Therefore, driver sequences could be abundant substrates for amplification after subtraction. This has not been found to cause any problems in the subtraction; in fact, we do not detect any amplification of driver post-subtraction. If desired, removal of adaptors and free primers could be carried out by *Hpa*II digest and gel extraction as described for the tester amplicon, but is likely to entail significant loss of material.

1. Digest ~3 µg tester amplicon with 30 U *Hpa*II in minimum volume for ≥2 h. Save one-tenth of the digest as recovery control.
2. Load the remainder into a 1-cm preparative well of a 2% LMP/1× TAE gel containing EtBr (0.1 µg/mL) against a size marker that will cover the size range of the amplicons (e.g., *Hae*III-digested φX174 DNA). Run at ~30 V until bromophenol blue has migrated 20–25 mm.
3. Cut off marker lane; photograph on transilluminator, and note position of the size markers that correspond to the size range of the amplicons. Excise the corresponding portion of the preparative lane and check that the primers are in the remainder of the gel.
4. Weigh the gel slice, divide into 300- to 350-µL volumes and place each into separate 1.5-mL microcentrifuge tubes (dividing each again if necessary).
5. Add 0.10 vol 10× β-agarase buffer and allow gel pieces to melt completely at 68°C (*see* **Note 7**).
6. Equilibrate to 40°C, add β-agarase at 2 U/100 µL, and mix well. (This is an excess of enzyme, but it takes into account the high percentage of agarose, and that the gel slices are not equilibrated in two changes of buffer before digestion.) Incubate at 40°C for >1 h.
7. Add sodium acetate to 0.3 *M* and chill on ice >5 min. Centrifuge 7 min to remove debris and transfer supernatant to fresh tube.
8. Precipitate supernatant with 2 vol isopropanol, on ice >1 h. Centrifuge 5 min. Remove supernatant and wash pellets with 70% ethanol (*see* **Note 8**).
9. Take up pellet 20 µL TE or H_2O.
10. Run 2 µL on 1.2% 0.5× TBE gel, with starting digest and concentration standards.
11. Set up annealing and ligation of 500 pmol each of adaptors M*Hpa*24 and 12 to 0.5–1 µg gel purified tester amplicon under the conditions described under **Subheading 3.1.2.**
12. Perform an analytical PCR to check that ligation was successful (e.g., analytical PCR in 20 µL using 1 ng ligation and 1 µ*M* primer M*Hpa*24). Ensure that the PCR products reflect the size range of the starting tester amplicon, and do not contain novel prominent bands.

3.3. Subtraction

The subtraction includes co-precipitation, denaturation, and hybridization of tester and driver amplicons. Amplification is then performed using tester-specific primers, single-stranded DNA is removed, and further amplification is carried out in order to obtain a kinetic enrichment of target sequences. Subtraction is performed in the minimum volume and over sufficient time to ensure that fragments corresponding to single-copy sequences are able to reassociate. The fact that the amplicons are greatly simplified representations of the starting genome also favors reassociation kinetics. It is important to ensure complete resuspension and denaturation of the DNA, since residual double-stranded tester molecules will be perfect templates for PCR.

3.3.1. Subtractive Hybridization

1. Mix 40 μg driver amplicon and 400 ng religated tester amplicon.
2. Extract with aqueous phenol/chloroform/IAA and chloroform/IAA. Precipitate with 0.3 *M* sodium acetate, pH 5.2, and 2.5 vol ethanol on ice (>1 h), then centrifuge 7 min to bring down pellet.
3. Remove supernatant, wash pellet in 70% ethanol, and allow to dry. Resuspend as well as possible in 4 μL 2.5× EE. Preequilibrate a 1.5-ml microcentrifuge tube with 50 μL paraffin oil in a heating block at 98°C.
4. Pipet the subtraction reaction under the oil. Denature 10 min at 98°C. Remove from heating block, add 1 μL of 5 *M* sodium chloride and mix well. Incubate >20 h at 67°C.
5. After subtraction, dilute with 95 μL TE. Ensure that any precipitate that may have formed during the hybridization is thoroughly resuspended.

3.3.2. Recovery of DPs by PCR

This first PCR after the subtraction/reassociation can be regarded as recovering the small quantities of DNA that are homoduplexes of tester amplicons. However, the PCR product will also contain relatively large amounts of annealed and unannealed driver amplicons, primers, and linearly amplified PCR products from the heteroduplex annealed component. The bulk of the linearly amplified products are single-stranded molecules, which may be removed by digestion with mung bean nuclease (MBN). PCR reactions are then carried out on the MBN-digested material, to amplify duplex tester DNA further. The resulting DPs should be enriched for the target sequences.

1. Set up a 200-μL PCR using 5-μL subtraction, 200 μ*M* each dNTP, 1× PCR buffer (appropriate to the amplicons used), and 4 U *Taq* DNA polymerase.
2. Incubate 5 min at 72°C, to fill in ends.
3. Add primer M*Hpa*24 to 1 μ*M*, and proceed with amplification. Incubate at 95°C for 1 min 30 s, then amplify through 10 cycles of 95°C for 30 s, 72°C for 2 min (final 72°C for 3 min). Recovery (equivalent amount of religated tester) and specificity (equivalent amount of driver) controls could be included as separate PCR reactions (*see* **Note 9**).

4. Extract PCRs with phenol/chloroform/IAA and chloroform/IAA. Precipitate with 0.3 *M* sodium acetate, pH 5.2, and 2.5 vol ethanol on ice >1 h, then centrifuge 7 min to pellet DNA.
5. Remove supernatant and wash precipitated DNA with 70% ethanol. Take up pellet in 20 µL of H_20.
6. Save 2 µL as an undigested control, and to the remainder add 2 µL of 10× MBN buffer and 10 U MBN. Incubate 30 min at 30°C.
7. Dilute with 100 µL of 50 m*M* Tris-HCl, pH 8.8. Inactivate MBN 5 min at 95°C. (These inactivation conditions are designed not to inhibit downstream reactions.)
8. Set up PCR of 2 µL MBN digest with 200 µ*M* each dNTP, 1 µ*M* primer M*Hpa*24, 1× PCR buffer, and 1 U *Taq* DNA polymerase in final volume of 20 µL.
9. Amplify through 95°C for 30 s, 72°C for 2 min (final 72°C, for 3 min 30 s) for an appropriate number of cycles. It might be advisable to set up analytical reactions for the second PCR, possibly multiple reactions for a "time course" (e.g., 14, 17, and 20 cycles).
10. Examine products on 1.2% agarose/0.5× TBE gel. You might also consider running the following samples on the gel: 200 ng driver amplicon, 200 ng tester amplicon, 200 ng equivalent of subtraction, 1 µL before MBN digestion, 5 µL after inactivation of MBN, recovery control, specificity control (*see* **Note 9**).
11. Expect to see a smear of DPs in the PCRs, against which discrete bands may be visible. The size range of the DPs should be within that of the amplicons.

3.3.3. Subsequent Rounds of Subtraction

A single round of RDA will give enrichment of target sequences, up to an estimated 15-fold. Additional rounds of subtraction are recommended to increase the enrichment of target sequences and to suppress repetitive sequences that are likely still to be present in the first-round DPs.

1. For a second round of subtraction, set up a preparative PCR (200 µL) from the inactivated MBN digest. The number of cycles used should be those determined as suitable in the analytical PCR (*see* **Subheading 3.3.2.**), and the reaction conditions should be scaled up directly from the analytical PCR. Precipitate DPs, digest with *Hpa*II, and gel-purify as before (**Subheading 3.2.**). Ligate J*Hpa*24/12 adaptors to 200 ng of gel-purified DPs as before (**Subheading 3.2.**). Check ligation with analytical PCRs as before.
2. Set up subtraction with 10–100 ng DPs and 40 µg driver amplicon. You could set up subtractions at two or more different ratios of DPs to driver, to ensure coverage of optimal subtraction conditions. Perform subtraction and amplification steps as before. The second postsubtraction PCR might require 20–30 cycles. Examine by gel electrophoresis; you should expect to see discrete bands in the DPs. The less background smear, the more complete the subtraction is likely to be, and the less need to undergo a third round of subtraction.

3. If a third round of subtraction is necessary, the adaptors on the DPs are replaced again with the M*Hpa*24/12 set. Subtraction is set up with 100–400 pg DPs and 40 µg driver amplicon.

3.4. Analysis of DPs

The DPs produced in a Me-RDA screen may be examined individually, by cloning out single fragments; alternatively, the DP population could be used, for example, as a probe in library screening. The success of the subtraction will have depended on the complexity of the tester and driver amplicons, and on the size of the target fragment population in the tester amplicon, and this will vary from experiment to experiment. It is important to assess the efficiency of the Me-RDA by determining the proportion of DPs that represent target sequences: in our experience, two rounds of subtraction will yield anything between <10% and 100%. For library screening, this assessment may give confidence that positives correspond to target sequences, and the assessment is also critical in determining whether to proceed through further rounds of subtraction. Excessive rounds of Me-RDA may reduce the complexity of the DPs, which may lead to the exclusion of some target sequences. It is recommended that a measure of the completeness of subtraction be obtained following the second and subsequent rounds of Me-RDA. This can be done by cloning the DPs, and testing whether individual clones correspond to fragments differentially represented in the starting amplicons (*see* **Fig. 3A**).

3.4.1. Cloning DPs

Cloning of DPs allows the separation of a complex collection of fragments into their constituent parts, which will allow DP fragments to be analysed individually.

1. Digest 1 µg DPs (from first, second, or third subtraction) with 10 U *Hpa*II and run on 1.5% LMP agarose/1× TAE gel stained with EtBr (0.1 µg/mL).
2. Excise the DPs and dilute the gel slice with water, or purify DNA from gel using β-agarase as described under **Subheading 3.2.1.**
3. Ligate 5–10 ng DPs into 200 ng of a suitable vector (*Hpa*II and *Hin*6I fragments may be cloned into the *Acc*I site of pBS+ [Stratagene]. Allow to ligate overnight at 16°C and transform into an appropriate *E. coli* host. The use of blue/white selection may aid identification of colonies containing plasmids with inserts.

3.4.2. Determining Efficiency of Subtraction

Cloned DPs can be tested by hybridization back onto filters of the starting tester and driver amplicons to verify the proportion corresponding to differences (*see* **Fig. 3A**). Testing ≥10 clones will give an impression of how effective

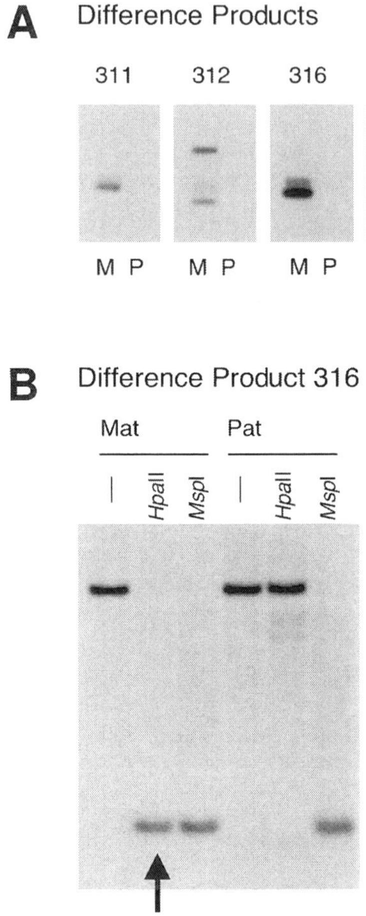

Fig. 3. Testing difference products from a subtraction involving DNAs from mouse embryos with paternal or maternal UPD for distal chromosome 2. **(A)** Cloned DPs are hybridized to filters containing *Hpa*II amplicons prepared from maternally disomic DNA (M) and paternally disomic (P) DNAs. The DPs correspond to *Hpa*II fragments unmethylated in maternally disomic DNA, and therefore present specifically in the M amplicon. A control clone (Con) detects a fragment in both M and P lanes. **(B)** Demonstration that DP clone 316 detects imprinted methylation by Southern blot analysis of genomic DNAs from maternal (Mat) and paternal (Pat) chromosome 2 disomy embryos. DNAs are digested with *Hind*III alone (–), or in combination with *Hpa*II or *Msp*I, and hybridized with probe 316. The arrow indicates a *Hpa*II restriction fragment with imprinted methylation, since it is detected specifically in the Mat DNA. The *Msp*I digests serve as a control to exclude the possibility that the *Hpa*II fragment difference is the result of a *Hpa*II RFLP between the two DNAs.

the subtractions have been, and whether there may be a need to undergo a further round of subtraction.

1. Pick individual colonies and lyse in 25 µL colony lysis buffer for 15 min at 55°C. Inactivate digest by heating 15 min at 80°C. Centrifuge briefly. The lysates can be stored at –20°C.
2. Use 10 µL colony lysate in 100 µl PCR using 200 µ*M* each dNTP, 1× PCR buffer, 0.5 U *Taq* DNA polymerase, and 0.2 µ*M* primers corresponding to M13–20/M13 reverse, T3/T7, or other appropriate primers depending on vector used (*see* **Note 10**).
3. Remove primers, nucleotides, and so on, from PCR product using QIAquick (QIAgen), or equivalent PCR purification kit, following manufacturer's protocols.
4. Examine yield on 1.6% agarose/0.5× TBE gel stained with 25 ng/mL EtBr, by comparing to concentration standards (e.g., varying amounts of commercially available molecular weight marker).
5. Label 25 ng DNA with 10 µCi [α^{32}P] dCTP by random priming. Clean probes by precipitation with 0.75 *M* ammonium acetate and 2.5 vol ethanol. Resuspend in H_2O and dilute with hybridization buffer.
6. Hybridize with Southern blots of tester and driver amplicons, wash, and expose to X-ray film.
7. Examine whether DP clones represent differences between starting amplicons (present only in the tester amplicon), and from this calculate the efficiency of subtraction.
8. DPs that are of interest may be sequenced directly from the purified PCR product.

3.4.3. Determining Differential Methylation of DP Clones

Testing DP clones on Southern blots of tester and driver amplicons will indicate whether the fragment represents a difference between the amplicons. This does not exclude the possibility that the DP derives from a *Hpa*II or *Hin*6I fragment that is polymorphic between tester and driver DNAs, or some other artefact of the amplicons (such as unequal representation or different size ranges). Whether the differences correspond to fragments derived from regions of allele-specific methylation may be tested by hybridizing the probes to Southern blots of DNAs digested with methylation-sensitive enzymes (e.g., *Hpa*II, *Hin*6I) and non-methylation-sensitive isoschizomers (e.g., *Msp*I) (*see* **Fig. 3B**). The DNAs used on the Southern blots may be from various sources that depend on the tester and driver DNAs used. For full details on the analysis of methylation by Southern blotting, refer to Chapter 14.

1. The most direct way of testing for allele-specific methylation is to examine the hybridization pattern of the probe on Southern blots of tester and driver

DNAs digested with methylation-sensitive restriction enzymes. This may not be practical if the DNA used as the tester and driver is limiting, such as when parthenogenetic and androgenetic embryos are used.

2. If the map location of the DP clone is known, allele-specific methylation may be examined in mice with appropriate uniparental disomy.
3. If neither of the above approaches is possible, an RFLP must be identified in/close to the sequence spanned by the clone, between two different species/subspecies (e.g., *Mus musculus domesticus* and *Mus musculus castaneus*). DNAs from these species and hybrids may then be prepared for Southern blotting by digestion with methylation-sensitive enzymes and the enzyme that identifies the RFLP, thus allowing the maternal and paternal alleles to be distinguished.

4. Notes

1. The adaptor oligonucleotides must be unphosphorylated and of high quality. It is important that no shorter synthesis products are present in the adaptors, as these could ligate illegitimately and destroy the *Hpa*II restriction site. The quality of the adaptors can be assessed by α^{32}P-end labeling and polyacrylamide gel electrophoresis. Adaptors can be purified by thin-layer chromatography or preparative polyacrylamide gel electrophoresis.
2. Annealing can be performed in a beaker of water (~250 mL) that is heated to 50°C and then placed in a refrigerator.
3. The concentration standard used is genomic DNA digested with a suitable restriction enzyme in order to give a smear of fragments of similar size range to the amplicons. *Sau*3A or *Msp*I restriction endonucleases may be used for this purpose.
4. Due to the extent of methylation, the DNA will still appear practically undigested. Digestion is indicated by a shift in the largest molecular-weight fragments and smearing of the DNA. If necessary, completeness of digestion could be determined by Southern blotting and hybridizing with a known unmethylated sequence.
5. Two different PCR buffers have been tried for this protocol: 10× STD (100 m*M* Tris-HCl pH 8.8 (at 25°C), 500 m*M* KCl, 15 m*M* $MgCl_2$) is typically that supplied with *Taq* DNA polymerase, and produces amplicons in the range of ~200–1000 bp. 10× TRI (300 m*M* Tricine, pH 8.4 (at 25°C), 20 m*M* $MgCl_2$) was designed for amplification of longer products ***(10)*** and produces amplicons in the range of ~240–2200 bp. The yield in TRI buffer is generally greater than in STD buffer. The same buffer used in the production of amplicons should be used in all subsequent amplification stages of the Me-RDA using these amplicons.
6. Expect to obtain 200–500 ng per 10 µL of PCR. Poorer yields and amplicons of a predominantly large size range might indicate that DNA, primers, or *Taq* DNA polymerase are limiting. For example, if the original *Hpa*II digest is incomplete, there might be insufficient ligated material in the size range for the PCR. The amplicons should appear as a smear on the gel; if prominent bands are present, they may be indicative that one of the above components is limiting.

7. As an alternative to β-agarase digestion, a variety of gel extraction kits are available. However, we have found the most consistent yields are obtained using β-agarase.
8. If residual agarose has been brought down (sometimes pellets are larger than expected for 3 μg DNA), take up pellets in a combined volume of 90 μL H_2O, add 10 μL 10× buffer, melt, and equilibrate as before. Then add a further 4 U β-agarase and digest as before. Precipitate as before, and take up finally in 20 μL TE or H_2O.
9. The specificity control (an appropriate amount of driver amplicon taken through the postreassocation steps) indicates whether the amplification is specific to the religated tester amplicon. If there is amplification of the driver amplicon, this may indicate that a large quantity of free primer has been carried over into the subtraction, which should be repeated using gel-purified driver amplicon. The recovery control (an appropriate amount of tester amplicon ligated to M*Hpa*24&12 adaptors and taken through the postreassocation steps) serves to indicate whether it is possible to amplify from the very small amounts of DNA likely to exist as tester–tester homoduplexes.
10. The primers chosen represent standard priming sites on the vector, and flank the cloning site. Primers farther from the cloning site allow high-quality sequencing of the complete insert from the PCR product. A large amount of polylinker in the PCR product sequence may cause problems using these fragments as probes, for example, in library screening. We have encountered no problems using M13 primers and the pBS+ (Stratagene) vector.

Acknowledgments

R. J. Smith is a student of the Biotechnology and Biological Sciences Research Council. G. Kelsey is a Senior Fellow of the Medical Research Council.

References

1. Lisitsyn, N., Lisitsyn, N., and Wigler, M. (1993) Cloning the differences between two complex genomes. *Science* **259,** 946–951.
2. Constância, M., Pickard, B., Kelsey, G., and Reik, W. (1998) Imprinting mechanisms. *Genome Res.* **8,** 881–900.
3. Li, E., Beard, C., and Jaenisch, R. (1993) Role for DNA methylation in genomic imprinting. *Nature* **366,** 362–365.
4. Tycko, B. (1997) DNA methylation in genomic imprinting. *Mut. Res.* **386,** 131–140.
5. Bartolomei, M. S., Webber, A. L., Brunkow, M. E., and Tilghman, S. M. (1993) Epigenetic mechanisms underlying the imprinting of the mouse *H19* gene. *Genes Dev.* **7,** 1663–1673.
6. Plass, C., Shibata, H., Kalcheva, I., Mullins, L., Kotelevtseva, N., Mullins, J., Kato, R., Sasaki, H., Hirotsune, S., Okazaki, Y., Held, W. A., Hayashizaki, Y., and Chapman, V. M. (1996) Identification of *Grf*1 on mouse chromosome 9 as an imprinted gene by RLGS-M. *Nature Genet.* **14,** 106–109.

7. Hayashizaki, Y., Shibata, H., Hirotsune, S., Sugino, H., Okazaki, Y., Sasaki, N., Hirose, K., Imoto, H., Okuizumi, H., Muramatsu, M. Komatsubara, H., Shiroishi, T., Moriwaki, K., Katsuki, M., Hatano, N., Sasaki, H., Ueda, T., Mise, N., Takagi, N., Plass, C., and Chapman, V. M. (1994) Identification of an imprinted U2af binding protein related sequence on mouse chromosome 11 using the RLGS method. *Nature Genet.* **6,** 33–40.
8. Peters, J., Wroe, S. F., Wells, C. A., Miller, H. J., Bodle, D., Beechey, C. V., Williamson, C. M., and Kelsey, G. (1999) A cluster of oppositely imprinted transcripts at the *Gnas* locus in the distal imprinting region of mouse chromosome 2. *Proc. Natl. Acad. Sci. USA* **96,** 3830–3835.
9. Sambrook, J., Fritsch, E. F., and Maniatis, T. (1989) *Molecular cloning: A Laboratory Manual*, 2nd ed., Cold Spring Harbor Laboratory Press, Cold Spring Harbor, NY.
10. Ponce, M. R. and Micol, J. L. (1992) PCR amplification of long DNA fragments. *Nucleic Acids Res.* **20,** 623.

9

Ribonuclease Protection

Joanne L. Thorvaldsen and Marisa S. Bartolomei

1. Introduction

The ribonuclease protection assay (RPA) is a sensitive technique for the analysis of total cellular RNA. It involves generating a specific antisense riboprobe, hybridizing the probe to total RNA, removing unprotected RNA by RNases, and finally isolating and analyzing the protected RNA on a denaturing gel. Although the RPA is somewhat more labor-intensive than Northern analysis, it has the advantage of being more sensitive (as little as 0.1 pg of target RNA can be detected with ideal hybridization conditions). RPAs are also more tolerant of partially degraded RNA (provided the area that is protected is intact). Although RPAs are not as sensitive as polymerase chain reaction (PCR)-based RNA analyses, the target RNA is analyzed directly; a reverse transcription step is not required. Finally, the RPA is quantitative as long as the probe is in excess. More important for the study of imprinted genes, the RPA can be designed to detect allele-specific expression of the target gene of interest.

1.1. Design of Probes

There are several points to consider when designing a riboprobe for the typical RPA. The probe template is prepared by cloning the sequence of interest into a transcription vector such that the 3′ end of the fragment is adjacent to bacteriophage promoter *(1)*. The template can be cloned from cDNA or genomic DNA. The recommended probe size is 100–500 bases. In addition, the probe should be designed to be larger than the protected RNA fragment for ease of discriminating between RNA:RNA vs RNA:DNA hybridization products, the latter resulting from DNA contamination. Further useful considerations

From: *Methods in Molecular Biology, vol. 181: Genomic Imprinting: Methods and Protocols*
Edited by: A. Ward © Humana Press Inc., Totowa, NJ

are presented by others *(2,3)*. We subsequently focus on the design of probes for analysis of imprinted genes.

1.2. Allele-Specific Analysis

The RPA is an excellent tool for the analysis of allele-specific expression of imprinted genes, since the methodology can be used to detect RNA polymorphisms between different (sub)species of mice and their F1 hybrid offspring. It was shown that RNase A is capable of cleaving base-pair mismatches *(4)*. Consequently, the RPA can be used to detect single base changes between a probe from one species and RNA from a divergent species. Thus the products from each parental allele can be distinguished in an F1 hybrid and can thereby be independently assayed. The RPA developed to detect allele-specific expression of the mouse *H19* gene was used to prove that *H19* is expressed exclusively from the maternal allele *(5)*.

To design a probe to detect allele-specific expression by an RPA, it is critical to identify polymorphisms within the transcribed region between the alleles of hybrid species. For *H19*, the 5th exon contains the most divergent sequence between mouse subspecies *(5)*. **Figure 1** illustrates the template used to generated the allele-specific *H19* RPA probe and the position of polymorphisms between the *Mus musculus domesticus (M. domesticus)* and *Mus musculus castaneus (M. castaneus)* subspecies. A probe has also been designed to detect allele-specific expression of the imprinted mouse *Igf2* gene in F1 hybrids generated from the same intersubspecific cross *(6)*. Specifically, the *Igf2* probe expression vector contains a PCR generated fragment of 3′ untranslated *Igf2* sequence.

We typically analyze *H19* as well as *Igf2* imprinted expression in F1 hybrids generated between C57BL/6 and B6(CAST-*H19*), which has *M. castaneus* sequence at the distal portion of mouse chromosome 7, the region bearing *H19* and *Igf2 (7)*. Normally, *H19* is expressed exclusively from the maternal allele and *Igf2* is expressed primarily from the paternal allele in neonatal tissues *(5,6)*. Allele-specific RPAs of *H19* *(5)* and *Igf2* *(6)* have been instrumental in detecting the disruption of normal *H19* and/or *Igf2* imprinted expression in mice with targeted mutations at the *H19* locus *(6,8–11)*. We generated a targeted deletion of important imprinting regulatory sequences 5′ to *H19* that, when inherited from the paternal allele, results in biallelic expression of *H19* and when inherited from the maternal allele results in biallelic expression of *Igf2*. In our study, expression was assayed in F1 hybrids generated from reciprocal crosses of heterozygous mutants (maintained in a C57BL/6 background) with B6(CAST-*H19*). The targeted mutations were generated on a 129Sv/J allele, which results in the same *H19* and *Igf2* protected fragments

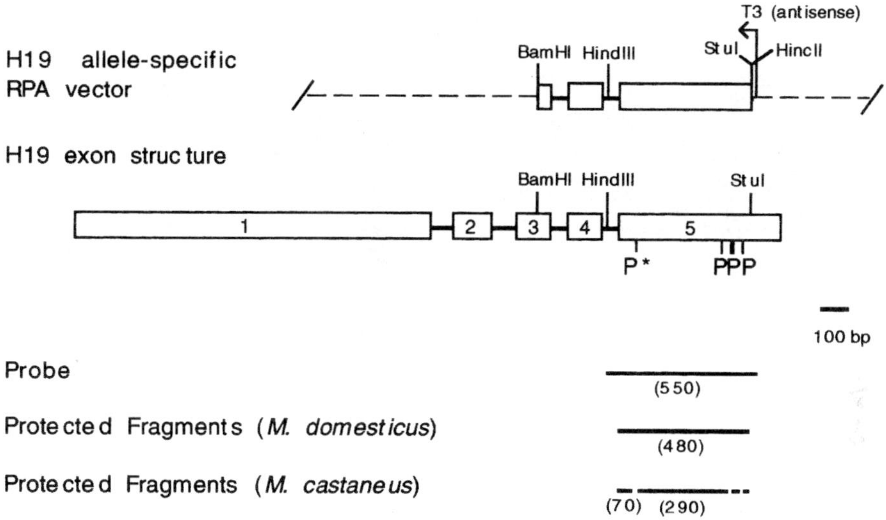

Fig. 1. Schematic of the allele-specific RPA for *H19*. The top line depicts the RPA vector, which was constructed by cloning a 754-bp *Bam*HI/*Stu*I M. domesticus H19 genomic DNA fragment into the *Bam*HI and *Hinc*II sites of Bluescript II KS plasmid (Stratagene) ***(5)***. The T3 transcription start site used to generate the *H19* antisense riboprobe is 3′ of the *Hinc*II cloning site. The *H19* DNA fragment includes the exon 3 through exon 5 sequence (open boxes) as defined in the *H19* exon structure. We generally use the *Hin*dIII site in intron 4 to linearize the RPA vector. In-vitro transcription of *Hin*dIII-digested vector with T3 RNA polymerase results in a probe that is 550 nt long. Using this probe we detect one *M. domesticus* fragment (480 nt) and two *M. castaneus* fragments (one ~290 nt and the other ~70 nt). Our assay conditions are such that we do not detect the shorter *castaneus*-specific protected fragments. Polymorphisms (P) have been mapped near the *Stu*I site in exon 5.

as seen in the assay of C57BL/6 RNA. **Figures 2A,B** present the allele-specific expression analysis of *H19* and *Igf2*, respectively.

1.3. Quantitative Analysis

RNA may be assayed by the RPA with more than one probe simultaneously, thereby allowing for quantitation of relative RNA abundance among independent samples. When total RNA expression levels are to be assayed for imprinted genes, it may be useful to assay with probes that do not detect allele-specific expression. Leighton et al. ***(8)*** used *H19* and *Igf2* RPA probes simultaneously to determine the effect of the targeted deletion of the endodermal enhancer located 3′ to the *H19* gene on levels of expression of both *H19*

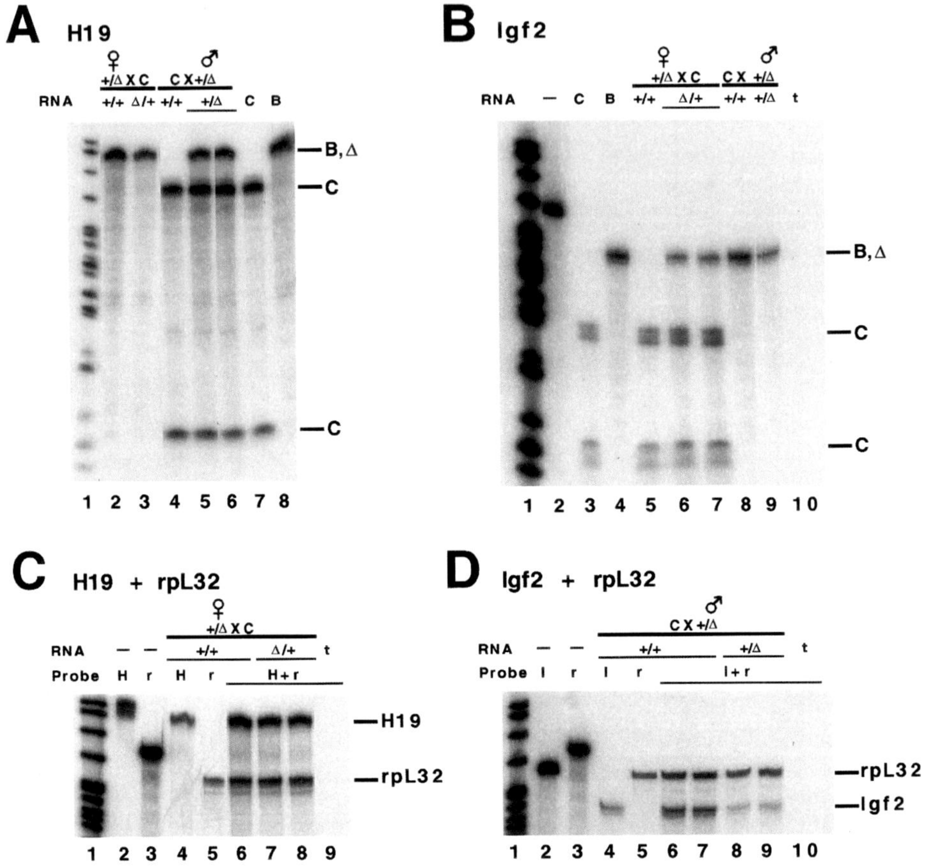

Fig. 2. Ribonuclease protection analysis of *H19* and *Igf2* in mouse neonatal liver RNA. The examples presented are taken from our allele-specific and quantitative analyses of *H19* and *Igf2* in $H19^{\Delta DMD}$ heterozygous mutant mice ***(11)***. Lane 1 in each panel is radiolabeled marker *Msp*I digested pBR322 DNA. (DNA is estimated to run 5–10% faster than RNA in a denaturing gel.) The allele-specific riboprobes for *H19* (H) and *Igf2* (I), as described elsewhere, are utilized in the allele-specific assays (**A** and **B**, respectively) and, as indicated, in combination with the riboprobe for *rpl32* (r) in the quantitative assays (**C** and **D**, respectively). The RNA assayed in a given lane is designated above the gels. Control RNAs include yeast tRNA (t) and neonatal liver RNA isolated from B6(CAST-*H19*) (C) or C57BL/6 (B) mice. Experimental RNAs were isolated from neonatal livers of wild-type (+/+) and mutant (+/Δ or Δ/+) littermates generated from reciprocal crosses of B6(CAST-*H19*) (C) and F_1 $H19^{\Delta DMD}$ heterozygotes (+/Δ) . The female parent is noted first in each cross; only the sex of the heterozygous mutant parent is designated in the figure, to emphasize the parental origin of the inherited mutant allele. (A) Allele-specific expression of *H19*. Normally, the maternal *H19* allele is expressed while the paternal allele is repressed. The control

and *Igf2*. Although the probes used in this assay were not allele-specific ***(8,12)***, the unique size of the *H19* and *Igf2* protected fragments allowed for comparison of their relative RNA abundance. To analyze the relative expression levels of *H19* and *Igf2* independently of each other, a riboprobe for the nonimprinted gene *rpl32* ***(9,13)*** has been used in the RPAs of *H19* or *Igf2* ***(9–11)***. This allows for comparative quantification of *H19* and *Igf2* levels in normal vs mutant mice. As illustrated in **Fig. 2C** (*H19* and *rpl32*) and in **Fig. 2D** (*Igf2* and *rpl32*), altered levels of *H19* and *Igf2* expression can be detected from the respective mutant allele.

Having designed the probe(s) for allele-specific or quantitative analysis of the imprinted gene of interest, there are numerous variations to the RPA methods that one can use. The protocol presented here is a modification of that described by Gilman *(2)*.

2. Materials

In the preparation and handling of all materials required for this protocol, it is essential to avoid RNase contamination and degradation of samples. Work surfaces and pipetman must be cleaned before use and gloves changed often.

Fig. 2. *(continued)* RNAs include *M. castaneus* RNA (C, lane 7) and C57BL/6 RNA (B, lane 8). The distinct allele-specific *M. castaneus* (C), C57BL/6 (B) and mutant allele (Δ) fragments are noted to the side of the gel. The experimental lanes 2–6 indicate that when the targeted *H19* mutant allele is inherited from the mother, only the maternal mutant allele is expressed (Δ/+, lane 3), as expected, and when the mutant allele is inherited from the father, the normally silent paternal *H19* is activated (+/Δ, lanes 5 and 6). (B) Allele-specific expression of *Igf2*. Normally, the paternal *Igf2* allele is expressed while the maternal allele is silent. The control samples include the *Igf2* probe (—, lane 2), *M. castaneus* RNA (C, lane 3), C57BL/6 RNA (B, lane 4), and yeast tRNA (t, lane 10). When the targeted *H19* allele is inherited from the mother, the normally silent maternal *Igf2* allele is activated (Δ/+, lanes 6 and 7) and when the mutation is inherited from the father, only the paternal *Igf2* allele is expressed (+/Δ, lane 9). (C) Quantitative analysis of *H19*. The *H19* (H) and *rpL32* (r) riboprobes are assayed individually (lanes 2–5), and the distinct protected fragments are noted to the right. Although H runs as doublet, this does not interfere with the assay. Quantitative analysis of *rpl32* and *H19* indicate that *H19* RNA levels are slightly reduced in the mutant samples (lanes 7 and 8) compared to the wild-type sample (lane 6). (D) Quantitative analysis of *Igf2*. The *Igf2* (I) and *rpL32* (r) riboprobes are also assayed individually (lanes 2–5). Quantitative analysis of *rpl32* and *Igf2* indicate that *Igf2* RNA levels are significantly reduced in the mutant samples (lanes 8 and 9) compared to the wild-type samples (lanes 6 and 7).

Pipet tips and solutions are reserved for RNA use only. These precautions, together with the proper sterilization and storage of solutions, eliminates the need to treat solutions with the RNase inhibitor diethylpyrocarbonate (DEPC).

2.1. Preparation of Probe

2.1.1. In Vitro Transcription of Riboprobe (see **Note 1**)

1. 5× transcription optimized buffer.
2. 0.1 *M* Dithiothreitol (DTT).
3. RNasin® ribonuclease inhibitor: 40 U/µL.
4. Nucleotide mixture: 2.5 m*M* ATP, 2.5 m*M* GTP, and 2.5 m*M* UTP (prepared from 10 m*M* nucleotide stocks and nuclease-free H_2O).
5. 0.1 m*M* CTP (prepared from 10 m*M* CTP and nuclease-free H_2O stocks).
6. Template DNA: linearized plasmid at 0.2–1.0 µg/µL in H_2O (*see* **Note 2**).
7. [α-^{32}P]CTP, 10 mCi/mL, 400–800 Ci/mmol (New England Nuclear Life Science Products).
8. Bacteriophage RNA polymerase (SP6, T3 or T7): 10–20 U/µL.
9. Nuclease-free H_2O.
10. RQ1 RNase-free DNase: 1 U/µL.

2.1.2. Purification of Probe by Gel Filtration

1. G-50 Sephadex columns for radiolabeled RNA purification (Boehringer Mannheim Biochemicals, Item 1 274 015) (*see* **Note 3**).
2. Nuclease-free H_2O.
3. Hybridization buffer: 40 m*M* PIPES, pH 6.4, 400 m*M* NaCl, 1 m*M* ethylenediaminetetraacetic acid (EDTA), 80% deionized formamide freshly prepared from stocks described under **Subheading 2.2.**

2.2. Hybridization of Probe and RNA

1. 5× Hybridization buffer: 200 m*M* PIPES, pH 6.4, 2 *M* NaCl, 5 m*M* EDTA (filter-sterilized, stored at 4°C).
2. Deionized formamide: prepared as described by Sambrook et al. ***(14)*** and stored in aliquots at –80°C.
3. Hybridization buffer: 1× hybridization buffer in 80% deionized formamide, freshly prepared.
4. RNA probe (*see* **Note 4**).
5. Total cellular RNA (*see* **Note 5**).

2.3. Ribonuclease Digestion

1. Ribonuclease digestion buffer: 10 m*M* Tris-HCl, pH 7.5, 300 m*M* NaCl, 5 m*M* EDTA (sterilized by autoclaving).
2. RNase A: 4 mg/mL in 10 m*M* Tris-HCl, pH 7.5; 15 m*M* NaCl: boiled 15 min to remove DNase; stored in aliquots at –20°C (*see* **Note 6**).

3. RNase T1: 100,000–500,000 U/mL, 4°C (Boehringer Mannheim Biochemicals) (*see* **Note 7**).
4. 20% sodium dodecyl sulfate (SDS), filter sterilized.
5. Proteinase K: 20 mg/mL in 50 m*M* Tris-HCl, pH 8.0; 50% glycerol; stored at –20°C.
6. Phenol/chloroform/isoamyl alcohol, 25/24/1, equilibrated in 0.1 *M* Tris-HCl, pH 7.5.
7. Yeast tRNA: 10 mg/mL.
8. Ethanol.

2.4. Analysis of Ribonuclease Protected Fragments

1. Denaturing polyacrylamide-urea gel: 7% acrylamide, 7 *M* urea, 1× TBE.
2. 20× TBE: 1.78 *M* Tris-Borate, 0.08 *M* EDTA.
3. RNA loading buffer: 80% (vol/vol) deionized formamide; 1 m*M* EDTA, pH 8.0; 0.1% bromophenol blue; 0.1% xylene cyanol.
4. Radiolabeled marker.
5. Gel fixative: 10% methanol; 10% glacial acetic acid.

3. Methods

As the protocol involves working with radioactive materials, appropriate laboratory safety procedures should be followed to prevent the exposure and spread of radioactivity.

3.1. Preparation of Probe

3.1.1. In Vitro Transcription of Riboprobe

1. Add the following ingredients in order to a microfuge tube at room temperature: 4 μL 5× transcription optimized buffer, 2 μL DTT, 0.8 μL RNasin® ribonuclease inhibitor, 4.0 μL ribonucleotide mix, 2.4 μL CTP, 1.0 μL linearized template, 5.0 μL [α-^{32}P]CTP, 0.5 μL RNA polymerase, 0.3 μL RNase-free water to a final volume of 20 μL (*see* **Note 8**).
2. Incubate reaction for 1 h at 37°C.
3. Add 1 μL of RQ1 RNase-free DNase and incubate the reaction for 15 min at 37°C (*see* **Note 9**).
4. Place probe on ice.

3.1.2. Purification of Probe by Gel Filtration (*see* **Note 3**)

1. Gravity drain the storage buffer from the G-50 Sephadex column into a microfuge tube, as per instructions of the kit manufacturer (Boehringer Mannheim Biochemicals). Place the column in a collection tube provided by the manufacturer. Insert this tube and column into a capped tube (e.g., Falcon 2059) (to prevent the spread of contamination) and spin at 1000*g* for 2 min in a centrifuge (e.g., Beckman GS-6R) that is prechilled to 4°C. Discard collection tube with

eluate and then place the column in the second collection tube provided by the manufacturer.

2. Add 50 μL RNase-free H_2O to probe, then load onto column. Again insert the collection tube and column in the capped tube to prevent the spread of contamination. Centrifuge at 1000*g* for 4 min at 4°C. Transfer purified probe from collection tube into a new microfuge tube containing 100 μL of hybridization buffer.
3. Measure the ^{32}P incorporation in a scintillation counter (*see* **Note 10**).
4. Store the probe at –20°C until ready for use.

3.2. Hybridization of Probe and RNA

1. Determine the amount of RNA to be assayed. For abundant RNA species, 1–3 μg of total RNA is typically assayed. For rare RNA species, as much as 50 μg of total RNA may be assayed. The maximum volume of RNA that can be incubated with the hybridization reaction is 7 μL. If necessary, ethanol precipitate the required amount of RNA (*see* **Note 11**).
2. Determine the total amount of probe that is needed for the hybridization of n + 1 samples. Each sample should be assayed with 50,000–500,000 cpm of probe (we typically use 50,000 cpm/sample). A negative control sample with 10 μg tRNA should be included to verify that the ribonuclease digestion is complete. A second tRNA control sample may be included that will not be exposed to RNase, to analyze the integrity of the probe after hybridization. In addition, a range of RNA concentrations for a given sample should be assayed to verify that the probe is in excess. When analyzing allele-specific expression in RNA isolated from F1 hybrid tissue, it is important to include RNA samples isolated from the parental mouse strain, to ensure that allele-specific fragments will be detected (*see* **Fig. 2**).
3. Prepare the amount of hybridization buffer that is required for n + 1 samples (30 μL/sample, *see* **Subheading 2.2.**). Add the appropriate amount of probe to the hybridization buffer and place on ice.
4. Add 30 μL of the hybridization buffer and probe mix to n number of microfuge tubes. Add the appropriate amount of RNA (not to exceed 7 μL) and mix by pipetting up and down. Alternatively, if a sample has been precipitated, add hybridization buffer and probe mix to the precipitated RNA. Pipet up and down to ensure that the pellet is dissolved.
5. To denature the RNA, incubate tubes in a hot block at 85°C for 10 min (*see* **Note 12**).
6. Rapidly transfer the tubes to a water bath set at the appropriate hybridization temperature for the specific RPA probe (30–65°C). Hybridize for 4 h to overnight, as required (*see* **Note 13**).

3.3. Ribonuclease Digestion

1. Prepare the required amount of ribonuclease digestion buffer (350 μL/reaction) containing 40 μg/mL RNase A and 1 U/μL RNase T1 (*see* **Notes 6** and **7**).

2. Remove the hybridization reactions from the water bath and spin briefly. Add 350 µL ribonuclease digestion buffer mix to each reaction.
3. Incubate at optimum temperature (room temperature, 37°C) for 60 min (*see* **Note 14**).
4. To terminate the ribonuclease digestion, add 10 µL of 20% SDS and 2.5 µL of 20 mg/mL proteinase K.
5. Incubate at 37°C for 15 min.
6. Extract once with 400 µL of phenol/chloroform/isoamyl alcohol and remove the aqueous phase to a clean microfuge tube containing 10 µg of yeast tRNA on ice. Vortex and then spin 5 min in microcentrifuge.
7. Add 1 mL of ethanol and precipitate on dry ice for 15 min or at –20°C for at least 2 h.

3.4. Analysis of Ribonuclease-Protected Fragments (*see* Note 15)

3.4.1. Preparation of Denaturing Polyacrylamide-Urea Gel

1. Pour a 7% acrylamide/7 *M* urea/1× TBE gel and allow to polymerize (30 min–1 h). (A lower percent acrylamide may be used, particularly if analyzing long protected fragments.)
2. Assemble gel in gel apparatus, add 1× TBE buffer to upper and lower chambers, then carefully remove comb. Immediately flush the wells with buffer solution and, if necessary, adjust wells with syringe and needle.
3. Pre-run gel at least 30 min at ~250 V.

3.4.2. Analysis of RPA Products on Gel

1. Spin samples at the maximum speed for 15 min, preferably at 4°C.
2. Carefully remove supernatant with a pipetman and dry samples in speed vac.
3. Resuspend samples in 7 µL of loading buffer. In a separate tube, combine 0.5–2.0 µL (depending on specific activity) radiolabeled marker and 7 µL of loading buffer.
3. Immediately before loading, heat samples (including marker) at 85°C for 3 min and place on ice.
4. As samples are being heat-denatured, flush wells again from the prerun gel to remove urea that has leached out of the gel.
5. Load samples on gel, and run at ~250 V for appropriate length of time for the size of the protected RNA fragments.
6. Fix gel for 15 min, then dry before either exposing to X-ray film or a phosphorimage screen (Molecular Dynamics).
7. Quantitate relative RNA levels with densitometry of the X-ray film or by using Image Quant (Molecular Dynamics) to analyze the phosphorimage scan.

4. Notes

1. All of the reagents required for the synthesis of the RNA probe, except for the template DNA and the [α-^{32}P]CTP, are obtained from Promega Corporation and

stored at –20°C. The reagents are available individually or as a kit (Item P1420 for SP6, P1430 for T3, P1440 for T7).

2. It is important to generate the probe template from purified plasmid devoid of nucleases. We use DNA isolated from a CsCl-ethidium bromide gradient. To prepare an RPA template, we linearize 20 μg of plasmid with an appropriate restriction enzyme and verify that digestion is complete on an agarose gel. The DNA is subsequently phenol/chloroform-extracted, chloroform-extracted and ethanol-precipitated. The pellet is dissolved in RNase-free H_2O at a final concentration of 0.2–1 μg/μL. The plasmid may be digested to generate multiple fragments. However, digests that generate a 3′ overhang should be avoided, as 3′ overhangs serve as templates for bacteriophage RNA polymerase ***(15,16)***.
3. We routinely use gel filtration to purify the probe by removing unincorporated nucleotides. Alternatively, the probe can be phenol/chloroform/isoamyl (25/24/1) extracted, ethanol-precipitated with 10 μg tRNA, and resuspended in 100 μL of hybridization buffer ***(2)***. If partial length products are routinely synthesized by the in-vitro transcription assay, then gel isolation of the full-length probe is recommended.
4. The RPA probes are labeled to high specific activity and are therefore prone to rapid radiolysis. Although it is best to use a freshly made probe, probes can be used for up to a week after synthesis. It is suggested that the integrity of the probe be analyzed on a gel prior to each use. If degraded, the probe should be discarded.
5. Total cellular RNA prepared from tissues by extraction with LiCl-urea ***(17)*** is used for RPA analysis. The RNA is resuspended in TE (10 m*M* Tris-HCl, 1 m*M* EDTA), pH 7.5, and stored at –80°C.
6. Caution should be exercised when preparing the RNase A stock solution so that the surrounding work area is not contaminated. As the specific activity can vary from stock to stock, it is suggested that optimal RNase A concentration for the RPA be determined empirically each time a new stock is prepared.
7. The RNase T1 stock solution (1–2 mg/mL) that was previously provided by Sigma is no longer available. We currently use a Boehringer Mannheim Biochemicals stock solution that is provided at a concentration of 100,000 or 500,000 U/mL. We use 1 U/μL RNase T1 for the ribonuclease digestion. This is roughly equivalent to the recommended 2 μg/mL of the Sigma solution, as determined from the specific activity.
8. Multiple probes may be used simultaneously in an RPA to detect more than one RNA product in a sample. However, if the relative abundance of the RNA products varies, it is advisable to alter the specific activities of the probes. For example, when using a probe to detect an abundant RNA product, the "cold" CTP concentration should be increased and the [α-^{32}P]CTP concentration should be decreased; vice versa for the probe monitoring the rare RNA.
9. Removal of DNA template from the probe is very important. Contaminating DNA may be a source of background on a RPA gel that results from RNA : DNA hybrid formation. Contaminating DNA will also reduce the amount of probe

available for RNA:RNA hybridization. If the protected RNA product is smaller than the probe, full-length DNA:RNA hybrids will be readily distinguishable from the protected RNA fragment.

10. If the probe activity is less than 100,000 cpm, incorporation of the radionucleotide may not have been sufficient. To verify the integrity of the probe before use, we suggest that 0.5 µL of the probe be electrophoresced on a polyacrylamide/urea gel (*see* **Subheading 3.4.**). The wet gel covered with plastic wrap can then be exposed to a phosphorimage screen for 15 min or X-ray film for 1 h to analyze the integrity of the probe: a single discrete band should be observed.
11. The quantity of total RNA to be analyzed depends on the abundance of the target RNA in the tissue of interest. For *H19* and *Igf2*, 1–3 µg of total neonatal liver and skeletal muscle RNA is normally analyzed by RPA. In tissues where these mRNAs are less abundant, 10 µg (neonatal kidney) to 50 µg (neonatal brain) of the total RNA is analyzed.
12. It is important to eliminate secondary structure of the probe and RNA by denaturing the sample before the hybridization step. To avoid stable secondary structure, immediately transfer the denatured sample from the 85°C heating block to a water bath preset at the appropriate hybridization temperature.
13. Hybridization conditions are dependent on the melting temperature (T_m) of the probe and should be determined empirically. Specifically, for allele-specific analysis of *H19*, we hybridize samples at 65°C with a probe generated from the DNA template constructed by Bartolomei et al. *(5)*. For allele-specific analysis of *Igf2*, we hybridize the samples at 55°C with a probe generated from the DNA template constructed by Leighton et al. *(6)*. In general, longer hybridization times are required for less abundant targets. We typically perform all RPA incubations overnight.
14. To detect a single-base-pair polymorphism between alleles of a gene it may be necessary to incubate the ribonuclease reaction at an elevated temperature. For example, we perform the ribonuclease digestion at 37°C for the allele-specific detection of *H19* RNA and we perform the digestion at 30°C for the allele-specific detection of *Igf2* RNA. Thus, the optimum ribonuclease digestion temperature and the hybridization temperature should be determined empirically. Alternatively, the nonspecific RNase I (cleaves the bond of single-stranded RNA 3′ of any ribonucleotide) provided by Boehringer Mannheim Biochemicals may be utilized.
15. The gel running conditions are specific for 220 × 165 × 1 mm vertical gels.

Acknowledgments

The authors wish to thank Tamara Davis and Mellissa Mann for helpful suggestions in preparing this document.

References

1. Melton, D. A., Krieg, P. A., Rebagliati, M. R., Maniatis, T., Zinn, T., and Green, M. R. (1984) Efficient *in vitro* synthesis of biologically active RNA and RNA

hybridization probes from plasmids containing a bacteriophage SP6 promoter. *Nucleic Acids Res.* **12,** 7035–7056.
2. Gilman, M. (1993) Ribonuclease protection assay, in *Current Protocols in Molecular Biology* (Ausubel, F. M., et al., eds.), Wiley, New York, pp. 4.7.1–4.7.8.
3. Belin, D. (1998) The use of RNA probes for the analysis of gene expression, in *Methods in Molecular Biology, Vol. 86* (Rapley, R. & Manning, D. L., eds.), Humana, Totowa, NJ, pp. 87–102.
4. Myers, R. M., Larin, Z., and Maniatis, T. (1985) Detection of single-base substitutions by ribonuclease cleaveage at mismatches in RNA:DNA duplexes. *Science* **230,** 1242–1246.
5. Bartolomei, M. S., Zemel, S., and Tilghman, S. M. (1991) Parental imprinting of the mouse *H19* gene. *Nature* **351,** 153–155 .
6. Leighton, P. A., Ingram, R. S., Eggenschwiler, J., Efstratiadis, A., and Tilghman, S. M. (1995) Disruption of imprinting caused by deletion of the *H19* gene region in mice. *Nature* **375,** 34–39.
7. Tremblay, K. D., Saam, J. R., Ingram, R. S., Tilghman, S. M., and Bartolomei, M. S. (1995) A paternal-specific methylation imprint marks the alleles of the mouse *H19* gene. *Nature Genet.* **9,** 407–413.
8. Leighton, P. A., Saam, J. R., Ingram, R. S., Stewart, C. L., and Tilghman, S. M. (1995) An enhancer deletion affects both *H19* and *Igf2* expression. *Genes Dev.* **9,** 2079–2089.
9. Webber, A. L., Ingram, R. S., Levorse, J. M., and Tilghman, S. M. (1998) Location of enhancers is essential for the imprinting of *H19* and *Igf2* genes. *Nature* **391,** 711–715.
10. Jones, B. K., Levorse, J. M., and Tilghman, S. M. (1998) *Igf2* imprinting does not require its own DNA methylation or *H19* RNA. *Genes Dev.* **12,** 2200–2207.
11. Thorvaldsen, J. L., Duran, K. L., and Bartolomei, M. S. (1998) Deletion of the *H19* differentially methylated domain results in loss of imprinted expression of *H19* and *Igf2*. *Genes Dev.* **12,** 3693–3702 .
12. Brunkow, M. E. and Tilghman, S. M. (1991) Ectopic expression of the *H19* gene in mice causes prenatal lethality. *Genes Dev.* **5,** 1092–1101 .
13. Dudov, K. P. and Perry, R. P. (1984) The gene encoding the mouse ribosomal protein L32 contains a uniquely expressed intron containing gene and an unmutated processed gene. *Cell* **37,** 457–468.
14. Sambrook, J., Fritsch, E. F., and Maniatis, T. (1989) *Molecular Cloning, A Laboratory Manual*, 2nd ed., Cold Spring Harbor Laboratory Press, Cold Spring Harbor, NY.
15. Schenbon, E. T. and Mierendort, R. C. (1985) A novel transcription property at SP6 and T7 RNA polymerases: dependence on template structure. *Nucleic Acids Res.* **13,** 6223–6234 .
16. Krieg, P. A. and Melton, D. A. (1987) *In vitro* RNA synthesis with SP6 RNA polymerase. *Methods Enzymol.* **155,** 397–415.
17. Auffray, C. and Rougeon, F. (1980) Purification of mouse immunoglobulin heavy-chain messenger RNAs from total myeloma tumor RNA. *Eur. J. Biochem.* **107,** 303–314.

10

Quantitative RT-PCR-Based Analysis of Allele-Specific Gene Expression

Judith Singer-Sam and Chunguang Gao

1. Introduction

F1 hybrids resulting from intercrosses of inbred strains have provided an invaluable tool for the study of imprinting. The hybrids can be used to analyze parent-of-origin differences in expression of any gene, provided sequence differences exist between the two parental alleles. Methods used to detect allele-specific expression include ribonuclease protection assays *(1)* and allele-specific RNA *in situ* hybridization *(2)*, as well as a number of reverse transcriptase polymerase chain reaction (RT-PCR)-based assays (*see*, for example, **refs.** *3* and *4*). We describe here two such assays that are quantitative and require only single base differences between the two alleles. Both assays rely on the amplification of the RNA of interest by RT-PCR using primer sets that flank the sequence polymorphism, a method shown previously to yield amplicons whose allelic ratio is proportional to the ratio in the starting material, regardless of the number of cycles of amplification *(5)*.

The first assay, single nucleotide primer extension (SNuPE), is adapted from a method first described for the detection of allelic variants in DNA *(6,7)*. Briefly, it involves the following steps: (1) The gel-purified RT-PCR product is used as a template for the enzymatic extension by one radioactive nucleotide of a primer whose 3′ terminus is adjacent to the position of the allelic difference. Each allele is probed separately, by use of the appropriate ^{32}P-labeled dNTP. (2) Following denaturing gel electrophoresis, the contribution of each allele is determined from the ratio of the two primer-extension products. The method is highly reproducible, and allows the detection of an allelic variant in a 99-fold excess of the other.

From: *Methods in Molecular Biology, vol. 181: Genomic Imprinting: Methods and Protocols*
Edited by: A. Ward © Humana Press Inc., Totowa, NJ

The second assay combines automated fluorescent-based sequencing of the gel-purified RT-PCR products with software allowing quantitative analysis of the signal from each allele. There have been previous reports of automated sequencing of allelic variants *(8–12)*; the method as outlined here does not require specific tagged primers or a size difference between the two PCR products, and, rather than tracings, raw sequencing data is used to determine peak heights. **Figure 1** demonstrates the quantitative nature of the assay. PCR products differing as shown (two A/G mismatches and a G/T mismatch) were mixed in proportions ranging from 1:9 to 9:1, then subjected to automated sequencing with fluorescent-tagged dideoxynucleoside triphosphates. The ratio of the peak heights varies with the input ratio; the relative proportion of a given allelic variant can thus be determined by comparison with a standard curve specific for the mismatch being measured. This rapid method has the following advantages: (1) As part of the automated sequencing reaction, one learns the identity and quality of the surrounding sequence, to verify the accuracy of the measurement. (2) Where there are multiple mismatches, as in the sequence shown in **Fig. 1**, the allelic ratios obtained can be averaged. (3) The technique is rapid, nonradioactive, and amenable to the analysis of multiple samples. (4) The data appear to be as reproducible as the SNuPE assay, the major drawback being that, depending on the strength of the signal for a given base, ratios below 1:10 may be undetectable above background. Thus, in cases where detection of alleles expressed at relatively low levels is desired, the quantitative SNuPE assay is still preferable.

1.1. Analysis of Single Cells

Imprinting is not always an all-or-none process, in many cases varying with developmental state and tissue type (reviewed in **ref.** ***13***). For this reason, it is sometimes desirable to analyze allele-specific expression in individual cells or cell types. While conditions for amplification of each gene must be determined empirically, we and others have found that even genes expressed at relatively low levels can be amplified from single cells, and describe here the method we used to amplify Ncam RNA from individual neurons *(14)*.

2. Materials

2.1. Purification of RNA

1. Presiliconized RNase/DNase-free microcentrifuge tubes (0.65 mL) (Sorenson Bioscience).
2. RNAzol (Tel-Test, Friendswood, TX).
3. Mussel glycogen, 20 mg/mL in DEPC-treated water (Sigma, St. Louis, MO).
4. Chloroform.

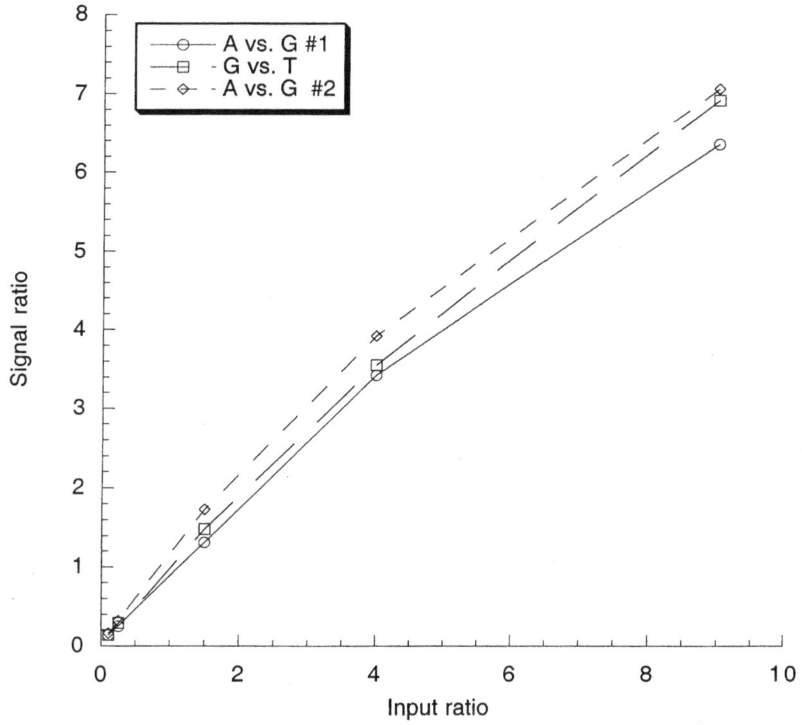

Fig. 1. Quantitative measurement of allele-specificity by direct sequencing. A region of the *Snrpn* gene containing two G/A mismatches and one G/T mismatch between *Mus spretus* and *Mus musculus* C57BL/6 DNA was amplified from DNA of the two strains. The PCR products were mixed in the proportions indicated (*x* axis), and the signal ratio determined by the combined use of automated sequencing and of a program for peak analysis (*y* axis). See text for details.

5. 2-Propanol.
6. 75% ethanol.

2.2. RT-PCR

1. DEPC-treated water.
2. 10× buffer: 100 m*M* Tris-HCl, pH 8.3, 500 m*M* KCl.
3. Magnesium chloride, 25 m*M*.
4. Upstream and downstream gene-specific primers with T_m of ~60°C, 20 μ*M*.
5. dNTPs (BoehringerMannheim, Germany), diluted to 10 m*M*.
6. RNasin, 40 U/μL (Promega, Madison, WI).
7. Moloney murine leukemia virus reverse transcriptase (MMLV-RT), 200 U/μL (Gibco BRL, Gaithersburg, MD).

8. AmpliTaq DNA Polymerase, 5 U/μL (Perkin Elmer).
9. Mineral oil (not necessary if thermal cycler is equipped with a hot bonnet).

2.3. Gel Purification

1. 50 m*M* TBE buffer: Dilute from 1 *M* solution: 121.1 g Tris base, 51.4 g boric acid, 3.7 g disodium ethylenediaminetetraacetic acid (EDTA) in 1 L, pH 8.3 ***(15)***.
2. Agarose, ultrapure (GibcoBRL).
3. 10 mg/mL ethidium bromide.
4. GeneClean (Bio 101, Vista CA).

2.4. SNuPE Assay

1. 10× buffer: 100 m*M* Tris-HCl, pH 8.3, 500 m*M* KCl, 20 m*M* $MgCl_2$, 0.01% gelatin (optimum pH and $MgCl_2$ concentration should be determined empirically for each primer).
2. SNuPE primer, 18-mer or longer, with 3′ end just 5′ to allelic base difference (20 μ*M*).
3. [^{32}P]dNTPs corresponding to bases that differ, 3000 Ci/mmol, diluted to 1 μCi/μL in H_2O just before incubation.
4. 12% polyacrylamide–7 *M* urea gel in 100 m*M* TBE, pH 8.3 (*see* **Subheading 2.3.**).
5. Sequencing gel-loading buffer: 98% deionized formamide, 10 m*M* EDTA, pH 8.0, 0.025% xylene cyanol FF, 0.025% bromphenol blue ***(15)***.
6. AmpliTaq polymerase, 5 U/μL (Perkin Elmer).

2.5. Direct Sequencing

1. ABI Prism 377 DNA Sequencer (Perkin Elmer).
2. BigDye™ Terminator Cycle Sequencing Kits with AmpliTaq DNA Polymerase, FS, including dRhodamine Matrix Standards Kit and protocol #4303237 (Perkin Elmer).
3. GeneScan 2.1 (Perkin Elmer).

3. Methods

3.1. Purification of RNA

1. Add 5 μL chloroform to each cell dissolved in 50 μL RNAzol in 0.65-mL microcentrifuge tube. Vortex vigorously. Keep on ice 5 min.
2. Centrifuge at 12,000*g*, 15 min, 4°C.
3. Transfer aqueous phase to fresh microcentrifuge tube.
4. Add 2 μL glycogen solution. Vortex.
5. Add an equal volume of 2-propanol.
6. Keep at –20°C overnight (*see* **Note 1**).
7. Centrifuge for 30 min at 12,000*g* 4°C.

8. Wash one time with 75% ethanol without mixing.
9. After decanting as much of ethanol as possible, air dry for 5 min.
10. Use sample immediately for RT-PCR.

3.2. RT-PCR

For each gene analyzed, a primer set should be selected that flanks the base(s) differing between alleles, and primers chosen that give a single strong band after amplification. The optimal concentration of $MgCl_2$ in both RT and PCR steps should be determined empirically for each primer set. In some cases the use of oligo($dT_{30}N_2$) or random primers in the RT step in place of the downstream primer improves the RT-PCR signal. Following is the procedure that we have used to amplify Ncam RNA from single mouse hippocampal neurons ***(14)***.

1. Prepare pre-RT master mix: Mix together (per sample) 2 μL 10× buffer, 4 μL $MgCl_2$, 2.5 μL downstream primer, and 6.5 μL DEPC-treated water. Suspend each sample in 15 μL of the mix and add mineral oil, 50 μL. Subsequent incubations are all done using a thermal cycler.
2. Incubate at 64°C for 10 min, then cool slowly at ~5 s/°C.
3. Prepare RT master mix: Mix together (per sample) 2 μL dNTPs, 0.25 μL MMLV-RT, 0.5 μL RNasin, and 2.25 μL DEPC-treated H_2O. Add 5 μL to each sample, and incubate at 42°C for 1 h.
4. Inactivate the MMLV RT at 99°C for 10 min, then keep samples at 55°C while preparing PCR master mix below (maximum time of incubation, 30 min).
5. Prepare PCR mix: Mix together (per sample) 8 μL 10x buffer, 0.5 μL Amplitaq polymerase, 2.5 μL upstream primer, and 69 μL H_2O. Add 80 μL to each sample (*see* **Note 2**).
6. Amplify using PTC-100 Thermal Cycler (MJ Research) programmed as follows: 95°C, 2 min, 1 cycle; 95°C, 1 min, *N*°C, 1 min, 72°C, 1.5 min, 2 cycles, where *N* is successively 60, 57, 54, 51, 49, and 47°C; 95°C, 1 min, 45°C, 1 min, 72°C, 1.5 min, 40 cycles; 95°C, 2 min, 72°C 5 min.

3.3. Gel Purification

1. Electrophorese each sample (up to 100 μL) on 1.5% agarose gel in 50 m*M* TBE.
2. Excise band from agarose gel using GeneClean, following manufacturer's instructions.
3. Elute DNA with water.
4. To determine the concentration of each sample, run serial twofold dilutions on an agarose gel stined with ethidium bromide (use 10 μL of 10-mg/mL ethidium bromide per 100 mL of agarose solution), and compare with known quantities of DNA .

3.4. SNuPE Assay

1. Dilute each sample to 2 ng/μL and distribute 5 μL (10 ng) to each of two microcentrifuge tubes.
2. Prepare master mix: Mix (per tube) 1 μL 10× buffer, 0.5 μL SNuPE primer, 0.15 μL AmpliTaq Polymerase, and 1.35 μL H_2O. Add 3 μL of master mix to each tube.
3. For each sample, add 2 μL (2 μCi) of one allele-specific [^{32}P]dNTP into one tube, and 2 μL of the second [^{32}P]dNTP into another (*see* **Note 3**).
4. Incubate the samples in a thermal cycler for one round of denaturation, annealing, and synthesis: 95°C, 1 min, 42°C, 2 min, 72°C, 1 min. Place on ice.
5. As controls for background and maximum incorporation, incubate 10 ng of amplified product from each allele with the "incorrect" and appropriate [^{32}P]dNTP, respectively. Initially, controls should also include various ratios of the two alleles, mixed before amplification, to establish a standard curve (*see* **Note 4**).
6. After addition of sequencing gel-loading buffer (10 μL), incubate each sample at 90°C for 1–2 min, and place on ice. Load 10 μL of each sample on a 1-mm-thick 12% polyacrylamide-urea gel (18.5 cm long × 16.5 cm wide) and electrophorese at 30 mA/gel until the bromphenol blue reaches two-thirds of the way to the bottom of the gel (~40 min).
7. Cut out the portion of the gel between the bromphenol blue and xylene cyanol markers. Wrap the gel slice in Saran wrap, and determine the amount of radioactivity above lane background in each (n + 1) band (n = length of SNuPE primer) by use of a radioisotope scanning system, such as the PhosphorImager (Molecular Dynamics, Sunnyvale, CA) (*see* **Notes 5** and **6**).
8. For each sample, calculate the ratio of cpm incorporated with each of the two [^{32}P]dNTPs. Correct for the following: (1) background incorporation of the incorrect [^{32}P]dNTP, calculated from the ratio of the signals seen when the control templates for each allele are incubated with the incorect vs correct [^{32}P]dNTP—the ratio should be ≥1%; (2) different efficiency of incorporation with each [^{32}P]dNTP, determined from the ratio seen when equal amounts of each control template are incubated with the corresponding [^{32}P]dNTP.

3.5. Direct Sequencing

1. Perform automated DNA sequencing using 10 ng of PCR product for each 100 bp in length, and the protocol supplied with the BigDyeTM Terminator Cycle Sequencing Kit. Use dRhodamine Matrix Standards and filter E when performing acrylamide gel electrophoresis.
2. For each sample, open the corresponding sequencing file in the GeneScan program.
3. In the Settings menu, under Preferences, vary (1) Results Dye Scales, to optimize ability to select minor peaks, and (2) Dye Indicators, so that the color correspondence of the bases on the screen is the same as that of the sequencing program. This window also provides a reference for the default color correspondence used by GeneScan in tabulating peak heights of the bases.

Then, under Auto-Analysis Defaults, select GS2500.1 as the size standard, and R as the dye.
4. In the Sample menu, under Install New Matrix, open the dRhodamine matrix (this must be done each time a sample is analyzed). Then select Analyze Sample.
5. Once the results of quantitative analysis appear in the window, use the magnification tool to locate the base mismatch(es) of interest. Select the bases, then in the View menu, select View Only Selected Rows.
6. Copy resulting table to Microsoft Excel or other spreadsheet.
7. To determine allelic ratio, compare ratio of peak heights to standard curve obtained as in **Fig. 1**.

4. Notes

1. While overnight precipitation in 2-propanol has given the most consistent results, in some cases a 2-h incubation at –20°C is sufficient.
2. Use of variants of AmpliTaq Polymerase such as AmpliTaq Gold (Perkin Elmer) and HotStarTaq together with Q-Solution (Qiagen) often result in increased RT-PCR signals. While we have not yet confirmed quantitative allele-specific amplification from single cells using these polymerases, they may ultimately prove to be superior to AmpliTaq.
3. It is best to use [^{32}P]dNTPs within several days of arrival, and to store templates and SNuPE primers at 4°C or frozen in aliquots, to avoid repeated freezing and thawing.
4. The sensitivity of the SNuPE assay depends on the particular mismatch. While C/A can be measured down to the 0.001% level, G/C, C/G, and G/A mismatches have a higher limit of ~1%. C/T, T/C, and A/G mismatches give backgrounds similar to C/A.
5. When the base 3′ to the incorporated [^{32}P]dNTP is identical to it, (n + 2) bands of varying intensity will be obtained, where n is the length of the SNuPE primer. These products should be included in the calculation along with the (n + 1) products.
6. Occasional minor bands of incorrect size may be seen. These can be ignored as long as control experiments show that the expected standard curve is obtained by calculations using only the SNuPE product of the expected size(s).

Acknowledgments

We thank Mr. Dan Darcy for data analysis and intelligent questions, and Dr. Gerald Forrest and Mr. Hector Rivera for their invaluable help with the GeneScan program. We also thank Drs. Michael E. Barish, Jeanne M. LeBon, Jeffrey R. Mann, Michael Reed, Arthur D. Riggs, Louise Shively, and Piroska Szabo for their kind help.

References

1. Bartolomei, M. S., Zemel, S., and Tilghman, S. M. (1991). Parental imprinting of the mouse *H19* gene. *Nature* **351,** 153–155.

2. Adam, G. I. R., Cui, H., Miller, S. J., Flam, F., and Ohlsson, R. (1996). Allele-specific in situ hybridization (ASISH) analysis: a novel technique which resolves differential allelic usage of H19 within the same cell lineage during human placental development. *Development* **122,** 839–847.
3. Nozari, G., Rahbar, S., and Wallace, R. B. (1986). Discrimination among the transcripts of the allelic human β-globin genes, β^A, β^S and β^C using oligodeoxy-nucleotide hybridization probes. *Gene* **43,** 23–28.
4. Giddings, S. J., King, C. D., Harman, K.W., Flood, J. F., and Carnaghi, L.R., (1994). Allele specific inactivation of insulin 1 and 2, in the mouse yolk sac, indicates imprinting. *Nature Genet.* **6,** 310–313.
5. Gilliland, G., Perrin, S., Blanchard, K., and Bunn, H. F. (1990). Analysis of cytokinine mRNA and DNA: detection and quantitation by PCR. *Proc. Natl. Acad. Sci. USA* **87,** 2725–2729.
6. Kuppuswamy, M. N., Hoffmann, J. W., Kasper, C. K., Spitzer, S. G., Groce, S. L., and Bajaj, S. P. (1991). Single nucleotide primer extension to detect genetic diseases: experimental application to hemophilia B (factor IX) and cystic fibrosis genes. *Proc. Natl. Acad. Sci. USA* **88,** 1143–1147.
7. Singer-Sam, J., LeBon, J. M., Dai, A., and Riggs, A. D. (1992). A sensitive, quantitative assay for measurement of allele-specific transcripts differing by a single nucleotide. *PCR Methods Appl.* **1,** 160–163.
8. Gibbs, R. A., Nguyen, P. N., McBride, L. J., Koepf, S. M., and Caskey, C. T. (1989). Identification of mutations leading to the Lesch-Nyhan syndrome by automated direct DNA sequencing of in vitro amplified cDNA. *Proc. Natl. Acad. Sci. USA* **86,** 1919–1923.
9. Porcher, C., Malinge, M.-C., Picat, C., and Grandchamp, B. (1992). A simplified method for determination of specific DNA or RNA copy number using quantitiative PCR and an automatic DNA sequencer. *BioTechniques* **13,** 106–114.
10. Hattori, M., Shibata, A., Yoshioka, K., and Sakaki, Y. (1993). Orphan peak analysis: a novel method for detection of point mutations using an automated fluorescence DNA sequencer. *Genomics* **15,** 415–417.
11. Leren, T. P., Rodningen, O. K., Rosby, O., Solberg, K., and Berg, K. (1993). Screening for point mutations by semi-automated DNA sequncing using sequenase and magnetic beads. *BioTechniques* **14,** 618–623.
12. Kwok, P.-Y., Carlson, C., Yager, T. D., Ankener, W., and Nickerson, D. A. (1994). Comparative analysis of human DNA variations by fluoresence-based sequencing of PCR products. *Genomics* **23,** 138–144.
13. Kelsey, G. and Reik, W. (1998). Analysis and identification of imprinted genes. *Methods* **14,** 211–234.
14. Shen, S. I., Gao, C., and Singer-Sam, J. (1998). Use of a reverse transcriptase-polymerase chain reaction assay to analyze allele-specific expression in individual hippocampal neurons. *Mol. Genet. Metab.* **63,** 96–102.
15. Sambrook, J., Fritsch, E. F., and Maniatis, T., eds. (1989). *Molecular Cloning*, Cold Spring Harbor Laboratory Press, Cold Spring Harbor, NY.

11

Allele-Specific *In Situ* Hybridization (ASISH)

Rolf Ohlsson, Kristian Svensson, Hengmi Cui, Helena Malmikumpu, and Gail Adam

1. Introduction

An unexpected outcome of the diploid genome is that evolutionary strategies have evolved to express only one of the alleles *(1)*. The rapidly expanding list of genes that are expressed monoallelically fall into three main categories: random inactivation, allelic exclusion, and genomic imprinting. These categories are distinguished by whether the expressed and inactivated alleles are maintained from one cell division to the next (random inactivation occurs with each cell division, compared to stable propagation through subsequent cell divisions as seen for allelic exclusion and genomic imprinting) and whether allele inactivation or expression is determined by parent of origin of inheritance (this differentiates allelic exclusion from genomic imprinting). Genomically imprinted genes are currently more numerous than members of the other categories. This fact is likely to reflect that the persistent (in)activation of one allele in a parent of origin-specific manner has facilitated the detection of monoallelic expression patterns in RNA extracted from homogenized tissue. This crude approach does not take into account, however, different imprinted states within a tissue, or, indeed, random allelic (in)activation. It is not surprising, therefore, that well-known genes, such as *Il2* *(2)*, were only recently found to be monoallelically expressed in a random manner *(1)*.

In order to avoid the problems posed by analysing tissue homogenates, allele-specific expression needs to be examined at the individual cell level. The need to visualize complex allelic expression patterns has promoted the application of two complementary techniques. The first is a fluorescent *in situ* hybridization (FISH)-based visualization of nuclear transcripts still attached

From: *Methods in Molecular Biology, vol. 181: Genomic Imprinting: Methods and Protocols*
Edited by: A. Ward © Humana Press Inc., Totowa, NJ

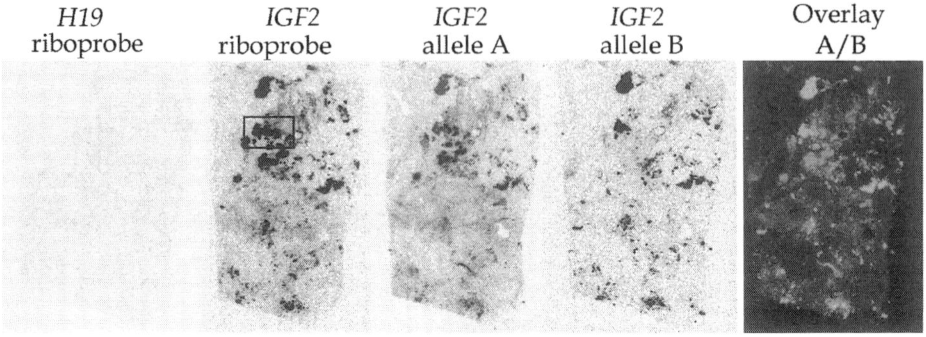

Fig. 1. Heterogeneous allelic expression patterns of the *IGF2* gene in a Wilms' tumor as determined by the ASISH method. Panels show expression analysed using *H19* and IGF2 riboprobes, as well as oligomers for the A and B alleles, as indicated in the figure (*see* **Fig. 2A** for further information). The rightmost panel is a false-colored overlay of X-ray images of allelic expression patterns. This image serves mainly to illustrate the overall heterogeneity in allelic usage of the *H19* gene in this tissue. Original magnification is 1.3-fold for all panels.

to the corresponding genes ***(3)***. This approach has been successfully applied to examine mechanisms underlying the random silencing of the *Xist* gene during X-inactivation in the soma, for example ***(4)***. The second technique, allele-specific *in situ* hybridization (ASISH), is based on oligomers that are designed to span polymorphic sites in exons of genes of interest ***(5–10)***.

By controlling hybridization conditions, allele-specific oligoprobes can anneal specifically to expressed sequences from opposite alleles based on single nucleotide differences. Although its resolution cannot currently be matched with that of RNA FISH, it has two unique advantages. First, it provides an excellent overview of allelic expression patterns in individual tissues or whole embryos (**Fig. 1**). Second, in contrast to RNA FISH techniques, it allows a determination of the parental origin of the expressed sequence ***(5–10)***.

1.1. Choice of Allele-Specific Probe

The oligo probe selected for the ASISH analysis depends on the availability of polymorphic sequences within exons. Although these can initially be narrowed down by restriction fragment length polymorphism (RFLP) or single-stranded conformational polymorphism (SSCP) analysis, both alleles of the region of interest must be sequenced not only at each polymorphic site, but also in the region surrounding the polymorphism that will be encompassed by the oligonucleotide probes. Once the polymorphic site has been identified, it is prudent to design oligomers with the mismatch or sequence difference

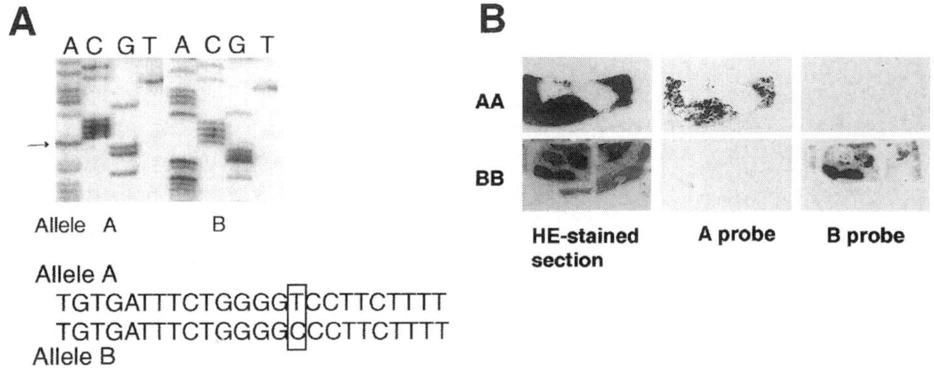

Fig. 2. Fidelity of the allele-specific *in situ* hybridization (ASISH) technique. **(A)** Example from tissue material homozygous for the AA (*Apa* I noncutting) or BB (*Apa* I cutting) alleles of *IGF2*, respectively (arrow indicates polymorphism within the *Apa* I recognition sequence). The oligomers that were used to discriminate allelic *IGF2* expression patterns *in situ* (the sequence polymorphism is boxed) are also depicted. **(B)** Control *in situ* hybridization analysis of Wilms' tumor and neuroblastoma (homozygous for the A and B alleles, respectively). The left panels of (B) are hematoxylin (HE)-stained images of sections immediately adjacent to each other to indicate the areas analyzed for *H19* expression.

positioned in the center. The overall size of the oligomer should not exceed 35–40 bases, as it may otherwise be difficult to find hybridization and posthybridization wash conditions that prevent longer oligo probes from interacting with opposite allele sequences. The base composition of the sequences flanking the polymorphic site also plays a role: As a rule, the internal stability profiles of each oligo probe should be balanced across the polymorphic sequence; for example, GC-rich sequences may be shorter and AT-rich sequences longer to enable optimal allele-specific annealing. It should also be noted that the length of the oligo is inversely proportional to the specific activity of the oligo probe (see below). We have found it suitable, therefore, to limit the size of the probe to around 30 bases, which is sufficiently long to ensure specific annealing to the sequence of interest, but sufficiently short to enable allele-specific annealing of a probe with optimal specific radioactivity. One such example (for *IGF2*) is shown in **Fig. 2A**

1.2. Specificity of the Hybridization Signal

Given that the ASISH analysis exploits minor sequence differences, it is important to define carefully the conditions that allow discrimination of allelic expression patterns. This is performed by examining tissue specimens which are homozygous for either allele type (see **Fig. 2B**, for example). Using such specimens, hybridisation conditions where allele A probe does not anneal to allele B sequences and vice versa can be established. In our experience, however, the annealing conditions alone may be insufficient to establish allele-specific hybridization signals satisfactorily. Indeed, in some cases, particularly where only a single nucleotide mismatch is available for analysis, we have found it helpful to anneal allele-specific probes at a relatively stringent but allele-nondiscriminative temperature and focus on the posthybridization washes.

Although the mere fact that an allele-specific hybridization signal can be obtained depending on a single nucleotide difference is indicative of a high degree of specificity, there is a need to rule out that one of the allelic probes does not generate a false signal. The specificity of the hybridization signal can be assessed in several ways. First, it is imperative that homozygous control specimens accompany each ASISH analysis. Second, it is essential to examine adjacent sections with an antisense riboprobe covering a larger portion of the same expressed region of interest and compare the signal patterns obtained by the riboprobe and oligoprobe. Third, it is vital to reproduce the hybridization signal in series of sections. A suitable strategy is to anneal allele-specific probes A and B, for example, and antisense riboprobes to a series of adjacent slides in the following order: A, B, antisense riboprobe, A, B (*see* **Fig. 3**) *(9)*. It is also advised that this strategy includes a sense control riboprobe. Fourth, we have often found it helpful to include a severalfold excess of unlabeled oligomer to reduce nonspecific hybridization of the oligo probe. Ideally, the unlabeled oligomer employed as a cold competitor is derived from the opposite allele. In this instance, any hybridization signal originating from the opposite allele should be reduced. It should be noted, however, that if the allele-specific discriminative step in the ASISH protocol is dependent on the stringency of posthybridization washes rather than hybridization temperature, the allele-specific cold competitor may fail to differentiate between the allelic sequences during hybridization. If so, the inclusion of cold oligomers representing sequences of the opposite allele may run the risk of reducing the allele-specific hybridization signal without attaining increased specificity.

1.3. Sensitivity of Allele-Specific Probes

The low sequence complexity of the oligomers reduces the sensitivity of the ASISH analysis. As a rule of thumb, we generally obtain a signal that is an order of magnitude lower using the oligoprobes compared to the hybridization signal obtained by using a 400- 500-nucleotide antisense RNA probe (covering the same expressed region). Although nonisotopic detection systems, such as alkaline phosphatase detection systems, provide higher resolution and might

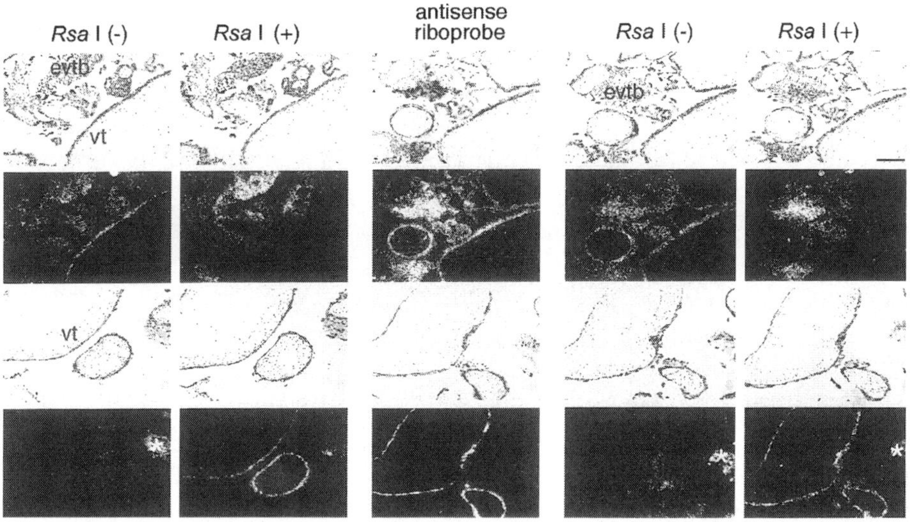

Fig. 3. Sequential ASISH analysis of allelic *H19* expression patterns in dispermic complete hydatidiform moles, illustrating the importance of verifying the reproducibility of the allele-specific hybridization signal in series of adjacent sections. Two different areas (upper and lower rows, respectively) were examined in sections 200 μm apart. For details of the *H19* ASISH protocol, *see* **Subheading 3.3.4.**). The asterisks indicate areas that generate false dark-field signals. Abbreviations: evtb = extravillous cytotrophoblasts; vt = villous cytotrophoblasts. The bar corresponds to 0.2 mm.

even allow simultaneous analysis of allelic expression patterns, their sensitivity is currently inferior to isotopic detection systems. The success of the ASISH approach depends, therefore, on the generation of radioactive oligo probes with an optimal specific activity. We have found it useful to end-label the oligomers at their 3′-end using ^{35}S-dATP and terminal transferase. The resulting product has a poly(A) tail, which can be exploited in the purification process using oligo dT columns ***(11)***. Ideally, the purified oligomer has a poly (A) tail with a mean size of 20–40 bases, which in practical terms means that every second base is labeled, on average. The specific radioactivity of the probe is limited, therefore, by the specific activity of the ^{35}S-dATP, the performance of the terminal transferase used in the labeling process, and the initial length of the oligomer. The poly (A) tail generated in the labeling process does not normally generate a background problem, since the hybridization conditions used in our protocol prevent the formation of any oligo (U)/poly (dA) hybrids (but *see* **Subheading 3.4.4.**). Although we have so far failed to see any significant difference with or without cold poly (A) competitor, we nevertheless routinely include both poly (A) and poly (dA) in our hybridization cocktails.

Following posthybridization washes, it is essential to assess the result and estimate the time that the slides need to be exposed with autoradiographic emulsion. We routinely expose the hybridized and washed slide, therefore, to Kodak Biomax films for 1–7 days. The image produced can be very helpful in analyzing the overall allelic expression patterns and signal intensity. The slides are then dipped in photographic emulsion and exposed for a time period that may vary from a few days to a few weeks, depending on signal intensity. As a rule of thumb, to estimate the time needed to expose the slides to photographic emulsion, we multiply the time needed to generate an acceptable image on the X-ray film approximately fivefold. An emulsion background control slide should always be included when slides are intended for long-term emulsion exposures.

2. Materials

2.1. Slide Preparation

2.1.1. Glass Slides and Rubber Cement

1. SuperFrost*/Plus (Menzel-Gläser) glass slides.
2. Rubber cement (Fastik, AB Thure Bunger, Sweden).

2.1.2. Removal of Paraffin

1. Coplin jars: oven-baked at 165°C overnight.
2. Xylene.
3. A series of 70%, 90%, 95%, and 99% ethanol.

2.1.3. Pretreatment

1. 0.2 *M* Tris-HCl, 2 m*M* CaCl, pH 7.2.
2. 0.1× SSC (20 × SSC: 3 *M* NaCl, 0.3 *M* Na citrate).
3. 0.2 *M* HCl.
4. 2× SSC.
5. Proteinase K (Merck); 10 mg/mL in 10 m*M* Tris-HCl, pH 7.4.
6. Acetic acidanhydride (BDH).
7. 0.1 *M* triethanolamine (BDH), pH 8.0.
8. A series of 70%, 90%, and 95% ethanol.

2.2. Labeling and Purification of Oligomers

1. Double-distilled water (ddH_2O).
2. 5× TdT buffer (Boehringer-Mannheim).
3. 2 pmol oligo DNA (stored in 5 µL 2×10^{-6} *M* aliquotes at –70°C).
4. 50 µCi ^{35}S-dATP (Amersham SJ1334).
5. Terminal transferase (TdT, Boehringer-Mannheim).
6. OligoTex kit (Quiagen).

2.3. Hybridization Cocktails

2.3.1. Prehybridization

Prehybridization mixture	Final conc.	Volume	Notes
Formamide (Fluka)	50%	600 µL	
SSC (20×SSC = 3 *M* NaCl and 0.3 *M* Na citrate)	3×	180 µL (from 20× SSC stock)	
Vanadyl ribonucleoside (Gibco BRL)	10 m*M*	60 µL (from 200 m*M* stock)	
ss Herring sperm DNA (Gibco BRL)	0.2 mg/mL	24 µL (from 10 mg/mL)	Denature at 95°C for 4 min
Yeast RNA (Sigma)	0.5 mg/mL	12 µL (from 50-mg/mL stock)	
Poly A (Sigma)	100 µg/mL	12 µL (from 10-mg/mL stock)	
Poly dA (Sigma)	50 µg/mL	2 µl (from 30-mg/mL stock)	
BSA (grade V, Sigma)	2 mg/mL	24 µL (from 100-mg/mL stock)	
Double-destilled water (ddH_2O)		286 µL	

Add 100 mg DTT (dithiotreitol, Sigma) immediately before use.

2.3.2. Hybridization

Hybridization mixture	Final conc.	Volume	Notes
Dextran sulfate (Pharmacia)	10%		See below
Formamide	50%	696 µL	See below
SSC	3×	180 µL (from 20× SSC stock)	
Vanadyl ribonucleosides	10 m*M*	60 µL (from 200 m*M* stock)	
ss Herring sperm DNA	0.2 mg/mL	24 µL (from 10 mg/mL stock)	Denature at 95°C for 4 min
Yeast RNA	0.5 mg/mL	12 µL (from 50-mg/mL stock)	
Poly A	100 µg/mL	12 µL (from 10-mg/mL stock)	
Poly dA	50 µg/mL	2 µL (from 30-mg/mL stock)	

1. Dextran sulfate–formamide mixture should be prepared in advance by thoroughly dissolving 12 g Dextran sulfate in 50 mL formamide at 60°C. The solution is then divided into 696-μL aliquots and stored at –20°C.
2. A few hours before use, place desired number of aliquots into a heat block (37°C for oligo probe hybridization cocktails).
3. Add 100 mg DTT immediately before use.
4. To make a 100-μL final hybridization cocktail, take 83 μL of hybridization mixture, add probe, and ddH_2O up to 98 μL. Denature at 95°C for 4 min, cool rapidly on ice, and add 2 μL BSA. Mix thoroughly.

2.3.3. Autoradiography

1. 25 mL of ddH_2O/0.5% glycerol.
2. Solid emulsion shreds of Kodak NTB2 emulsion.
3. Dipping chamber. Should have the same depth as the slides but sufficient volume to allow dipping of 10–15 slides without replenishing the final emulsion.
4. D19 developer.
5. 30% sodium thiosulfate.

3. Methods

3.1. Slide Preparation

3.1.1. Removal of Paraffin

1. Place slides in 60°C oven for 30 min to melt paraffin.
2. Cool to room temperature and dissolve remnant paraffin by immersing sections for 2 × 4 min in xylene.
3. Remove xylene by sequential treatment with 99%–95%–90%–70% ethanol (2 × 4 min in each).
4. Air dry.

3.1.2. Pretreatment

1. Place baked Coplin jars with 80 mL of 0.2 *M* Tris-HCl, 2 m*M* $CaCl_2$, pH 7.2, at 37°C.
2. Prewarm baked Coplin jars containing 80 mL of 0.1 × SSC at 56°C.
3. Incubate slides for 10 min in 0.2 *M* HCl in a baked Coplin jar.
4. Carefully transfer sections to a baked Coplin jar with 80 mL of 2 × SSC and incubate for 2 × 5 min. Make certain that the slides are completely immersed in the solution.
5. Transfer slides to a new baked Coplin jar (*see* **step 2**) and heat treat in 0.1 × SSC at 56°C for 30 min.
6. Transfer slides to a baked Coplin jar (*see* **step 1**) and permeabilize sections with 2-μg/mL proteinase K (16 μL of 10-mg/mL stock for 80 mL of solution) in prewarmed 0.2 *M* Tris-HCl, 2 m*M* CaCl, pH 7.2 at 37°C for 15 min.

7. Dehydrate and rehydrate sequentially in 70%–90%–95%–90%–70% ethanol (4 min in each).
8. Rinse in 2 × SSC, 2 × 5 min.
9. Place slides in a baked coplin jar and acetylate slides (to reduce background signals) by adding 200 µL of acetic acid anhydride and 80 mL of 0.1 *M* triethanolamine, pH 8.0, simultaneously. Vigorously rotate Coplin jar sideways for 10 s and incubate for 10 min. It is important that the acetic acid anhydride solution is rapidly distributed in the Coplin jar, due to its short half-life.
10. Rinse slides for 2 × 5 min in 2 × SSC.
11. Dehydrate slides in 70% ethanol and store in 70% ethanol at –20°C. Use slides within 2 wk.

3.2. Labeling and Purification of Oligo Probes

3.2.1. 3′-End Labeling

1. Reaction mixture contains 31 µL, 10 µL of 5 × TdT buffer, 2 pmole oligo DNA, 50 µCi of ^{35}S-dATP, 20–30 U of TdT, and ddH_20 to make a total volume of 50 µL.
2. Incubate for 1 h at 37°C

3.2.2. Purification (Using Quiagen Oligotex mRNA Mini Kit)

1. Prewarm binding buffer that contains SDS at 65°C for 1 h before use.
2. 10 min before use, prewarm oligotex at 37°C, vortex for 1 min, then store on ice.
3. Aliquot elution buffer into autoclaved microcentrifuge tube and prewarm at 75–95°C.
4. Add to the oligo labeling reaction mix the following: 150 µL ddH_2O, 200 µL binding buffer, and 10 µL Oligotex.
5. Vortex 10 min at room temperature.
6. Pulse-spin before transferring to spin columns.
7. Resuspend beads by gentle pipetting and transfer to spin column and spin for 30 s at 14,000 rpm in an Eppendorff centrifuge.
8. Resuspend beads by gentle pipetting, transfer spin column to fresh tube, and wash with 2 × 400-µL wash solution. Centrifuge as above.
9. Resuspend beads by gentle pipetting and transfer to fresh tube for elution. Elute 2 × 25-µL prewarmed elution buffer. Centrifuge for 1 min at 14,000 rpm.
10. Remove a 2-µL aliquot for scintillation counting.

3.3. Allele-specific hybridisation

3.3.1. Prehybridization

1. Remove slides from alcohol storage, dehydrate in 90%, 95%, 99% ethanol, and air dry.
2. Line tissue sections with rubber cement and allow this to dry while preparing prehybridization cocktail (see above). It is important that the rubber cement layer

is even and with a thickness of less than 1 mm. This can be facilitated by diluting the rubber cement solution 10–50% with xylene.

3. Add prehybridization mix to sections and prehybridize without coverslips in 50% formamide, 2 × SSC saturated atmosphere for 2–3 h at the appropriate hybridization temperature (see below).

3.3.2. General Hybridization Protocol

1. Prepare hybridization cocktail and prewarm to hybridization temperature.
2. Add DTT (10 mg per 100 µL hybridisation solution) and oligo probe (10,000–30,000 cpm/µL hybridization solution) immediately before use.
3. Remove as much prehybridization solution from tissue section as possible.
4. Add hybridization cocktail (between 25 and 50 µL/cm^2).
5. Cover sections with coverslips and hybridize overnight at the appropriate hybridization temperature (see below).

3.3.3. General Posthybridization Washes

1. Remove coverslips with a pair of fine forceps.
2. Wash sections 2 × 45 min at a temperature that is 5–15°C higher than that of the hybridization temperature in prewarmed 2 × SSC, 50% formamide, 20 m*M* DTT (approx. 150 mg DTT per Coplin jar). The optimal wash conditions have to be established independently for each allele-specific probe (see below, for example).
3. Transfer slides to 2 × SSC (2 × 15 min), followed by 1 × SSC (2 × 15 min at room temperature) and 2 × 1 min washes in cold tap water.
4. Dehydrate sequentially in 70%–90%–95%–99% ethanol.
5. Air dry (*see* **Notes 1–4**).

3.3.4. Specific Hybridization and Posthybridization Protocols

3.3.4.1. Human *H19*

1. Oligomers used for discrminating between *H19* allelic transcripts are based on an *Rsa* I polymorphism in exon 5 *(5)*.

   Rsa (–) 5′-CTC ACG CAC ACT CGC ACC GAG ACT CAA GGC C

   Rsa (+) 5′-CTC ACG CAC ACT CGT ACT GAG ACT CAA GGC C

2. Prehybridisation and hybridization temperature: 33°C.
3. Posthybridization washes: 38°C.

3.3.4.2. Human *IGF2*

1. Oligomers used for discriminating between *IGF2* allelic transcripts are based on an *Apa* I polymorphic site in exon 9 (*see also* **Fig. 2A**) *(5)*.

   IGF2 allele (A) 5′-TGT GAT TTC TGG GGT CCT TCT TTT CTC TT

   IGF2 allele (B) 5′-TGT GAT TTC TGG GGC CCT TCT TTT CTC TT

2. Prehybridization and hybridization temperature: 31°C.
3. Posthybridization washes: 38°C for the B probe and 45°C for the A probe.

3.3.5. Autoradiography

1. Expose to KODAK BioMax film for 1–3 d.
2. Develop to assess success of ASISH analysis and estimate time necessary for subsequent exposure to dipped emulsion (as a general rule of thumb, the exposure time to emulsion should be approximately fivefold that used to generate an acceptable image on X-ray film).
3. The emulsion should be made freshly each time. Use safe-light Kodak 6B or equivalent filters.
4. Prewarm 25 mL of ddH_2O/0.5% glycerol in a 50-mL polypropylene screw-cap Falcon tube to 43°C in a water bath.
5. Add solid emulsion shreds of Kodak NTB2 emulsion to the prewarmed ddH_2O/glycerol such that it displaces the liquid to the 50-mL mark.
6. Wrap the Falcon tube in aluminium foil and allow the emulsion to melt for 30–45 min at 43°C. Gently invert the mixture to produce a homogenous solution, but avoid creating air bubbles.
7. Transfer emulsion to a dipping chamber that is kept at 43°C in a water bath. The dipping chamber should be of the same depth as the glass slides.
8. Dip slides individually into the chamber and allow to drain for 5–10 min.
9. Transfer slides into a dark chamber that has capacity to hold 10–20 slides and plenty of space for silica gels.
10. Keep at +4°C for the desired length of time.
11. On the day of development, slide boxes are allowed to warm up to room temperature for 2–3 h.
12. Under safe-light conditions, slides are transferred into glass racks (Coplin jars, for example) and immersed into freshly prepared D19 developer for 2–3 min.
13. Transfer slides into a freshly prepared solution of 30% sodium thiosulfate for 2–3 min.
14. Transfer slides to several baths of ddH_2O and rinse for 10 min each time. Normal lights can be turned on at this point.
15. Allow slides to air dry.
16. Staining is optional, since tissue morphology can be visualized under water or glycerol coverslip without counterstain (using dark-field or Nomarski filters).

3.4. Multicolor Display of Allelic Expression Patterns

Allele-specific expression patterns can be illustrated in informative multicolour images (as, for example, in **Fig. 1**). These are based on scanned X-ray images in which the allelic hybridization signal has been normalized between allelic probes (by comparing hybridization signal in similar tissues homozygous for either genotype). Differences in allelic hybridization signals require differential exposure times (we analyze the hybridization signals in a phosphorimager to determine the duration of film exposure for each allelic probe used). The scanned images are then false-colored and overlayed in the computer according to the following protocol.

1. Scan the desired images of two adjacent sections (hybridized to allele-specific probes) at a high resolution (minimally 600 pixels per inch) and import into Adobe Photoshop software, version 4.0 or higher.
2. Crop the images to the desired size such that the hybridization signals from adjacent sections occupy approximately the same surface and have similar orientations.
3. Set the mode to RGB colors if the X-ray images were scanned in gray scale.
4. Set the foreground color to blue by adjusting the RGB color map (4 [R], 4 [G], 251 [B], for example).
5. Color one of the images blue using the "fill-in" command (found under the "edit" command). "Contents" should be set at "foreground color." The opacity is set at 100% and the mode of blending is "screen."
6. Set the foreground color to red by adjusting the RGB color map (251 [R], 4 [G], 4 [B], for example).
7. Color the image of the adjacent section red by repeating **step 5**.
8. Save both images and create duplicates.
9. Pick one of the duplicated images and enter the "apply image" command (found under the "image" command).
10. Select as source the duplicated image from the adjacent section. Set opacity at 100% and use the "difference" command under the "blending" command.
11. Adjust the orientation and position of the two images by selecting one of the layers.
12. Select the "rotate" command (found under the "layer" command, which is under the "transform" command).

3.5. Future Modifications

It is clearly of interest not only to perform simultaneous ASISH on the very same cells, but also to increase the resolution of the hybridization signal. Both of these current shortcomings of the ASISH technique could be resolved by including both allelic probes in the same hybridization cocktail and by establishing fluorescence-based detection of the hybridization signal. The technical problems that may arise from this approach depend on whether (1) the hybridization efficiency between the allelic probes is different at a given hybridization temperature, (2) the discrimination of the allelic signal is based on hybridization temperature or posthybridization washes, and (3) the sensitivity of the fluorescence detection systems can be readily normalized for the allelic signals. It is possible that the most optimal approach to resolve these issues requires that the allele-specific oligo probes detect different portions of the transcript. This would not only neutralize the possibility of competition for the same exon-specific sequences between allele-specific oligo probes during hybridization, but might also increase the flexibility in choosing the conditions for posthybridization washes.

Additional improvements will reflect more general advances in hybridization technology. For example, currently much mention is made of single nucleotide polymorphism (SNP) analysis, which demonstrates the need for the development of techniques to allow analysis at the individual allele level. This will be required for basic research purposes, such as the analysis of imprinted and nonimprinted monoallelic expression at individual loci, for rapid diagnosis of disease loci, and for disease treatment through improvements in individual patient treatment and in new drug developments.

In the context of single nucleotide polymorphism analysis, hybridization-based techniques have proved to be less straightforward in their applications than might have been hoped, with optimal conditions having to be developed for individual gene positions. Regarding future modifications to the ASISH technique, a number of contemporary developments may prove invaluable in the future. One example, which might overcome the relatively low sensitivity of the ASISH technique, is the adaptation of the "padlock" technique, which has a very high degree of sensitivity due to a rolling-circle-based detection method ***(12)***. Another example is the dynamic allele-specific hybridization (DASH) technique, which is based on advances in fluorescent labeling chemistry and where signal emission is determined by whether the probe is single-stranded in the hybridization solution or is bound to its target sequence ***(13)***. Advances in the area of nucleotide analogs may also prove promising. For example, locked nucleic acids (LNAs) have recently been described ***(14)***, in which incorporation of one or more LNA base analogs into an oligonucleotide probe substantially alters the temperature at which that oligonucleotide will hybridize to its target sequence. Utilization of this type of technology in conjunction with ASISH protocols may alleviate the need for individual allele hybridization and wash temperatures to be optimized and allow a more general protocol to be determined for a wider range of future applications, including a simultaneous detection of allelic transcripts in the same cell.

4. Notes

1. Allele-specific hybridization signals cannot be discriminated: if optimal hybridization/posthybridization conditions have not been found, it is generally a prudent strategy to optimise allele-specific signals by focusing on posthybridization wash conditions. The reason for this is that optimization of the hybridization conditions will tend to reduce hybridization efficiency, whereas posthybridization wash conditions may achieve allele specificity without compromising the signal intensity. In addition, the search for posthybridization wash condi-

tions is less time-consuming than finding optimal hybridization conditions.To increase the stringency of posthybridization wash conditions, begin by raising the temperature at which the initial posthybridization washes are performed.

2. In cases in which the allele-specific hybridization signals are not identical with the antisense riboprobe signal. This problem invokes two mutually exclusive explanations. First, both parental alleles could be differentially expressed, such that only their combined expression pattern is comparable with that of the riboprobe. Alternatively, one of the parental alleles could be expressed in only a subpopulation of cells. Second, the signal is unspecific, since it is found within a tissue that does not hybridize to the riboprobe. It is advised that the chosen oligo sequence be checked for the presence of repeat sequence by performing Blast analysis, for example. The Web address for the Blast service is: *http://www.ncbi.nlm.nih.gov/BLAST/*. Also consider the possibility that the riboprobe may be enriched in exons that are spliced in a tissue-specific manner. This possibility can be tested be examining a wider range of probes covering the gene of interest.
3. When there is excessive background, this can be traced to three main sources, acting alone or in combination. First, the hybridization cocktail may contain suboptimal components. Repeated freeze-thawing of the BSA stock, for example, precipitates protein. The quality of the formamide stock is important not only for hybridization stringency but also for the pH of the hybridization cocktail. It is advisable, therefore, that the formamide stock be purchased in a recrystallized form and immediately aliquoted and frozen at –70°C upon arrival. Second, repeated freeze-thawing can rapidly degrade oligomers. As a consequence, the reduced sequence complexity may interfer with the specific hybridization signal. This possibility is enhanced by the fact that the specific activity of degraded 3′-labeled oligomers will tend to be higher than that for the full-length oligoprobe. The oligo DNA should be aliquoted, therefore, and stored at –70°C upon arrival. Third, too little DTT in the hybridization cocktail or posthybridization wash may lead to nonspecific background hybridizations (remedied by increasing the DTT concentrations). Since DTT is hygroscopic, it is advisable that freshly purchased DTT be kept in aliquots at +4°C under vacuum.
4. We have occasionally observed that the oligo probe may light up the nuclei of some cells in a pattern that cannot explain expression of the gene in question. It is probable that this artefact is due to that long stretches of poly (T)n sequence annealing in a cooperative manner with oligo probes equipped with overlong poly (A)*n* tails. Neither inclusion of competitive cold oligomer or oligo (A) nor excess DTT will neutralize this problem. It is possible, however, to overcome this background problem by decreasing the length of poly (A)*n* tails by slightly increasing the number of oligo template molecules in the labeling reaction.

Acknowledgments

This work was supported by the Natural Science Research Council (NFR), the Swedish Cancer Research Foundation (RmC/CF), and the Swedish Pediatric Cancer Foundation (BCF).

References

1. Ohlsson, R., Tycko, B., and Sapienza, C. (1998) Monoallelic expression: "there can only be one." *Trends Genet.* **14,** 435–438.
2. Holländer, G., Zuklys, S., Morel, C., Mizoguchi, E., Mobisson, K., Simpson, S., Terhorst, C., Wishart, W., Golan, D., Bhan, A., and Burakoff, S. (1998) Monoallelic expression of the interleukin-2 locus. *Science* **279,** 2118–2121.
3. Wijgerde, M., Grosveld, F., and Fraser, P. (1995) Transcription complex stability and chromatin dynamics *in vivo*. *Nature* **377,** 209–213.
4. Marahrens, Y., Loring, J., and Jaenisch, R. (1998) Role of the *Xist* gene in X chromosome choosing. *Cell* **92,** 657–664.
5. Adam, G., Cui, H., Miller, S., Flam, F., and Ohlsson, R. (1996) Allele-specific *in situ* hybridization (ASISH) analysis: a novel technique which resolves differential allelic usage of *H19* within the same cell lineage during human placental development. *Development* **122,** 839–847.
6. Cui, H., Hedborg, F., He, L., Nordenskjöld, A., Sandstedt, Pfeifer-Ohlsson S., and Ohlsson, R. (1997) Inactivation of *H19*, an imprinted and putative tumor repressor gene, is a preneoplastic event during Wilms' tumorigenesis. *Cancer Res.* **57,** 4469–4473.
7. Li, Y.-M., Franklin, G., Cui, H. M., Svensson, K., He, X. B., Adam, G., Ohlsson, R., and Pfeifer, S. (1998) The *H19* transcript is associated with polysomes and may regulate *IGF2* expression in *trans*. *J. Biol. Chem.* **273,** 28247–2852.
8. Svensson, K., Mattsson, R., James, T., Wentzel, P., Pilartz, M., MacLaughlin, J., Miller, So., Olsson, T., Eriksson, U., and Ohlsson, R. (1998) The paternal allele of the *H19* gene is progressively silenced during early mouse development: the acetylation status of histones may be involved in the generation of variegated expression patterns. *Development* **125,** 61–69.
9. Ohlsson, R., Cui, H., Flam, F., Fisher, R., Miller, S., He, L., Pfeifer, S., and Adam, G. I. R. (1999) Absence of discriminative parental imprints is linked with epigenetic mosaicism of the *IGF2 H19* genes. *Development, Genes & Evol.* **209,** 113–119.
10. Ohlsson, R., Cui, H., He, L., Pfeifer, S., Jiang, S., Feinberg, A. P., and Hedborg, F. (1999) Mosaic allelic *IGF2* expression patterns reveal a link between Wilms' tumorigenesis and epigenetic heterogeneity. *Cancer Res.* **59,** 3889–3892.
11. Cui, H. and Ohlsson, R. (1996) An improved protocol for purification of ^{35}S-labeled oligonucleotide DNA probes for *in situ* hybridisation applications. *Trends Genet.* **12,** 506–507.
12. Baner, J., Nilsson, M., Mendel-Hartvig, M., and Landegren, U. (1998) Signal amplification of padlock probes by rolling circle replication. *Nucleic Acids Res.* **26,** 5073–5078.
13. Howell, W. M., Jobs, M., Gyllensten, U., and Brookes, A. J. (1999) Dynamic allele-specific hybridization. A new method for scoring single nucleotide polymorphisms. *Nature Biotechnol.* **17,** 87–88.
14. Kumar, R., Singh, S. K., Koshkin, A. A., Rajeanshi, V. K., Meldgaard, M., and Wengel, J. (1998) The first analogues of LNA (locked nucleic acids): phosphorothioate-LNA and 2′-thio-LNA. *Bioorg. Med. Chem. Lett.* **8,** 2219–2222.

12

RNA-FISH to Analyze Allele-Specific Expression

Giovanna Braidotti

1. Introduction

1.1. Gametic vs Random Imprinting vs Allelic Exclusion

One of the difficulties associated with the analysis of imprinted gene expression is the need to distinguish RNA synthesis occurring at the maternal vs the paternally inherited copy of the gene. Most of the techniques used to examine allele-specific expression exploit naturally occurring polymorphisms and measure steady-state levels of RNA isolated from a pool of cells. Hence, a restriction fragment length polymorphism (RFLP) an be exploited in a heterozygote, by a reverse transcriptase polymerase chain reaction (RT-PCR)-based procedure, to analyze maternal vs paternal gene expression. The human *IGF2R* gene was analyzed in this way. Smrzka et al. *(1)* were thus able to show that the *IGF2R* gene possesses a hemimethylated, intronic CpG island analogous to the mouse imprinting box. However, *IGF2R* mRNA was detected that possessed the RFLP from both the maternal and paternal alleles in all but one of the 70 lymphoblastoid samples. (The one monoallelic sample reactivated its paternal allele with continued cell culturing.) It was concluded that monoallelic expression of the human gene is a polymorphic trait occurring in a small minority of all tested samples (reviewed in **refs.** ***2,3***). Although this is a sound conclusion, the question remains: Is the human *IGF2R* gene imprinted?

The difficulty associated with this question reflects, to a large extent, the technical difficulties associated with the analysis of genes subject to epigenetic regulation. For example, useful parent-of-origin polymorphism in heterozygous states are not always readily available. Furthermore, procedures that measure steady-state levels of RNA can only determine the overall or average level

From: *Methods in Molecular Biology, vol. 181: Genomic Imprinting: Methods and Protocols*
Edited by: A. Ward © Humana Press Inc., Totowa, NJ

of expression for all the cells within the examined tissue. This is fine for biallelically expressed genes that rely on cis-acting DNA regulatory sequences. However, imprinted expression states can be limited to a subpopulation of cells within a tissue, as observed for the mouse *Igf2* ***(4)*** gene. These subpopulations of monoallelic cells are hard to detect in homogenates of tissues. Another difficulty is provided by genes that randomly select one allele and express it exclusively. These monoallelic genes can be difficult to detect in pooled RNA that possess transcripts that arose from either allele. Allelic exclusion has been reported for the mouse olfactory receptor cluster of genes ***(5)*** and for the interleukin-2 locus in mature thymocytes and T cells ***(6)***. And then there are the genes that combine strategies. For instance, the *XIST* gene in females is imprinted in placenta (associated with inactivation of the paternal X chromosome) but is expressed from either allele in embryonic tissues, resulting in random X inactivation ***(7)***. To complicate matters further, the silent alleles of imprinted genes are known to "leak" reduced amounts of RNA. This phenomenon has been observed at the mouse paternal *Igf2r* gene ***(8)***. It is extremely difficult to determine the mechanism underscoring this leaky silencing. It may be due to a silencing mechanism that reduces transcription at the paternal allele, leaving a very low level of expression in all cells. Alternatively, a subpopulation of cells that fail to silence the paternal allele may be responsible for all of the leaky expression. Although the former possibility seemed the more likely, direct visualization of RNA synthesis by fluorescent *in situ* hybridization (RNA-FISH) provided evidence for the latter, at least for the cell lines examined (unpublished data). Although no paternal signal was observed in the vast majority of nuclei, a few cells were seen to express paternal *Igf2r* at levels similar to the maternal allele. This leaky expression seemed to occur in a nonclonal manner, suggesting that monoalleleic expression can be dynamically established in a stochastic manner as opposed to requiring a stably maintained, silenced state.

1.2. Gene Expression at the Single-Cell Level

The use of a fluorescent *in situ* hybridization (FISH) procedure that detects RNA in intact, interphase cells can circumvent some of the difficulties associated with the analysis of epigenetically regulated genes [*see*, for example, work on the human ***(9)*** and mouse ***(10,11)*** *Xist* gene]. The procedure involves hybridization of labeled probes to nondenatured cells that have been fixed on a microscope slide, followed by visualization using standard immunocytochemical methods. Expression of mRNA is observed at the single-nucleus level as discrete spots or "transcription foci" that localize, in cis, on the transcribed gene. Gene expression can therefore be observed at the single-cell level and within a proper cellular context, permitting the correlation of gene expression with other cellular events. Furthermore, the subcellular localization of imprinted, noncoding transcripts can be determined within both nuclear and cytoplasmic compartments, as was done when the noncoding *Xist* transcript was shown to coat the inactivated X chromosome ***(9–11)***.

Although discrete spots have been observed by RNA-FISH in a variety of genes, the reason for the spatially restricted morphology of the fluorescent signal is unclear. The spot could be due to the presence of multiple RNA polymerases on the gene, which produce a nested set of nascent transcripts that are targeted by the FISH probe. Alternatively, a rate-limiting step in the posttranscriptional processing of full-length transcripts (such as intron splicing) may act to retain RNA at the gene transiently. The question, therefore, arises as to which step in RNA synthesis is detected by RNA-FISH to produce the spots. The answer seems to vary depending on which gene is analyzed. Attempts to detect nascent transcripts specifically (by using probes that target the short-lived introns rather than the exons) have been made successfully (*Xist*, **ref. *12***; β-globin, **ref. *13***). On the other hand, analysis of the inducible *hsp70* gene indicated that the predominant source of the signal was due to retention of full-length RNA at the gene ***(14)***. These differences should be kept in mind when setting up the procedure. It may be unwise to assume that the strength of the RNA-FISH signal is analogous to the transcription rate. Posttranscriptional events affecting RNA (or its "processing fate") may also determine whether a spot is observed. Having said this, it should be clear that only a combination of RNA-FISH with nuclear run-on assays can provide unequivocal data about transcription rates.

Using RNA-FISH to determine whether a particular cell line or tissue expresses a gene mono- or biallelically involves counting the number of fluorescent transcription foci within a nucleus and comparing it to the ploidy number for that cell. However, there may be some instances (possibly rare) when an average spot-per-nucleus number that is half the ploidy number is not an indication of imprinted expression. These would be instances where gene expression is subject to strong cell timing phenomena. RNA-FISH visualizes gene expression in cells that are at various stages of the cell cycle. However, expression is not continuous in cells. For example, a general inhibition of transcription occurs during the mitotic stages and is reinitiated in interphase of the next cycle. If the gene to be analyzed is transcribed during a narrow window of the cell cycle and the two alleles express at different stages, then this allelic cell timing difference can result in nuclei that predominantly show a fluorescent spot at just one allele (despite the absence of imprinted expression). If there is an overlap in the timing of transcription between the two alleles, then a subset of nuclei will show two spots (the overall percentage will depend on the duration of the overlap). Cell timing phenomenon has been observed with β-globin transgenes carrying an incomplete locus control region ***(15)***.

Conversely, imprinted genes can be mistakenly classed as biallelic by RNA-FISH if the imprinting mechanism does not involve transcriptional silencing of one allele. This would occur if a gene is transcribed biallelically but the transcripts from only one allele are stable enough to function. This has been observed for mouse *Xist* in differentiated embryonic stem cells ***(11)***, where use of different promoters at the two alleles produces transcripts with different half-lives. Transcripts are detectable at both alleles by RNA-FISH with intronic probes, although only one allele transcribes the stable form of the transcript that coats the inactive X chromosome. With the appropriate controls, the RNA-FISH procedure can distinguish between the above possibilities and produce an increasingly more sophisticated picture of gene expression.

Several variants of the RNA-FISH procedure have been developed. The primary difference between them is the way the fixed cells are permeabilized in order to permit the probe access to the nucleus. Fixed cells can be treated with a detergent such as Triton X-100 or digested with a proteinase such as pepsin. Most cell types analyzed by RNA-FISH will produce results with either procedure, although the exact conditions may vary slightly for different cell types. The procedure outlined here is a simplified version of the procedure established by Roeland Dirks ***(16)***. It works well with a variety of cell types, including cultured fibroblasts, embryonic stem cells, and cell suspensions from homogenized tissues such as fetal liver. Precautions required for preventing the degradation of RNA are discussed. The trickiest part of establishing RNA-FISH frequently involves the preparation of the haptenized probe. A novel, PCR-based procedure that reliably produces single-stranded probes is described, and troubleshooting hints are provided.

2. Materials (*see* Note 1)

2.1. Probe Preparation

1. Taq polymerase (such as Gibco BRL).
2. 10× PCR buffer: 200 m*M* Tris-HCl, pH 8.4, 500 m*M* KCl.
3. 15 m*M* $MgCl_2$.
4. 10× nucleotide solution: 2 m*M* each of dATP, dCTP, and dGTP, 1.9 m*M* dTTP (lower concentrations of dTTP can be tried in order to increase the incorporation of the haptenized or fluorochromized dUTP).
5. 1 m*M* digoxigenin-11-dUTP or biotin-16-dUTP (Boehringer Mannheim).
6. 3 *M* NaOAc, pH 5.5.
7. 70% and 100% ethanol.

2.2. Slide Preparation

1. Cell culture media and disposables.
2. Adherent cell lines can be grown on sterilized microscope objects slides soaked in 0.1% gelatin or poly-L-lysine (highest degree of polymerization). Alternatively, nonadherent lines and cell suspensions can be spotted directly onto Poly-Prep slides (Sigma).
3. 10× PBS (pH 7.4): 80 g NaCl, 2 g KCl, 14.4 g $Na_2HPO_4 \cdot 2H_2O$, and 2.4 g KH_2PO_4. Adjust pH to 7.4 and the volume to 1 L. Autoclave the solution and store at room temperature.
4. Fixatives: 4% formaldehyde, 5% acetic acid in phosphate-buffered saline (PBS); 1% formaldehyde in PBS.

2.3. RNA in Situ Hybridization

1. 10% pepsin (Sigma) prepared in water and stored frozen.
2. Biotin- or digoxigenin-labeled DNA probes.
3. Formamide.
4. 20× SSC: 3 *M* NaCl, 0.3 *M* Na citrate, pH 7.0.
5. 2 *M* Tris-HCl, pH 7.5 (do not diethyl pyrocarbonate [DEPC]-treat).
6. 10× saline: 85 g NaCl dissolved in 1 L of distilled water. Autoclave the solution.
7. Tris-saline solution: 145 m*M* NaCl, 100 m*M* Tris-HCl, pH 7.5.
8. Tris-saline-Tween solution: 0.1 *M* Tris-HCl, pH 7.5, 0.15 *M* NaCl, 0.05% Tween-20.
9. 2× hybridization buffer: 4× SSC, 20% dextran sulfate, and 2% bovine serum albumin (BSA).
10. Cot-1 DNA.
11. Herring sperm or salmon sperm DNA (Sigma).
12. Yeast tRNA (Sigma).
13. Biotin and/or digoxigenin detection reagents:
    a. Avidin-FITC or TRITC for detection and biotinylated goat anti-avidin for amplification of signal (Vector).
    b. Sheep anti-digoxigenin-FITC or TRITC (Boehringer) for detection and rabbit anti-sheep-FITC or TRITC and goat anti-rabbit-FITC or TRITC for amplification.
14. 4′,6-Diamidino-2-phenylindole (DAPI).
15. Vectashield embedding medium (Vector).
16. Graded series of 70%, 90%, and 100% ethanol.
17. Hybridization buffer: 50–65% deionized formamide, 2× SSC, 50 m*M* sodium phosphate, 5% dextran sulfate, pH 7.4.
18. Blocking buffer: 4× SSC, 0.1% Tween-20, 4 mg/mL BSA.
19. Detection buffer: 4× SSC, 0.1% Tween-20, 1 mg/mL BSA.
20. Epifluorescence microscope equipped with appropriate filter sets and image-capturing devices (such as a CCD camera and imaging software).

3. Methods

3.1. Probe Preparation

Procedures for labeling DNA with haptenized (such as biotin or digoxigenin) or fluorochromized dUTP have been published elsewhere ***(15)***. Kits that incorporate these nucleotides by nick translation and PCR are available commercially (for example, see the Boehringer catalog).When choosing a detection system, keep the following points in mind.

Fluorochromized nucleotides (i.e., directly labeled probes) generally produce weaker signals compared to haptenized probes. For genes expressed at low levels it is preferable to use a detection system in which amplification of the signal is possible (i.e., digoxigenin and biotin-labeled dUTP).

When using biotinylated probes, check the endogenous level of biotin in the fixed cell by hybridizing in buffer lacking probe and processing the slide through the detection system. In mammalian cells, digoxigenin routinely produces lower background signals compared to biotin, especially if amplification is required.

A convenient control for hybridization specificity is provided by co-hybridizing two different probes to the same target. Using different probe-detection systems (i.e., two different colors) will ensure that the probes can produce co-localized signals and that the signal is not due to spurious hybridization. Conversely, two probes to distinct targets can also be tested to ensure that distinct signals can be obtained. These controls necessitate using at least a two color probe-detection system. A third color (usually DAPI) is used to counterstain the entire nucleus. Alternatively, a three-color hybridization is possible by treating fluorescein as a hapten. The requirement for more than one detection system should be kept in mind when deciding what immunocytochemical method to adopt for detection of each probe, and therefore, what antibodies are purchased.

The probes synthesized by the above procedures will produce double-stranded DNA that can simultaneously detect sense and antisense transcripts. However, the discovery of noncoding, antisense transcripts at an increasing number of imprinted loci ***(17–19)*** produces the need for strand-specific probes. The procedure listed below is a fast, PCR-based method first developed for the antisense transcript present at the mouse *Igf2r* locus (manuscript in preparation) and subsequently applied to study *Xist* ***(11)***. It relies on PCR amplification using oligos complementary to just one DNA strand (linear PCR).

1. Select the DNA template to be used for probe synthesis. Restriction fragments from genomic clones, PCR products, or cDNA clones can all be used, but gel-

purify the required inserts beforehand. Design a primer complementary to the target strand (avoiding repetitive sequences). A cocktail of oligos distributed over several kilobases can also be used. Use approximately 10–50 ng of template DNA amplified with 1 μ*M* final concentration of oligo(s).

2. Linear PCR reactions are performed in 50 μL in the presence of fluorochromized or haptenized dUTP as follows: 5 μL of 10× PCR buffer, 5 μL of 15 m*M* $MgCl_2$ (although a range of concentrations ought to be tested in order to optimize each PCR reaction), 5 μL of 10× nucleotide solution, 5 μL of 1 m*M* digoxigenin-11-dUTP or biotin-16-dUTP, 5 units Taq polymerase (such as Gibco BRL).
3. Labeled product is generated by 200 PCR cycles of 94°C for 20 s, the annealing temperature for the oligo for 30 s, and 72°C for 1–2 min. Check the linear PCR products by visualization on a 1% agarose gel. (To test the strand specificity of the probe, Southern-blot the agarose gel and hybridise with radiolabeled sense and antisense oligos contained within the amplified sequence).
4. The hybridization mix is prepared by precipitating approximately 50 ng of labeled single-stranded DNA adding 0.05 vol 3 *M* NaOAc, 2.5 vol 100% ethanol, and 10 μg of carrier DNA (such as salmon sperm DNA) and 10 μg of yeast tRNA. Wash the precipitate several times in 70% ethanol and resuspend in 10 μL formamide (for a 50% final concentration). Increased hybridization specificity can be achieved by increasing the concentration of formamide up to 75%. Before hybridization, the mix is denatured at 75°C for 10 min, placed on ice, and diluted with 10 μL of 2× hybridization buffer. Cot-1 DNA (1 μg) can also be added to the hybridization mix, in which case probe denaturation should be followed by a 15-min incubation at 37°C.

3.2. Cell Culture and Fixation

1. Culture subconfluent cells in the appropriate medium, CO_2 atmosphere, and temperature. Trypsinize adherent cells and seed onto gelatin or poly-L-lysine-coated glass object slides. Grow cells overnight to allow cells to attach. Nonadherent cells and cell suspensions from organ homogenates can be cytospun onto slides or spotted directly onto Poly-Prep slides (Sigma).
2. If the slides have been in cell culture medium, rinse the slides briefly in PBS.
3. Fix (4% formaldehyde, 5% acetic acid in PBS) for 20 min at room temperature. Prepare the fixative just before using.
4. Wash three times in PBS for 5 min. Slides can be stored at this stage in 70% ethanol at –20°C.

3.3. RNA in Situ Hybridization

1. Transfer slides from ethanol storage to Tris-saline solution for 5 min at room temperature.
2. Just before use, thaw and dilute the 10% pepsin stock (1/100 to 1/1000, depending on the cell type) in 0.01 *M* HCl prewarmed to 37°C. The time of protease treatment should be optimized for each cell type. Optimal incubation times commonly fall between 2 and 10 min (*see* **Note 2**).

3. Rinse slides in water for 1 min.
4. Incubate in 1% formaldehyde in PBS for 5 min at room temperature.
5. Rinse slides in PBS for 10 min.
6. Dehydrate the cells by immersing successively in 70%, 90%, and 100% ethanol for 3 min each. Air-dry the slides. Apply the denatured probe to the slide (use 15–20 µL with a 24 × 24-mm^2 coverslip). Hybridize overnight at 37°C in a humidified container. The hybridization temperature can be increased if more stringent conditions are required.

3.4. Immunocytochemical Detection

1. Prewarm a solution of 2× SSC at 37°C. Wash the slides in 2× SSC for 5 min and ensure that the coverslips are released. Wash the slides a further three times for 10 min.
2. Wash slides in Tris-saline-Tween solution for 5 min at room temperature.
3. Add 100 µL of blocking buffer to each slide and incubate under a coverslip for 30 min at room temperature.
4. Wash with Tris-saline-Tween for 5 min.
5. Dilute each antibody required for hapten detection just before use in detection buffer. Apply 100 µL under a coverslip and incubate for 30 min at room temperature.
6. Wash with Tris-saline-Tween two times for 5 min between each antibody application in the detection system (*see* **Notes 3** and **4**).
7. Rinse slides in Tris-saline for 5 min.
8. Dehydrate the slides through 70%, 90%, and 100% ethanol. Air-dry the slides.
9. Mount in 60 µL of Vectashield containing 10 ng/µL DAPI and cover with 24 × 50 mm^2 coverslip.
10. Examine the slides with a fluorescence microscope equipped with appropriate excitation and emission filters for FITC, TRITC, and DAPI fluorescence (*see* **Notes 5–7**).

4. Notes

1. To ensure the preservation of *in situ* RNA, a few simple precautions can be adopted. The first is to wear gloves throughout the procedure, including when preparing pipet tips, making solutions, and handling slides. The second is to set aside chemicals reserved for RNA use only. These need not be the entire set of solutions required for the procedure, but need only include those chemicals and solutions that are likely to be contaminated if in general use. The third is to heat-sterilize glassware and autoclave solutions. If feeling particularly cautious about the need to preserve RNA, then water and solutions (with the exception of Tris) can be DEPC-treated by adding 200 µL of DEPC (diethyl pyrocarbonate; Sigma) to 1 L of solution, mix, and leave overnight before autoclaving.
2. When setting up the technique it will be necessary to verify that the signal obtained is due to hybridization to cellular RNA. A useful set of controls involves RNase A-treating two slides. One is hybridized in its native form. The other is heat-

denatured to make the DNA available for hybridization. This will produce a slide for RNA alone (no RNase treatment, nondenatured), a negative control (RNase treated, nondenatured), and a DNA control (RNase treated and denatured).

The RNase treatment can be performed after pepsin treatment by incubating the slide in 100 µL of 0.1-mg/mL RNase A in 2× SSC. Apply a coverslip and incubate at 37°C in a moist chamber. Wash the slide three times in 2× SSC and continue as normal. Make sure the RNase A used in this way does not become a source of contamination for the RNA-FISH.

To heat-denature and make DNA available for hybridization, incubate the slide containing the hybridization mix on a hot plate set at 80°C for 5–10 min before incubating overnight at 37°C.

3. If multiple antibodies are used to amplify hapten detection, they should be applied sequentially and not combined except when an antibody and fluorochrome-conjugated avidin are involved.
4. Once fluorochrome-conjugated components of the hapten-detection system are applied to the slides, precautions should be taken to reduce the amount of exposure to light, to help prevent quenching.
5. Failure to obtain any signal may be due to inadequate permeabilization of the cells and, therefore, failure of the probe to access the nucleus. This may be due to overly long fixation times combined with inadequate pepsin treatment. However, as tempting as it may be to blame the histological side of the procedure for the failure to obtain a signal, it is generally more likely that the source of the problem is the probe. Possible problems include:
   a. Use of excessively long probes that fail to access nuclear RNA. Probes of a length of around 300–600 bp are optimal. Try decreasing the length of the PCR-amplified insert. However, note that, in a low percentage of nuclei, targeting just one short stretch of the cellular RNA can result in weak signal. If longer sequences need to be targeted, then label a large insert and digest the linear PCR product with a 4-bp restriction enzyme that possesses single-stranded DNA digestion activity (e.g., Mnl I).
   b. Excessive competition between the probe and the salmon sperm and COT-1 DNA. Your probe may become "competed out" by the carrier DNA. This will happen to any DNA fragment if the amount of competitor is high enough. Try reducing the amount of carrier DNA used. If the signal obtained this way is not specific for the targeted RNA, then consider labeling a different fragment with a high content of unique sequence.
   c. Inadequate incorporation of hapten into the probe. Try reducing the amount of dTTP without altering the amount of haptenized dUTP in the PCR reaction. This may only be necessary if the gene of interest is expressed in very low amounts.
   d. The simplest solution, of course, is to try and increase the amount of probe used in the hybridization and decrease hybridization stringency until some signal (even if just background) is obtained. It is easier to clean up a hybridization than to work with no signal at all.

6. Too much signal and background problems: Having gotten past the no-signal stage, you are likely to experience too much signal as the next problem. You now need to reverse most of the advice given above:
   a. Increase the amount of competitor DNA.
   b. Increase hybridization stringency, either by increasing the amount of formamide in the hybridization solution or increasing the hybridization temperature to 42°C.
   c. Try including COT-1 DNA and decrease the amount of probe used.

   Having sent you round in a circle, you may now discover that you are in a no-win situation: either you have no signal or you have persistent background problems. There are two situations when background signal will persist to the point that eliminating it will also eliminate the signal.
   a. The first situation involves probes with high G+C content that can hybridize nonspecifically to ribosomal RNA in the nucleolus. I know of no procedure that will circumvent this problem, other than to choose a different sequence, with a lower G+C content, for your probe.
   b. The second situation involves persistent fluorescent spots on the surface of the nucleus that commonly appear as "edge spots" on the periphery of the nucleus. I suspect the cause of this problem is the presence of long strands of DNA that are unable to penetrate the nucleus but capture the probe, forming very intensely fluorescent spots. If the probe has been labeled by nick translation, then reduce the size of the template DNA either by increasing the concentration of DNase I or by using restriction enzyme digestion prior to nick translation. If the probe has been generated by the linear PCR procedure, then include a restriction enzyme digestion after amplification. Choose a combination of 4-bp cutters that will dramatically reduce the size of the unlabeled, double stranded template DNA and also digest single-stranded DNA. Heat-inactivate the reaction mixture before using.
7. Correcting problems of poor cellular morphology requires a balancing act between the fixation step and the pepsin treatment. You must balance the need to preserve cellular morphology and ensure probe accessibility to obtain an optimal signal intensity. Inadequately fixed samples that are overly pepsin digested can lead to loss of cellular structures, including the nucleus. Overly fixed cells with too little pepsin treatment may produce excellent morphology but too little or inconsistent signal. To correct this problems you will need to determine optimal pepsin concentrations and digestion times empirically for each cell type that causes problems.

References

1. Smrzka, O., Faé, I., Stöger, R., Kurzbauer, R., Fischer, G. F., Henn, T., Wieth, A., and Barlow, P. P. (1995) Conservation of a maternal-specific methylation signal at the human IGF2R locus. *Human Mol. Genet.* **4,** 1945–1952.

2. Efstratiadis, A. (1994) Parental imprinting of autosomal mammalian genes. *Curr. Op. Genet. Dev.* **4,** 265–280.
3. Barlow, D. P. (1995) Gametic imprinting in mammals. *Science* **270,** 1610–1613.
4. DeChiara, T. M., Robertson, E. J., and Efstratiadis, A. (1991) Parental imprinting of the mouse insulin-like growth factor II gene. *Cell* **64,** 849–859.
5. Chess, A., Simon, I., Cedar, H., and Axel, R. (1994) Allelic inactivation regulates olfactory receptor gene expression. *Cell* **78,** 823–834.
6. Hollander, G. A., Zuklys, S., Morel, C., Mizoguchi, E., Mobisson, K., Simpson, S., Terhorst, C., Wishart, W., Golan, D. E., Bhan, A. K., and Burakoff, S. J. (1998) Monoallelic expression of the interleukin-2 locus. *Science* **279,** 2118–2121.
7. Penny, G. D., Kay, G. F., Sheardown, S. A., Rastan, S., and Brockdorff, N. (1996) Requirement for XIST in X chromosome inactivation. *Nature* **379,** 131–137.
8. Lerchner, W. and Barlow, D. P. (1997) Paternal repression of the imprinted mouse Igf2r locus occurs during implantation and is stable in all tissues of the post-implantation mouse embryo. *Mech. Dev.* **61,** 141–149.
9. Clemson, C. M., McNeil, J. A., Willard, H. F., and Lawrence, J. B. (1996) XIST RNA paints the inactive X chromosome at interphase: evidence for a novel RNA involved in nuclear/chromosome structure. *J. Cell Biol.* **132,** 259–275.
10. Sheardown, S. A., Duthie, S. M., Johnston, C. M., Newall, A. E., Formstone, E. J., Arkell, R. M., Nesterova, T. B., Alghisi, G. C., Rastan, S., and Brockdorff, N. (1997) Stabilization of Xist RNA mediates initiation of X chromosome inactivation. *Cell* **91,** 99–107.
11. Johnston, C. M., Nesterova, T. B., Formstone, E. J., Newall, A. E. T., Duthie, S. M., Sheardown, S. A., and Brockdorff, N. (1998) Developmentally regulated Xist promoter switch mediates initiation of X inactivation. *Cell* **94,** 809–817.
12. Panning, B., Dausman, J., and Jaenisch, R. (1997) X chromosome inactivation is mediated by Xist RNA stabilization. *Cell* **90,** 907–916.
13. Wijgerde, M., Grosveld, F., and Fraser, P. (1995) Transcription complex stability and chromatin dynamics *in vivo*. *Nature* **377,** 209–213.
14. Jolly, C., Robert-Nicoud, M., and Vourc'h, C. (1998) Contribution of growing RNA molecules to the nuclear transcripts foci observed by FISH. *Exp. Cell Res.* **238,** 299–304.
15. Milot, E. M., Strouboulis, J., Trimborn, T., Wijgerde, M., de Boer, E., Langeveld, A., Tan-Un, K., Vergeer, W., Yannoutsos, N., Grosveld, F., and Fraser, P. (1996) Heterochromatin effects on the frequency and duration of LCR-mediated gene transcription. *Cell* **87,** 105–114.
16. Dirks, R. W., van de Rijke, F. M., Fujishita, S., van der Ploeg, M., and Raap, A. K. (1993) Methodologies for specific intron and exon RNA localization in cultured cells by haptenised and fluorochromised probes. *J. Cell Sci.* **104,** 1187–1197.
17. Wutz, A., Smrzka, O. W., Schweifer, N., Schellander, K., Wagner, E. F., and Barlow, D. P. (1997) Imprinted expression of the *Igf2r* gene depends on an intronic CpG island. *Nature* **389,** 745–749.

18. Moore, T., Constancia, M., Zubair, M., Bailleul, B., Feil, R., Sasaki, H., and Reik, W. (1997) Multiple imprinted sense and antisense transcripts, differential methylation and tandem repeats in a putative imprinting control region upstream of mouse Igf2. *Proc. Natl. Acad. Sci. USA* **94,** 12509–12514.
19. Rougeulle, C., Cardoso, C., Fontés, M., Colleaux, L., and Lalande, M. (1998) An imprinted antisense RNA overlaps *UBE3A* and a second maternally expressed transcript. *Nature Genet.* **19,** 15–16.

13

Flow Cytometry and FISH to Investigate Allele-Specific Replication Timing and Homologous Association of Imprinted Chromosomes

Janine LaSalle and Marc Lalande

1. Introduction

Chromosome replication banding studies show that homologous regions on a pair of autosomes generally replicate at the same time in S phase ***(1)***. Izumikawa et al. first observed that this was not the case for the imprinted chromosomal region 15q11-q13 ***(2)***. This observation has been confirmed in other replication banding studies ***(3)*** as well by the fluorescence *in situ* hybridization (FISH) replication assay ***(4–9)***. The latter technique has also been used to observe DNA replication asynchrony in association with allelic inactivation of genes such as those encoding olfactory receptors and the cytokine, interleukin 2 ***(10,11)***. The latter genes are not imprinted but display random silencing of an allele in individual cells. In imprinted regions, DNA replication was generally observed to occur earlier on the paternal homologue ***(5,6,9,12,13)***. The patterns of allele-specific replication in the cells of Prader-Willi (PWS) and Angelman syndrome (AS) patients, however, have generally been synchronous ***(5,6,14)***. Furthermore, an investigation of the kinetics of allele-specific replication timing in the GABRB3/A5 cluster on 15q11-13 revealed that cells from PWS and AS have lost the strict replication timing observed on the parental chromosomes of normal cells ***(12)***. These results suggested the requirement of a biparental contribution for the regulation of replication asynchrony and lead to the hypothesis that allelic cross-talk, perhaps via pairing of homologous chromosomes, might play a role in the imprinting process. This possibility was tested directly using three-dimensional (3-D) FISH in spatially intact nuclei and confocal laser scanning microscopy to

From: *Methods in Molecular Biology, vol. 181: Genomic Imprinting: Methods and Protocols*
Edited by: A. Ward © Humana Press Inc., Totowa, NJ

determine the distance between homologous chromosomes in cell cycle fractionated human lymphocytes *(15)*. The results demonstrated that, in normal cells, the chromosome 15 homologs pair during the late S phase. The most tightly paired region was observed to be 15q11-q13, consistent with the hypothesis that allelic cross-talk could be involved in the normal pattern of imprinting in chromosome 15q11-q13 *(15)*.

In this chapter, we detail the protocols used to perform flow cytometry, to fractionate cells on the basis of DNA content, before FISH analyses of DNA replication and homologous chromosome pairing. Although FISH detection of allele-specific replication can be performed on unsorted cycling cells, comparisons between replication timing of loci in different cell lines (i.e., patient and normal) are imprecise due to differences in cell growth. The kinetic approach to investigating both allele-specific replication and homologous pairing outlined here was essential in determining the biparental requirement for both normal replication timing and homologous pairing *(12,15)* that had been overlooked by previous studies of unsorted cells *(5,6,16)*. In addition, the protocol for culturing primary human T lymphocytes to obtain optimal numbers of cycling cells is included, as it varies from that of standard cytogenetic methods. Immortalized human cell lines such as lymphoblasts may be used instead of primary cultures, but epigenetic and cytogenetic alterations may occur as a result of long-term culture of these lines. FISH methods are given for both the standard 2-D preparation for allele-specific replication timing (**Subheading 3.3.**) and the 3-D preparation for investigating homologous associations (**Subheading 3.4.**). Normally, well-spread 2-D nuclei are the best target for FISH replication studies because separated sister chromatids can be readily observed *(4)*. Conversely, 3-D preparations in which the spatial organization of nuclei is maintained are ideally suited for investigations of nuclear chromosome organization *(17)*. If a large number of cells in 2-D FISH preparations are scored, however, 3-D organization can be extrapolated *(18,19)*. In the future, an automated approach to detecting homologous pairing on 2-D FISH preparations may be possible by the new technology of laser scanning cytometry *(20)*.

2. Materials

2.1. Culture of Primary Human T Lymphocytes

1. 20–60 mL of human peripheral blood collected in heparinized tubes (*see* **Note 1**).
2. PBS: Phosphate-buffered saline without Ca^{2+} and Mg^{2+}, sterile.
3. Ficoll-Paque^{PLUS} (Pharmacia).
4. RPMI 1640 cell culture medum: Roswell Park Medical Institute media.
5. Human sera, pooled male AB (Sigma). Prescreen sera samples for T-cell growth and purchase enough sera to complete experiments (*see* **Note 2**).

6. Penicillin, streptamycin, and L-glutamine.
7. 24-well culture clusters (Costar).
8. PHA-P: Phytohemagglutinin P (HA17, Remmel).

2.2. Vital Staining for DNA Content

1. RPMI, prewarmed to 37°C in tissue culture incubator.
2. Hoechst 33342 (Molecular Probes), 1-mg/mL stock in water, store in dark at 4°C.
3. PBS, 4°C.

2.3. Two-Dimensional FISH for Allele-Specific Replication

1. 0.075 *M* KCl.
2. Fixative 3/1 methanol/glacial acetic acid, make fresh, 4°C.
3. 70%, 90%, 100% ethanol.
4. Clean glass microscope slides in 100% methanol, stored at –20°C.
5. Biotin- or digoxigenin-labeled DNA probes (*see* **Note 3**).
6. Human placental DNA (Sigma) or Cot-1 DNA (Gibco).
7. Deionized formamide. Store in aliquots at –20°C and use within 2 wk of thawing.
8. Hybridization solution (2×): 40% dextran sulfate, 4× SSC. Make fresh, vortex to a uniform viscosity.
9. Denaturation solution: 70% deionized formamide, 2× SSC, make fresh.
10. Rubber cement.
11. Wash A: 50% formamide (inexpensive grade), 2× SSC.
12. Wash B: 0.1× SSC.
13. Wash C: 4× SSC, 0.5% Tween-20.
14. Blocking solution: 3% bovine serum albumin (BSA), 4× SSC, 0.1% Tween-20.
15. Detection solution: 1% BSA, 4× SSC, 0.1% Tween-20.
16. FITC-avidin (Vector Laboratories) and/or Rhodamine-anti-digoxigenin (Roche Diagnostics).
17. Mounting media: Vectashield (Vector Laboratories) plus 1 µg/mL DAPI (Molecular Probes).
18. Cover glasses, No. 1, 18 × 18 mm, 20 × 30 mm, and 20 × 50 mm.
19. Glass Coplin jars (at least 4).

2.4. Three-Dimensional FISH for Homologous Association

Most solutions and materials are the same as in **Subheading 2.3.**, except *see* **Note 4** and the following list:

1. Poly-D-lysine, MW 30–70 kDa (Sigma), 100 µg/mL in water, store aliquots at –20°C.
2. Microscope slides that have been wiped clean with 100% methanol (not cold as under **Subheading 2.3., step 4**).
3. Fixative: 4% paraformaldehyde/1% methanol in PBS. 10× stock, keep for 1 mo, make up as follows: to 4 g paraformaldehyde (Kodak), add 6 mL water plus

2 drops 4 *N* NaOH, heat slowly while stirring and do not allow to boil. Heat until in solution. When solution is cool, add 1 mL methanol and bring volume to 10 mL if needed. Filter and store protected from light. Make fresh dilution in PBS before each use.
4. Postfixation wash and storage solution: 0.3 *M* glycine in PBS, dilute 1/10 in PBS from 3 *M* glycine stock, add 0.04% sodium azide from 100× stock. Make fresh each time.
5. Permeabilization solution: 0.5% Tween-20 in 0.2 *N* HCl. Make fresh each time.
6. Wash solution: 0.5% Tween-20 in PBS.
7. Nail polish.
8. Cytofuge and parts (StatSpin Technologies, Norwood, MA).

3. Methods

3.1. Culture of Human T Cells

1. Dilute peripheral blood 1:1 with sterile PBS in 50-mL conical tubes with a final volume of less than 30 mL per tube.
2. Underlay Ficoll (at least 1/3 blood volume) by placing a pipet at the bottom of the tube and slowly releasing the Ficoll until a lower gradient has formed. Handle tubes gently to maintain the gradient.
3. Centrifuge at 500*g* for 30 min with the rotor brake turned off.
4. Aspirate and discard the upper serum layer. The white blood cells are at the interface and should be carefully transferred with a pipet to a new 50-mL tube. Discard the Ficoll layer and red cell pellet.
5. Wash cells twice with 50 mL PBS. Centrifuge at 300*g* for 10 min for each wash. The centrifuge brake can be turned back on.
6. Prepare T cell growth media: 20% human sera in RPMI with 10-U/mL penicillin, streptomycin, and L-glutamine.
7. Resuspend the cell pellet in 10 mL T-cell growth medium and count a 1:10 dilution on a hemacytometer.
8. Add T-cell growth medium to bring cells to ~10^6/mL. Add 2 µg/mL PHA-P.
9. Seed cultures in 24-well plates at 1–2 mL per well. Incubate for at least 72 h at 37°C in 5% CO_2. Cells should form large clumps from PHA-P stimulation after 24 h.

3.2. Vital Staining for DNA Content and Flow Sorting

1. Transfer cells from culture to a 50-mL tube, centrifuge at 300*g* for 10 min.
2. Make up staining buffer, use ~1 mL buffer per 10^6 cells, dilute Hoechst 33342 1/100 in prewarmed RPMI 1640 (10-µg/mL final concentration).
3. Resuspend cell pellet in Hoechst 33342 stain buffer, place in a tissue culture incubator for 30–45 min.
4. Centrifuge cells, resuspend in PBS at 10^7/mL and keep on ice prior to sort.
5. Sort cells into 4 or 6 fractions based on DNA content as shown in **Fig. 1**. *See* **Note 5** for details of flow cytometry setup.

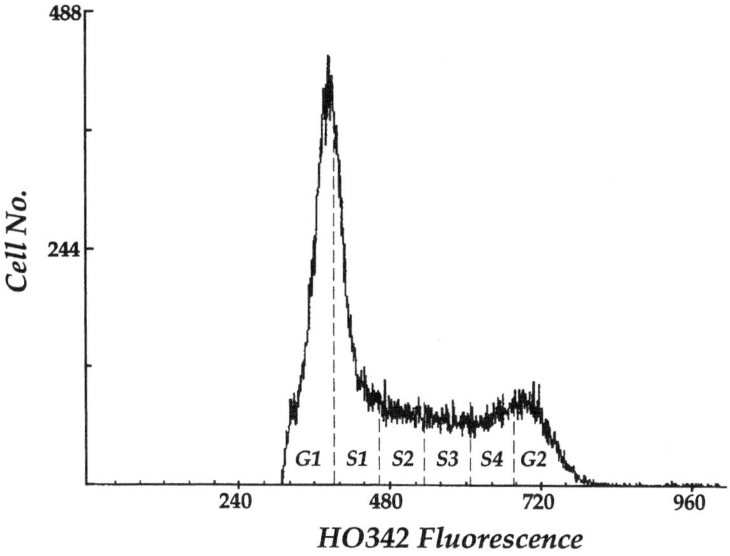

Fig. 1. Flow sorting cells on the basis of DNA content. PHA-stimulated T lymphocytes were stained with Hoechst 33342 (HO342) to allow flow cytometric analysis of DNA content in vitally stained cells. A representative histogram shows the distribution of DNA content (determined by HO342 fluorescence) in the cell population. Fluorescence-activated cell sorting gates were set at positions represented by the dotted lines. For kinetics of DNA replication, six fractions were collected and labeled G1, S1, S2, S3, S4, or G2 as shown. For detection of homologous pairing, four fractions were collected so that S1 and S2 were early S phase and S3 and S4 were late S phase.

3.3. Two-Dimensional FISH for Allele-Specific Replication

3.3.1. Fixation of Cells

1. Centrifuge cells from cell cycle sort (0.1 to 1.0×10^6) at 300*g*.
2. Resuspend in 1 mL of 0.075 KCl.
3. Incubate at 37°C for 13 min. Centrifuge cells.
4. Aspirate, leaving some fluid, resuspend cell pellet in 1 mL fixative and incubate for 20 min on ice.
5. Centrifuge cells and resuspend in 20–100 µL fixative, depending on pellet size (roughly 10^7/mL). Store cells at –20°C.
6. Drop 10 µL of cell fixation solution on slides that have been immediately removed from methanol and wiped dry. The drop should spread and dry quickly (within 5–15 s) for optimal spreading (*see* **Note 6**).
7. Air-dry slides overnight. Dehydrate in a 70%, 90%, 100% ethanol series for 10 min each. Store desiccated for up to 1 mo.

3.3.2. Hybridization and Washing

1. Prepare probe: combine 20–100 ng of labeled probe, 3 µg of human competitor DNA, and 7 µg of salmon sperm DNA per slide. Lyophilize DNA in a Speed-Vac. Resuspend DNA in 5 µL per slide in deionized formamide. Vortex for 5–10 min using a tube attachment head for the vortex machine. Add 5 µL per slide hybridization solution. Vortex for a further 5–10 min.
2. Prewarm slides for hybridization in a 55°C oven.
3. Prepare denaturation solution with sufficient volume to cover sample area in a Coplin jar. Cover jar, place in water bath, and heat bath to ~72°C.
4. Prepare postdenaturation ethanol series (70%, 90%, 100% ethanol) in Coplin jars and chill on ice. Prepare an extra 70% ethanol jar for first quick wash.
5. Test temperature inside jar of denaturation solution. Once it has reached 70 ± 2°C, add 3–4 slides. Denature for 3 min, monitoring the time exactly. Transfer slides immediately to ice-cold 70% ethanol quick wash for 10 s. Repeat as needed for additional slides.
6. Wash slides for 5 min each in a cold ethanol series. Air-dry and prewarm slides in a 37°C oven.
7. Denature probe for 5 min at 72°C. Incubate at 37°C for 10 min to preanneal competitor DNA.
8. Add 10 µL denatured probe to slide. Spread the drop with a 18 × 18 mm coverslip, using care to exclude and remove bubbles (*see* **Note 7**).
9. Seal the coverslip with rubber cement. Place slides in a prewarmed moist chamber at 37°C. Incubate overnight.
10. Prewarm wash solutions A and C to 45°C and wash solution B to 60°C.
11. Carefully remove the rubber cement seal with forceps to avoid removing or dragging the coverslip. Place slides in Coplin jar with wash A at 45°C and agitate to remove coverslips for 5 min. Transfer slides to new wash A jar and wash twice for 5 min each at 45°C.
12. Wash 3 times in wash B for 5 min each at 60°C.
13. Blot excess liquid from the back and sides of slide. Add 200-µL blocking solution per slide. Spread solution with a 20 × 50 mm coverslip, being careful to exclude bubbles. The coverslip should float very loosely on slide. Do not allow slides to dry.
14. Incubate in a moist chamber for 30 m at 37°C.
15. Prepare fluorescent detection solution: dilute FITC-avidin 1/400 and Rhodamine-anti-digoxigenin 1/100 in detection solution. Centrifuge for 5 min to remove fluorescent debris. Protect from light.
16. Tilt slides to remove the coverslip and excess block solution. Pipette 200 µL of the fluorescent detection solution onto slide and spread with a 20 × 50 mm coverslip.
17. Incubate in a moist chamber for 30 min at 37°C in the dark.
18. Tilt slides to remove coverslip and excess detection solution. Place slides in Coplin jar with 45°C wash C solution for 5 min. Wash twice more in wash C solution.

19. Blot excess liquid from the back and sides of slide. Pipette 30 µL of mounting medium per slide. Spread solution with a 20 × 30 mm coverslip, being careful to exclude bubbles.
20. Store slides upright in a covered slide box at 4°C.

3.3.3. Fluorescence Microscopy and Scoring of Replication Events

Examine slides on a fluorescent microscope with a mercury light source (100-W bulb) and a 100× oil objective with 1.4 numerical aperture. The microscope should be equipped with filter sets appropriate for DAPI, FITC, and Rhodamine/Texas red. A triple-bandpass filter is also useful for scoring replication events, but brightness of signals may be compromised by this approach. If FISH signals are dim and nuclear background is low, a cooled CCD-based imaging system may be useful for merging multiple colors. A Photometrics Sensys CCD camera and IPLab Spectrum software were used to obtain the three-color images in **Fig. 2**.

The FISH signals used to determine allele-specific replication events (green signals in **Fig. 2**) should be scored as singlets or doublets on each allele in interphase nuclei. Fine focus adjustment should be performed while scoring, to ensure that all FISH signals are counted. Doublets should be two closely spaced but discernable signals. For determining hybridization efficiency, *see* **Note 8**. A second FISH probe that detects a polymorphism can be used to distinguish parental alleles (red signals in **Fig. 2**). In this case, the FISH replication signal closest to the FISH polymorphism signal should be scored as being on the same chromosome. The final data analysis of kinetic allele-specific replication is performed as in **ref. *12***.

3.4. Three-Dimensional FISH for Homologous Association

3.4.1. Fixation of Cells

1. Centrifuge cells ~300*g* from the cell cycle sort (0.1–1.0 × 10^6 cells).
2. Resuspend in 1 mL fixative at 4°C and incubate for 20 min on ice.
3. Centrifuge cells and resuspend in 2–4 mL of 0.3 *M* glycine/PBS. Repeat once.
4. Resuspend cells at ~10^6/mL in 0.3 *M* glycine/PBS (with addition of sodium azide, cells can be stored for at least 2 mo at 4°C.

3.4.2. Slide Preparation

1. Polylysine coat slides: assemble cytospin gaskets and chambers (*see* **Note 9**). Pipette 20 µL of Poly-D-lysine per chamber, coat 5 m, wash twice with water and then air-dry.
2. Cytospin: pipet 100–200 µL of fixed cell suspension (~10^5 cells) into each chamber and centrifuge for 4 min at 55*g*. Aspirate solution from chamber. Without allowing slides to dry, disassemble chamber and place in a Coplin jar containing permeabilization solution for 10 min.

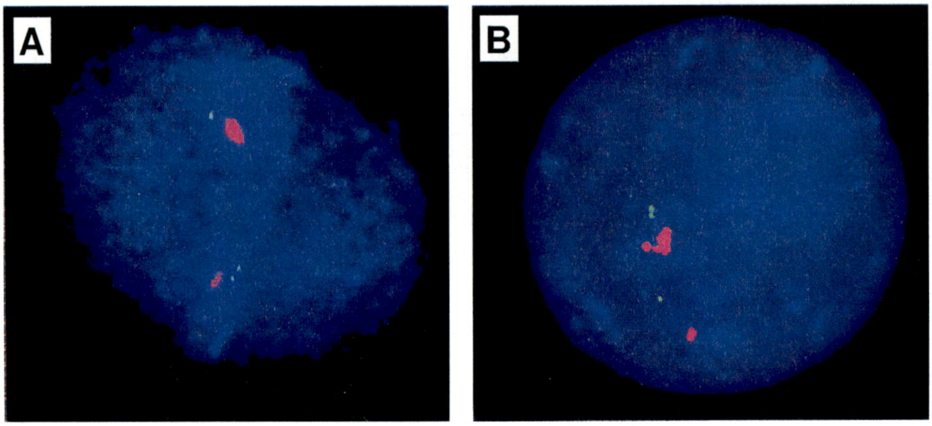

Fig. 2. Two-dimensional FISH detection of allele-specific replication. FISH detection of allele-specific replication was performed on interphase nuclei from a normal subject heterozygous for the chromosome 15 satellite size polymorphism detected by digoxigenin-labeled D15Z1 (red fluorescence). The chromosome 15 with the larger D15Z1 signal is maternally derived and the shorter D15Z1 signal is paternal. Biotin-labeled plasmid DNA was simultaneously hybridized to the nuclei to detect replication events (green fluorescence). Plasmid DNA signals appeared as either a singlet (unreplicated) or doublet (replicated) on each parental chromosome. Representative nuclei demonstrating the predominant parental allele specificity of asynchronous replication (singlet-doublet) are shown for two probes. **(A)** Probe p386 from the 5′ end of GABRB3 displayed earlier replication on the paternal chromosome 15, as the doublet is associated with the shorter D15Z1. **(B)** In contrast, probe p438 from the GABRB3/GABRA5 intergenic domain detected earlier replication on the maternal chromosome 15 shown by the proximity of the doublet to the larger D15Z1 signal.

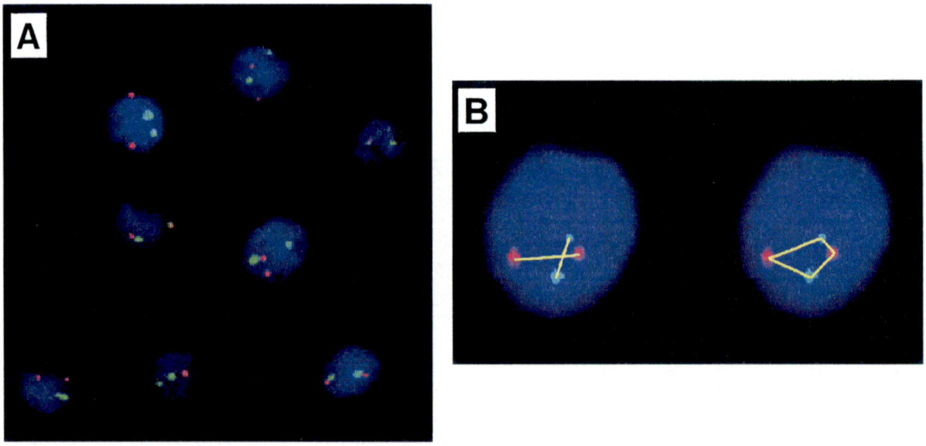

3. Wash twice for 5 min each in 2× SSC at 45°C. Slides can be held in the last wash prior to denaturation.

3.4.3. Hybridization and Washing

1. Prepare probe, as under **Subheading 3.3.2., step 1**.
2. Prepare denaturation solution, as under **Subheading 3.3.2., step 3**.
3. Prepare postdenaturation washes (two of PBS/0.5% Tween, two of 2× SSC) in Coplin jars and chill on ice.
4. Denaturation: as under **Subheading 3.3.2., step 5**, except first transfer slides to PBS/0.5% Tween and wash after denaturation.
5. Wash slides for 5 min each in a cold wash series. Do not air-dry. Slides may be held in the last wash.
6. Denature and preanneal probe, as under **Subheading 3.3.2., step 7**.
7. Wipe excess liquid from around the cell sample, leaving 1–3 µL of fluid remaining over cells. Do not allow cells to dry.

8–19. Same as under **Subheading 3.3.2., steps 8–19**, except *see* **Note 10**.

20. Seal coverslip with nail polish.
21. Store slides in covered slide box at 4°C.

3.4.4. Confocal Microscopy and Three-Dimensional Distance Measurements

Slides should be imaged with a confocal laser scanning microscope equipped with a 60× objective with 1.4 numerical aperture and appropriate excitation and emission to detect fluorophores. Clusters of 5–30 nuclei can be imaged in a single field if cytospin preparations are sufficiently dense (*see* **Fig. 3A**). Capture optical sections of 0.4 µm *z* resolution through all focal planes containing FISH signals. Measure 3-D distances between FISH signals using appropriate software (such as Molecular Dynamics ImageSpace). For each nucleus in a data set of at least 50 nuclei, take six different measurements, as shown in **Fig. 3B**. The two distances between homologous chromosomes (red to red and green to green) and four heterologous distances (red to green) are measured. Measurement data is graphed as described previously ***(15,21)***.

Fig. 3. *(see previous page)* Three-dimensional FISH detection of homologous association. **(A)** A field of densely distributed nuclei hybridized with D15Z1 (green) and D12Z3 (red) was optically sectioned using a confocal laser scanning microscope. Images are three-dimensional projection images. **(B)** Three-dimensional distance measurements are determined using a mouse-driven cursor. Two homologous distances (red to red and green to green) and four heterologous distances are determined for each nucleus.

4. Notes

1. If patient blood is limiting (less than 10 mL), the T-cell population in culture can be doubled by adding 5% Human T-stim (Collaborative Research) to the culture at 24–32 h and harvesting cells 4 d after PHA stimulation.
2. Fetal calf sera may be used in place of human sera, but the percentage of cells cycling will generally be lower.
3. Probes may be obtained commercially or labeled by nick translation or random priming methods. Probes may be labeled directly with fluorophores, but this is usually possible only for repetitive target sequence (centromeric satellite DNA). For replication studies, target sequence should be single-copy, and the ideal probe size is generally 5–40 kb. Although larger probes are generally easier to detect, doublets may be difficult to distinguish with signals obtained from probes larger than 50 kb.
4. To control for pH variation, use MilliQ or double-distilled water for all solutions before and during hybridization. Posthybridization washes can be done in standard distilled water.
5. UV excitation with an argon-ion laser is required for cell sorting on the basis of Hoechst 33342 staining. For cell sorting we have used a Becton Dickonson FACSCaliber at 10 psi and a 70-μm nozzle tip, collecting two fractions at a time. More recently, sorting times have been reduced by simultaneous four-fraction sorting on a Cytomation MoFlo run at 60 psi with a 75-μm nozzle tip.
6. Slides should be examined under a phase-contrast light microscope following air drying of fixative. Interphase nuclei should be large and flat, with no cytoplasm on top. If cells are not well spread, try placing slides briefly on a warm surface for 1 min. If fixative dried too quickly and cells are not refractory to light, try breathing on the slides after dropping, to increase the humidity. These modifications are best practiced on more plentiful samples (unsorted lymphocytes), before dropping sorted fractions.
7. To avoid getting bubbles under the coverslip, hold the coverslip with forceps at a 45° angle adjacent to one side of the drop. Slowly lower the coverslip, allowing the drop's leading edge to exclude bubbles. If a few bubbles remain, they may be gently pushed to the edge of the coverslip with a pipet tip.
8. Hybridization efficiency of FISH probes should be tested on G1 nuclei prior to analysis of replication events on all sorted fractions. At least two discrete FISH signals should be observed in at least 90% of the nuclei. If hybridization efficiency is low but background is also low, try increasing the amount of probe or decreasing wash stringency. If nuclear background is high, resulting in too many nonspecific signals, try reducing the amount of labeled probe, increasing wash stringency, or flattening nuclei further.
9. StatSpin Cytofuge 3-well cell concentrator gaskets and use one 4 mm × 12 mm chamber closest to the middle of slide.
10. If probes labeled directly with fluorophores are used, skip **Subheading 3.3.2., steps 13–18**.

References

1. Drouin, R., Holmquist, G. P., and Richer, C. L. (1994) High-resolution replication bands compared with morphologic G- and R- bands. *Adv. Hum. Genet.* **22,** 47–115.
2. Izumikawa, Y., Naritomi, K., and Hirayama, K. (1991) Replication asynchrony between homologs 15q11.2: cytogenetic evidence for genomic imprinting. *Hum. Genet.* **87,** 1–5.
3. Drouin, R., Boutouil, M., Fetni, R., Holmquist, G. P., Scott, P., Richer, C. L., and Lemieux, N. (1997) DNA replication asynchrony between the paternal and maternal alleles of imprinted genes does not straddle the R/G transition. *Chromosoma* **106,** 405–411.
4. Selig, S., Okumura, K., Ward, D., and Cedar, H. (1992) Delineation of DNA replication time zones by fluorescence in situ hybridization. *EMBO J.* **11,** 1217–1225.
5. Kitsberg, D., Selig, S., Brandeis, M., Simon, I., Keshet, I., Driscoll, D. J., Nicholls, R. D., and Cedar, H. (1994) Allele-specific replication timing of imprinted gene regions. *Nature* **364,** 459–463.
6. Knoll, J. H. M., Cheng, S.-D., and Lalande, M. (1994) Allele-specificity of DNA replication timing in the Angelman/Prader-Willi syndrome imprinted chromosomal region. *Nature Genet.* **6,** 41–46.
7. Smrzka, O. W., Fae, I., Stoger, R., Kurzbauer, R., Fischer, G. F., Henn, T., Weith, A., and Barlow, D. P. (1995) Conservation of a maternal-specific methylation signal at the human IGF2R locus. *Hum. Mol. Genet.* **4,** 1945–1952.
8. Brown, K. W., Villar, A. J., Bickmore, W., Clayton-Smith, J., Catchpoole, D., Maher, E. R., and Reik, W. (1996) Imprinting mutation in the Beckwith-Wiedemann syndrome leads to biallelic IGF2 expression through an H19-independent pathway. *Hum. Mol. Genet.* **5,** 2027–2032.
9. Greally, J. M., Starr, D. J., Hwang, S., Song, L., Jaarola, M., and Zemel, S. (1998) The mouse H19 locus mediates a transition between imprinted and non- imprinted DNA replication patterns. *Hum. Mol. Genet.* **7,** 91–95.
10. Chess, A., Simon, I., Cedar, H., and Axel, R. (1994) Allelic inactivation regulates olfactory receptor gene expression. *Cell* **78,** 823–834.
11. Hollander, G. A., Zuklys, S., Morel, C., Mizoguchi, E., Mobisson, K., Simpson, S., Terhorst, C., Wishart, W., Golan, D. E., Bhan, A. K., and Burakoff, S. J. (1998) Monoallelic expression of the interleukin-2 locus. *Science* **279,** 2118–2121.
12. LaSalle, J. M. and Lalande, M. (1995) Domain organization of allele-specific replication within the GABRB3 gene cluster requires a biparental 15q11-13 contribution. *Nature Genet.* **9,** 386–394.
13. Gunaratne, P. H., Nakao, M., Ledbetter, D. H., Sutcliffe, J. S., and Chinault, A. C. (1995) Tissue-specific and allele-specific replication timing control in the imprinted Prader-Willi syndrome region. *Genes Dev.* **9,** 808–820.
14. White, L. M., Rogan, P. K., Nicholls, R. D., Wu, B. L., Korf, B., and Knoll, J. H. (1996) Allele-specific replication of 15q11-q13 loci: a diagnostic test for detection of uniparental disomy. *Am. J. Hum. Genet.* 423–430.

15. LaSalle, J. and Lalande, M. (1996) Homologous association of oppositely imprinted chromosomal domains. *Science* **272,** 725–728.
16. Dietzel, S., Weilandt, E., Eils, R., Munkel, C., Cremer, C., and Cremer, T. (1995) Three-dimensional distribution of centromeric or paracentromeric heterochromatin of chromosomes 1, 7, 15 and 17 in human lymphocyte nuclei studied with light microscopic axial tomography. *Bioimaging* **3,** 121–133.
17. Manuelidis, L. and Borden, J. (1988) Reproducible compartmentalization of individual chromosome domains in human CNS cells revealed by in situ hybridization and three-dimensional reconstruction. *Chromosoma* **96,** 397–410.
18. Emmerich, P., Loos, P., Jauch, A., Hopman, A. H. N., Wiegant, J., Higgins, M. J., White, B. N., van der Ploeg, M., Cremer, C., and Cremer, T. (1989) Double in situ hybridization in combination with digital image analysis: a new approach to study interphase chromosome topography. *Exp. Cell Res.* **181,** 126–140.
19. Trask, B. J., Allen, S., Massa, H., Fertitta, A., Sachs, R., van den Engh, G., and Wu, M. (1993) Studies of metaphase and interphase chromosomes using in situ hybridization. *Cold Spring Harb. Symp. Quant. Biol.* **58,** 767–775.
20. LaSalle, J. and Lagdamen, C. (1999) Automated FISH spot counting by laser scanning cytometry detects homologous pairing of imprinted chromosomal domains (Abstr. #374) *Am. J. Hum. Genet.* **65**(suppl.), A74.
21. Weiner, B. M. and Kleckner, N. (1994) Chromosome pairing via multiple interstitial interactions before and during meiosis in yeast. *Cell* **77,** 977–991.

14

Southern Analysis Using Methyl-Sensitive Restriction Enzymes

Tom Moore

1. Introduction

Methylation of the cytosine base in cytosine-guanine (CG) dinucleotides of genomic DNA is likely to be one of the primary epigenetic "imprints" that results in parental allele-specific expression of certain mammalian genes *(1)*. Despite the increasing popularity of bisulphite sequencing *(2)*, ligation-mediated polymerase chain reaction (PCR) *(3)*, the availability of antibodies to 5-methylcytosine (^{5m}CG) *(4)*, and more recent developments in DNA array technology *(5)*, Southern analysis using methyl-sensitive DNA restriction enzymes remains one of the more popular methods for the detection and analysis of mammalian DNA methylation. However, it has two significant disadvantages. First, relatively large amounts of genomic DNA are required for Southern blotting, thereby reducing its usefulness in the study of DNA methylation in the female germline and preimplantation embryos. Second, the analysis is limited to CGs that occur in a DNA sequence context that is recognized by a methyl-sensitive restriction enzyme. However, despite these drawbacks, the fact that the technique is still widely used is testament to its robustness, reproducibility, and relative simplicity.

1.1. Sources of DNA for Imprinting-Related Methylation Studies

In general, 10–20 µg of good-quality genomic DNA is required for Southern blotting. Such amounts are readily extracted from postimplantation mouse embryos, human blood samples, tissue (e.g., tumor) biopsies, and cultured cells. Difficulties arise when particular cell types from tissues are to be analyzed in isolation (requiring customized purification or contamination-monitoring

From: *Methods in Molecular Biology, vol. 181: Genomic Imprinting: Methods and Protocols*
Edited by: A. Ward © Humana Press Inc., Totowa, NJ

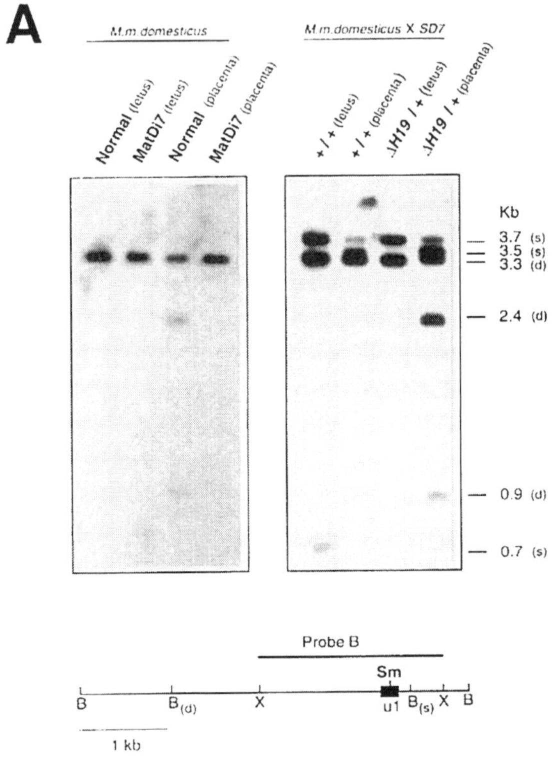

Fig. 1. Parental allele-specific DNA methylation analysis of the mouse *Igf2* upstream region reproduced from *(8)*, with permission. **(A)** Typical data from Southern analysis of mouse genomic DNA using a probe spanning a methyl-sensitive *Sma*I (Sm) site. The panel on the left shows the results of analysis using embryos maternally disomic for distal chromosome 7. The panel on the right shows the results of analysis of the same region using a *M. spretus BamHI* polymorphism in (*M. domesticus* X *SD7*) F_1 hybrid embryos. In this example, the *SD7* genotype is derived from the repeated backcrossing of the *M. spretus Igf2-H19* gene region to the C57BL/6 genetic background; the Δ*H19* genotype represents the targeted deletion of the *H19* gene upstream region. Subscripts in parentheses, (d) and (s), refer, respectively, to *M. domesticus*- and *M. spretus*-specific RE sites and fragments.

procedures), or when the amounts of starting material are very limited, as with female germ cells and preimplantation embryos. The latter problem has been a significant complication in imprinting studies, and the use of PCR-based alternatives can be problematic (*see* **ref. *6*** for discussion). Cells cultured in vitro provide ready access to large amounts of DNA. However, experiments in vitro must be carefully controlled because DNA methylation at imprinted loci may be susceptible to cell culture-induced changes *(7)*.

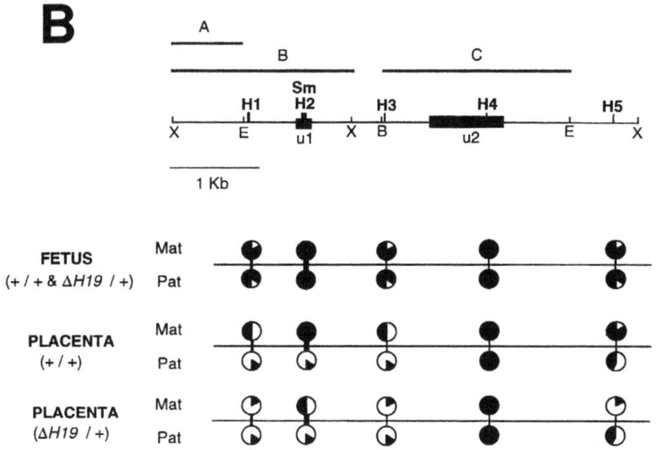

Fig. 1. *(continued)* **(B)** Graphical representation of analysis of parental allele-specific DNA methylation in the mouse *Igf2* upstream region in a range of tissues and genotypes, using probes A, B, and C. Parental alleles (maternal or paternal) are shown separately, above and below the horizontal line; degree of methylation of particular *Hpa*II (H) sites is represented by degree of circle filling. "u1" & "u2" are *Igf2* "upstream" exons 1 and 2, respectively. (Copyright ® 1997 National Academy of Sciences, U.S.A.)

Imprinting studies, by definition, require that the two parental alleles be discriminated. This requires the identification of suitable DNA restriction fragment length polymorphisms (RFLPs) within the region to be studied. In mice, this generally involves the use of interspecific hybrids (usually between *Mus spretus* or *Mus castaneus*, and an inbred *Mus musculus* laboratory strain). However, an alternative that does not require the identification of RFLPs is the comparison of normal tissues with tissues derived from individuals that are uniparentally disomic for the region of interest (e.g., **ref.** ***8***; **Fig. 1**).

In the study of DNA methylation at imprinted loci in the human, the use of samples from accessible tissues such as blood and term placenta is relatively common. However, the occurrence of tissue-specific differences in parental allelic expression and DNA methylation must be considered, particularly when extrapolating results to other tissues.

1.2. Analytical Strategies and Controls

1.2.1. HpaII and MspI

The simplest and most reliable approach to DNA methylation analysis with methyl-sensitive restriction enzymes involves the use of the isoschizomer pair, *Hpa*II and *Msp*I, which recognize the sequence CCGG. *Hpa*II, but not *Msp*I, restriction activity is inhibited by methylation of the 5-carbon of the cytosine ring in the CG dinucleotide, such that the sequence $C^{5m}CGG$ is not cut. *Msp*I therefore serves as an excellent positive control for DNA quality and completeness of digestion. (Note that most mammalian genomic DNA is methylated at CCGG sequences, therefore *Hpa*II and *Msp*I genomic DNA digests will look very different on agarose gel electrophoresis.) Other equivalent isoschizomer pairs with a more restricted distribution of recognition sites may be used to analyze specific sites within a region that contains many *Hpa*II sites, for example, *Xma*I–*Sma*I (recognition sequence: CCCGGG) (*see* **Note 1**).

1.2.2. Parental Allele-Specific Polymorphisms and Probe Placement

RFLPs between the parental alleles are most reliably identified by sequencing genomic DNA from a range of samples (by direct sequencing or by subcloning and sequencing PCR products), or by restricting genomic DNA with a range of enzymes, followed by Southern blotting and probing with cloned DNA fragments from the region of interest.

Generally, the genomic DNA sample to be analyzed is restricted with an enzyme that produces a previously characterized parental allele-specific RFLP, and then in parallel with either member of a chosen isoschizomer pair (e.g., *Msp*I and *Hpa*II). Following gel electrophoresis and Southern blotting, the membrane is hybridized to a radiolabeled DNA probe positioned between the parental allele-specific polymorphic RE site and the RE site being analyzed for presence of DNA methylation. However, an optimum strategy for the analysis of each individual region or methyl-sensitive RE site must be designed empirically.

2. Materials

2.1. Genomic DNA Extraction and Quantification

1. Tris-buffered saline (*see* **Note 2**): Dissolve 8 g NaCl, 0.2 g KCl, and 3 g Tris base in 800 mL H_2O, pH to 7.4 with HCl, add H_2O to 1 L and autoclave ***(9)***.
2. DNA extraction (lysis) buffer: 10 m*M* Tris-HCl pH 8.0, 50 m*M* NaCl, 50 m*M* ethylenediaminetetraacetic acid (EDTA), pH 8.0, 100-μg/mL proteinase K, 0.5% sodium dodecyl sulfate (SDS). (Proteinase K stock solution at 20 mg/mL in H_2O. Store –20°C).
3. Phenol containing 0.1% 8-hydroxyquinoline and equilibrated with 0.1 *M* Tris-HCl (pH 8.0) and 0.2% β-mercaptoethanol. Store at 4°C.
4. Chloroform/isoamyl alcohol (24/1).
5. 5 *M* NaCl.
6. 100% and 70% isopropanol.
7. Tris-EDTA (TE) buffer, pH 8.0: 10 m*M* Tris-HCl, pH 8.0, 1 m*M* EDTA pH 8.0.
8. UV spectrophotometer.

2.2. Digestion of Genomic DNA with Restriction Enzymes

Restriction enzymes and 10× reaction buffers are widely available commercially. Suitable standard buffer conditions for *Hpa*II and *Msp*I are 10 m*M* Tris-acetate (pH 7.5), 10 m*M* magnesium acetate, and 50 m*M* potassium acetate. *Sma*I requires a special buffer consisting of 20 m*M* KCl, 10 m*M* $MgCl_2$, 10 m*M* Tris-HCl (pH 8.0), and 1 m*M* dithiothreitol (DTT).

2.3. Gel Electrophoresis

1. Molecular biology-grade agarose.
2. Tris-borate-EDTA (TBE) electrophoresis buffer (10× stock/liter): 54 g Tris base, 27.5 g boric acid, 20 mL of 0.5 *M* EDTA, pH 8.0.
3. Ethidium bromide (10 mg/mL in H_2O).
4. 6× gel loading buffer: 0.25% bromophenol blue, 0.25% xylene cyanol FF, 40% (w/v) sucrose in H2O. Store at 4°C.
5. DNA molecular-weight markers (e.g., bacteriophage λ *Hind*III-digested DNA).

2.4. Southern Blotting

1. Gel denaturing buffer: 1.5 *M* NaCl, 0.5 *N* NaOH.
2. Gel neutralizing buffer: 1 *M* Tris (pH 7.4), 1.5 *M* NaCl.
3. 20× SSC: Dissolve 175.3 g NaCl and 88.2 g sodium citrate in 800 mL of H_2O, pH to 7.0 with 10 *N* NaOH, add H_2O to 1 L, and autoclave ***(9)***.
4. 20× SSPE: Dissolve 175.3 g NaCl, 27.6 g NaH_2PO_4, and 7.4 g EDTA in 800 mL H_2O, pH to 7.4 with 10 *N* NaOH, add H_2O to 1 L, and autoclave ***(9)***.
5. 0.2 *N* HCl.
6. 0.4 *N* NaOH (Store as 4 *N* stock solution) (*see* **Note 3**).
7. Membrane neutralizing buffer: 0.5 *M* Tris-HCl (pH 7.2), 1 *M* NaCl (*see* **Note 3**).
8. Charged nylon membrane (e.g., Nytran SuperCharge, Schleicher & Schuell). (Refer to manufacturer's instructions.)
9. Whatman 3MM chromatography paper.

2.5. Probe Labeling

1. 5× "oligolabeling" buffer ***(9)***: 250 m*M* Tris-Cl (pH 8), 25 m*M* $MgCl_2$, 5 m*M* β-mercaptoethanol, 2 m*M* each dATP, dGTP, dTTP, 1 *M* HEPES (pH 6.6), 1-mg/mL random oligonucleotides.
2. Bovine serum albumin (Fraction V; Sigma), 10 mg/mL.
3. DNA polymerase (Klenow fragment): 5–10 U/μL.
4. [α-^{32}P]-dCTP (3,000 Ci/mmol).
5. Buffer A ***(9)***: 50 m*M* Tris-Cl (pH 7.5), 50 m*M* NaCl, 5 m*M* EDTA (pH 8), and 0.5% SDS.
6. Sepharose G-25 column (*see* **Note 4**).

2.6. Blot Hybridization and Washing

1. Church hybridization buffer *(10)*: 1 m*M* EDTA, 0.5 *M* Na_2HPO_4/NaH_2PO_4, pH 7.2, 7% SDS. (For 1 L, mix 1 *M* Na_2HPO_4 and 1 *M* NaH_2PO_4 in the approximate ratio 10:3 (~385:115 mL), until pH 7.2 is achieved. Add 500 mL of this solution to 70 g of SDS and 2 mL of 0.5 *M* EDTA. Add deionized water to 800 mL. Stir to homogeneity on a heating block and add water to 1 L).
2. Church wash buffer *(10)*: 1 m*M* EDTA, 40 m*M* Na_2HPO_4/NaH_2PO_4 pH 7.2, 1% SDS. (For 5 L, mix 1 *M* Na_2HPO_4 and 1 *M* NaH_2PO_4 in the approximate ratio 10:3 (~192:58 mL), until pH 7.2 is achieved. Add 250 mL of this solution to 50 g of SDS and 10 mL of 0.5 *M* EDTA. Add deionized water to 3 L. Stir to homogeneity in a 5-L beaker on a heating block and add water to 5 L).
3. Medical X-ray film (e.g., Fuji RX or Kodak BioMax) and autoradiography film cassette.

3. Methods

3.1. Genomic DNA Extraction and Quantification

1. Disrupt soft tissues or pellets of cultured cells in 2–5 mL TBS using a hand-held or powered (e.g., Polytron) homogenizer.
2. Centrifuge the homogenate at 1500g for 10 min at 4°C.
3. Pour off the supernatant and add 0.5–1 mL of lysis buffer.
4. Cap the tube, vortex briefly to resuspend the pellet, and place in a (preferably, shaking) waterbath at 50°C for 4 h or overnight. (Loosely fitting caps may blow off and must be fixed in place with tape or other restraint.)
5. In a fume hood, add an equal volume of phenol to the lysate, cap, and shake gently for 10 min.
6. Centrifuge at 1500g for 10 min at room temperature.
7. Carefully remove the (DNA-containing) upper aqueous phase using a 1-mL Gilson or other pipet and transfer to a clean tube.
8. Add an equal volume of chloroform/isoamyl alcohol, cap the tube, and shake gently for 10 min.
9. Centrifuge, then remove the upper aqueous phase as in **steps 6** and **7** above.
10. Add an equal volume of 100% isopropanol, and 5 *M* NaCl in the ratio of 50 µL/mL of supernatant. Cap and invert the tube to ensure even mixing, and stand at room temperature for 10 min.
11. Using a plastic spatula or Gilson pipet tip, transfer the white DNA precipitate to an Eppendorf tube containing 1 mL of 70% isopropanol for 10 min. (This washing step removes excess salts.)
12. As in **step 11**, transfer the DNA to a 1-mL Eppendorf tube containing a suitable (50- to 500-µL) volume of nuclease-free water or TE buffer.

3.2. Digestion of Genomic DNA with Restriction Enzymes

1. Add 10–20 µg genomic DNA, 5 µL of 10× RE buffer, 10–20 units (1–2 µL) of each of appropriate enzymes (e.g., *Hpa*II or *Msp*I, and flanking RE), and H_2O to

50 µL total volume. (When digesting multiple samples, it is convenient to make up a stock solution of H_2O, RE buffer, and restriction enzymes).

2. Incubate for at least 3 h (or overnight) at 37°C to ensure complete digestion. This can be monitored by removing small aliquots of the reaction and running on an agarose minigel).
3. Following digestion, samples can be used immediately or stored at 4°C.

3.3. Gel Electrophoresis

1. Digested genomic DNA is usually run in 0.8% agarose gels. Add appropriate quantity of agarose and 1× TBE buffer to a 500-mL flask and heat to boiling in a microwave oven.
2. Allow agarose solution to cool to ~60°C and add ethidium bromide to a final concentration of 0.5 µg/mL.
3. Swirl the flask contents and pour into gel tray, allowing 30 min to set.
4. Remove comb, and flood gel with 1– TBE buffer.
5. Add 8 µL of 6× gel loading buffer to each sample and gently pipet up and down to mix.
6. Load up to 2 µg of λ-*Hind*III DNA molecular weight marker, skip one lane, and load the samples in the desired order (e.g., *Msp*I and *Hpa*II digests of each sample adjacent, etc.).
7. Run the gel slowly (at ~1 V/cm) to enhance resolution. (Running a 0.8% gel at 40 V overnight will give good resolution and retention of a wide range of DNA fragment sizes; this can also be monitored by viewing the gel under UV.)
8. Place a fluorescent ruler or other marker alongside the gel and photograph the gel using a UV light source.
9. The gel can now be blotted, or stored flat at 4OC in Saran wrap or cling film for up to 24 h.

3.4. Southern Blotting

3.4.1. Neutral Transfer (*see* **Note 3**)

1. Soak the gel for 30–60 min in gel denaturing buffer.
2. Rinse in distilled or deionized water for 1 min.
3. Soak in gel neutralizing buffer for 30 min.
4. Set up capillary blot using 10× SSC or SSPE in the reservoir and transfer for the desired period, usually overnight (**Fig. 2**).
5. Dismantle the blot apparatus and wash the membrane in 2× SSC for 15 min to remove adherent agarose.
6. Place the membrane between two pieces of 3MM paper to dry for approximately 30–60 min.
7. Fix the DNA to the membrane using a UV cross-linker. (Refer to manufacturer's instructions.)

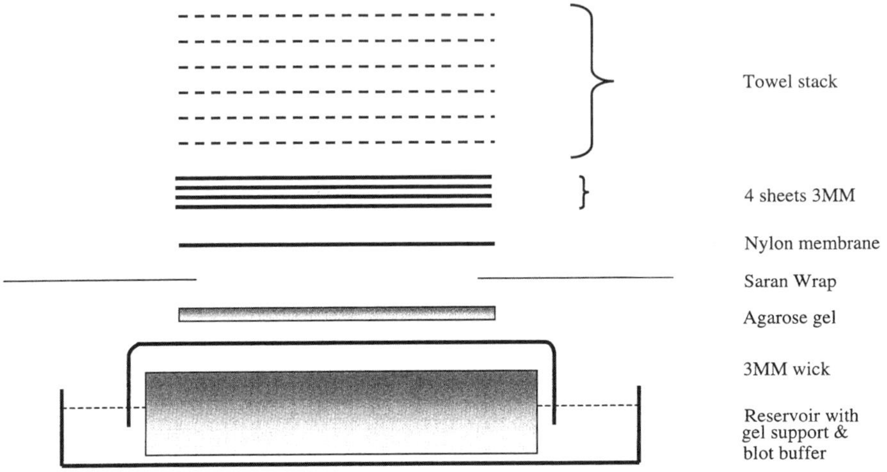

Fig. 2. Methods of transferring DNA to membranes (usually charged nylon) vary considerably; however, the widely used capillary method ***(9)*** is relatively straightforward. The gel rests on a wick consisting of four layers of 3MM paper, which contact the reservoir of transfer buffer at both ends. The membrane is placed directly on the gel and overlaid with two exactly fitting sheets of 3MM prewetted in transfer buffer, two sheets of dry 3MM, a stack of dry towels, a glass sheet, and (optionally) a ~500-g weight to exclude air and to ensure even blotting pressure. (Some protocols include a layer of Saran wrap between the gel and the membrane, in which the Saran overlaps the edges of the gel by ~0.5 cm and overhangs the sides of the reservoir: this will prevent "short-circuiting" of the blotting process.)

3.5. Probe Labelling

The following protocol and several commercially available kits (e.g., Pharmacia) are based on the technique of "oligolabeling" ***(9,11,12)***. The protocol produces probes of high specific activity (>10^9 dpm/µg).

1. Prepare probe DNA by heating 50–100 ng in a 500-µL Eppendorf tube to 100°C for 5 min to denature.
2. Place tube on ice for 2–5 min, then centrifuge briefly to collect condensate.
3. Add 10 µL 5× "oligolabeling" buffer, 2 µL BSA stock, 50–100 ng DNA in 32 µL H_2O, 5 µL [α-^{32}P]dCTP, 1 µL (5–10 U) Klenow fragment.
4. Incubate at room temperature for 3–12 h.
5. Add 200 µL buffer A to the reaction.
6. Denature probe by heating to 100°C for 5 min, repeat **step 2**, and use immediately (*see* **Note 4**).

3.6. Blot Hybridization and Washing

1. Wet the membrane briefly in deionized water and place in a hybridization bottle or bag. Add ~10 mL preheated (65°C) hybridization buffer and incubate in a rotary hybridization oven for 1–2 h (If necessary, this prehybridization step may be extended to overnight.)
2. Toward the end of the prehybridization period, denature the double-stranded DNA probe by heating to 100°C for 5 min in a water bath or heating block.
3. Using heat-resistant gloves, pour off approximately half (5 mL) of the hybridization buffer and add the denatured probe to the remaining buffer (do not pipet the concentrated probe directly on to the membrane).
4. Carefully seal the hybridization bottle or bag and replace in 65°C oven overnight.
5. Toward the end of the hybridization period, add ~500 mL of wash buffer to a conical flask and heat to 65°C.
6. Remove the hybridization bottle from the oven and pour the radioactive hybridization buffer into a 20-mL polypropylene tube. (This can be discarded or stored at –20°C for future use).
7. Half-fill the hybridization bottle with 65°C wash buffer and return to the oven for 15 min.
8. Remove the membrane from the bottle and wash in a shallow covered dish in 100–200 mL of wash buffer at 65°C for 15 min. Repeat as required. (The severity of washing must be determined empirically; however, removal of unwanted background radioactivity can be monitored using a hand-held counter.)
9. Place the membrane between two sheets of Saran wrap. Place the covered blot in an autoradiography cassette with intensifying screens and tape to the back of the cassette to immobilize.
10. Expose the film overnight at –70°C.

4. Notes

1. A comprehensive list of methyl-sensitive restriction enzymes, their DNA recognition sequences, and isoschizomers is available ***(13,14)***.
2. Phosphate-buffered saline (PBS) is also suitable: Dissolve 8 g NaCl, 0.2 g KCl, 1.44 g Na_2HPO_4 and 0.24 g KH2PO4 in 800 mL H_2O, pH to 7.4 with HCl, add H_2O to 1 L, and autoclave ***(9)***.
3. Gels can also be blotted in alkaline conditions using 0.4 *N* NaOH in the blotting reservoir. Use membrane neutralizing buffer instead of 2× SSC (or SSPE) in **step 5**. (In some protocols the gel is soaked in 0.2 *N* HCl for ~30 min to nick the DNA and enhance blotting efficiency. If this is done, the gel must be soaked in 0.4 *N* NaOH before blotting.).
4. It is recommended that, before using probe, unincorporated dNTPs are removed by G-25 column chromatography. Sepharose G-25 columns, and other alterna-

tives for probe purification, are available commercially (e.g., NAP™5 columns, Pharmacia). However, detailed instructions for cheaply constructing and using Sepharose "spin columns" in 1-mL syringe barrels have been described *(9)*.

References

1. Brannan, C. I. and Bartolomei, M. S. (1999) Mechanisms of genomic imprinting. *Curr. Opin. Genet. Dev.* **9,** 164–170.
2. Olek, A., Oswald, J., and Walter, J. (1996) A modified and improved method for bisulphite based cytosine methylation analysis. *Nucleic Acids Res.* **24,** 5064–5066.
3. Gillis, A. J., Verkerk, A. J., Dekker, M. C., van Gurp, R. J., Oosterhuis, J. W., and Looijenga, L. H. (1997) Methylation similarities of two CpG sites within exon 5 of human H19 between normal tissues and testicular germ cell tumours of adolescents and adults, without correlation with allelic and total level of expression. *Br. J. Cancer* **76,** 725–733.
4. Rougier, N., Bourc'his, D., Gomes, D. M., Niveleau, A., Plachot, M., Paldi, A., and Viegas-Pequignot, E. (1998) Chromosome methylation patterns during mammalian preimplantation development. *Genes Dev.* **12,** 2108–2113.
5. Huang, T. H., Perry, M. R., and Laux, D. E. (1999) Methylation profiling of CpG islands in human breast cancer cells. *Hum. Mol. Genet.* **8,** 459–470.
6. McDonald, L. E., Paterson, C. A., and Kay, G. F. (1998) Bisulphite genomic sequencing-derived methylation profile of the xist gene throughout early mouse development. *Genomics* **54,** 279–286.
7. Dean, W., Bowden, L., Aitchison, A., Klose, J., Moore, T., Meneses, J. J., Reik, W., and Feil, R. (1998) Altered imprinted gene methylation and expression in completely ES cell-derived mouse fetuses: association with aberrant phenotypes. *Development* **125,** 2273–2282.
8. Moore, T., Constancia, M., Zubair, M., Bailleul, B., Feil, R., Sasaki, H., and Reik, W. (1997) Multiple sense and antisense transcripts, differential methylation and tandem repeats in a putative imprinting control region upstream of mouse Igf2. *Proc. Natl. Acad. Sci. USA* **94,** 12,509–12,514.
9. Sambrook, J., Fritsch, E. F., and Maniatis, T. (1989) *Molecular Cloning: A Laboratory Manual*, 2nd ed., Cold Spring Harbor Laboratory Press, Cold Spring Harbor, NY.
10. Church, G. M. and Gilbert, W. (1984) Genomic sequencing. *Proc. Natl. Acad. Sci. USA* **81,** 1991–1995.
11. Feinberg, A. P. and Vogelstein, B. (1983) A technique for radiolabelling DNA restriction endonuclease fragments to high specific activity. *Anal. Biochem.* **132,** 6–13.
12. Feinberg, A. P. and Vogelstein, B. (1984) A technique for radiolabelling DNA restriction endonuclease fragments to high specific activity. Addendum. *Anal. Biochem.* **137,** 266–267.

13. Nelson, M. and McClelland, M. (1989) Effect of site-specific methylation on DNA modification methyltransferases and restriction endonucleases. *Nucleic Acids Res.* **17**(Suppl.), 389–415.
14. McClelland, M., Nelson, M., and Raschke, E. (1994) Effect of site-specific modification on restriction endonucleases and DNA modification methyltransferases. *Nucleic Acids Res.* **22,** 3640–3659.

15

A PCR-Based Method for Studying DNA Methylation

Mira Ariel

1. Introduction

DNA methylation is a mechanism for regulation of gene expression in animals *(1–3)*. The addition of a methyl group at the 5-position of cytosine bases occurs exclusively at CpG dinucleotides. CpG dinucleotides in the vertebrate genome are underrepresented and amount to 1% of the genome *(4)*. However, in some regions of the genome, CpG residues amount to 6% or more of the dinucleotides in the genome. These regions, known as CpG islands, are usually associated with the promoter regions of housekeeping genes and, in contrast to CpGs throughout the genome, are unmethylated *(5,6)*. Methylation of CpG islands occurs only in silenced genes on the inactive X chromosome and in parentally imprinted genes *(7)*. In addition, CpG islands may become methylated upon oncogenic transformation. These alterations in the methylation profile are correlated with silencing of tumor suppressor genes such as p15, p16, Rb, VHL, e-cadherin, ER, and HIC1 *(8)*.

Mapping methylated regions in the genome and detection of methylation changes is important for understanding both normal and pathological gene expression events such as silencing by methylation of tumor-suppressor genes. Another purpose is to study the role of methylation in imprinting *(7)*. Although not completely proven, methylation has been proposed to be the imprint signal *(9–11)*. Over the years, studies of cytosine methylation have focused on the adult organism *(12)*. Only with the development of polymerase chain reaction (PCR) methods to amplify small amounts of DNA has it become possible to determine the state of DNA methylation at CpG sites in specific genes in the embryo *(13–15)*. The accurate mapping of DNA methylation at different stages of development is essential for understanding how specific methylation

From: *Methods in Molecular Biology, vol. 181: Genomic Imprinting: Methods and Protocols*
Edited by: A. Ward © Humana Press Inc., Totowa, NJ

patterns are established and maintained in imprinted as well as other genes *(9,13,14)*. An additional field that requires methylation analysis is the evaluation of the pattern of X-chromosome inactivation, especially for the detection of female carriers of X-linked diseases *(16)*. X-chromosome inactivation is random in the majority of females. However, in asymptomatic female carriers of X-linked diseases, there is a preferential selection of the normal X chromosome, resulting in nonrandom inactivation. This pattern is currently evaluated by assays of differential methylation between the active and the inactive X chromosome *(17)*.

1.1. General Methods for DNA Methylation Analysis

Until recently, mapping of methylated DNA regions relied mainly on Southern hybridization approaches *(18)* following digestion of the genomic DNA with methylation-sensitive restriction enzymes that cleave only unmethylated CpG sites (*19*; *see* Chapter 14). These analyses are limited to the available sites in the region studied, which account for a small proportion of the potentially methylated CpG sites. Another limitation of these techniques is the relatively large quantity of DNA (5–10 µg) necessary for each sample. Maxam and Gilbert *(20)* sequencing techniques have also been used to determine methylation status *(21)*, but these chemical cleavage reactions have two disadvantages with respect to determining 5-methylcytosine residues. First, 5-methylcytosine is identified by the lack of a band in all the tracks of a sequencing gel; any background cleavage ladder or close spacing of bands can cause difficulties in interpretation. Second, this method also requires a relatively large amount of DNA (50 µg per lane).

Another method of studying DNA methylation, which avoids the use of restriction enzymes, is the chemical modification of cytosine, but not methylated cytosine, to uracil, by bisulfite treatment (*22,23*; *see* Chapter 16). This method determines the methylation status of every CpG site. It involves amplification of the modified DNA, cloning of the amplified region, and sequencing individual clones. This method requires only small amounts of DNA, but it is technically difficult and rather laborious. Bisulfite sequencing has not been widely applied to the analysis of cytosine methylation in preimplantation embryos.

1.2. PCR-Based Methods of DNA Methylation Analysis

Because of the above-mentioned limitations, researchers have taken advantage of PCR techniques developed in the last decade. PCR-based methods for methylation analysis were developed for assaying small quantities of DNA available from embryo samples and other sources of DNA. These methods require prior information of the sequence being analyzed.

There are two main approaches in the PCR-based methods for determining DNA methylation. One approach relies on the link between DNA methylation and sensitivity to DNA restriction enzymes *(24,25)*. Methylation can be determined by digestion with a methylation-sensitive restriction enzyme followed by PCR amplification. This assay is sensitive and quantitatively applicable to small quantities of DNA available from as little as 100–200 cells. The assay is technically very simple and rapid.

A second approach is based on PCR amplification of bisulfite treated DNA by methylation-specific PCR *(26–28)*. This assay allows analysis of any group of CpG sites. It is especially useful for mapping methylation patterns of CpG islands, but it cannot be used to determine the methylation state of an individual CpG dinucleotide. The method suffers also from the disadvantage of having to treat the DNA with bisulfite, which renders it more laborious than the methylation-sensitive restriction enzyme-coupled PCR assay that is the focus of this chapter.

1.2.1. Methylation-Sensitive Restriction Enzyme-Coupled PCR Assay

Singer-Sam et al. *(24,25)* developed a sensitive technique in which digestion of genomic DNA with a methylation-sensitive enzyme is coupled with PCR. The amplified product of genomic DNA does not retain the genomic profile of cytosine methylation. Thus, restriction of the methylation-sensitive site studied must be done on the genomic DNA harboring its methylation pattern, before the PCR amplification. Following digestion with HpaII, or any other methylation-sensitive restriction enzyme, PCR is performed with primers that flank the methylation-sensitive site being assayed. Only intact DNA can serve as a template for PCR amplification. Unmethylated sites will be digested by the methylation-sensitive enzyme, thus destroying the template. However, if the site is protected by methylation, it will not be cleaved, the template will remain intact, and a PCR product will be produced (**Fig. 1**).

Similar to Southern-based approaches, this method can only detect CpG methylation if suitable restriction sites are available. Moreover, the cleavage of unmethylated DNA must be complete, since any uncleaved DNA will be amplified by PCR, yielding a false-positive result for methylation.

1.2.1.1. Application for Allele-Specific Methylation Determination

Methylation-sensitive restriction enzyme-coupled PCR assay can be used to determine allele-specific methylation in imprinted genes *(9,29,30)*. A differentially methylated site where only one allele is methylated will always give a PCR product. In order to determine which allele is methylated, one must identify a polymorphism located adjacent to the HpaII or to another

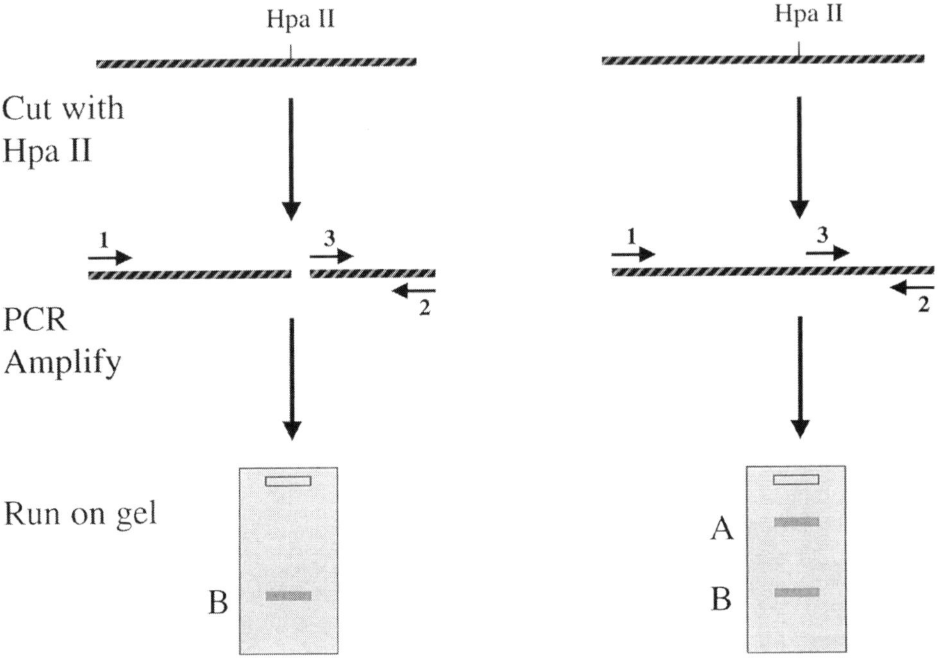

Fig. 1. Methylation-sensitive restriction enzyme-coupled PCR assay, unmethylated DNA on right, methylated DNA on right. Primers 1 and 2 flank the site being assayed. Their amplification product is A. Primers 2 and 3 are nested primers for internal control. Their amplification product is B.

methylation-sensitive enzyme site in the imprinted gene. By restricting the polymorphic site after PCR amplification, the methylation state of each allele can be determined. If the site is methylated on the allele with the polymorphic site, the PCR product will be cleaved by the enzyme. If the polymorphic allele is unmethylated and the template is destroyed by the HpaII restriction, the cleavage of the polymorphic site will not be visualized. The PCR product of the methylated allele will remain intact.

An example of an imprinted gene that has been studied in this manner is the *SNRPN* gene. *SNRPN* is a very well characterized gene from the 15q11-q13 cluster of imprinted genes ***(31)***, and is a useful diagnostic marker for Prader-Willi syndrome (PWS) and Angelman syndrome (AS) ***(32)***. These two disorders are characterized by deletions in a common region of chromosome 15q11-13, uniparental disomy for chromosome 15, and abnormal methylation of the imprinted genes in the region. Patients with PWS have a deletion in the paternal allele, while patients with AS have a deletion in the maternal allele. A rapid diagnostic test that was developed by Chotai and Payne ***(33)*** is based on the digestion of the unmethylated paternally expressed *SNRPN* sequence with the methylation-sensitive enzyme NotI or digestion of the methylated maternally repressed *SNRPN* sequence with the methylation-requiring nuclease McrBC followed by PCR amplification of the *SNRPN* promoter and first exon. This test enables determination of the methylation state of the paternal or the maternal allele in order to diagnose PWS and AS. PWS is expected to give an SNRPN product only after digestion with NotI, because PWS patients have only a maternal methylated *SNRPN*, which is not cleaved by NotI. AS, on the other hand, is expected to give a PCR product only after McrBC digestion because AS patients have only a paternal unmethylated *SNRPN*, which will not be digested by the McrBC. Normal control DNA is expected to yield a product after both NotI and McrBC digestion.

Imprinted genes may also play a role as tumor suppressors. *NOEY2* ***(34)***, for example, is a maternally imprinted tumor-suppressor gene that loses its expression in ovarian and breast cancer because of deletion of the paternally expressed allele. The maternal allele can be distinguished from the paternal allele on the basis of a TA repeat polymorphism. Methylation can be determined by digesting the genomic DNA with the methylation-sensitive restriction enzyme BstUI and PCR amplification of the DNA. Only the methylated allele will be amplified.

1.2.1.2. Application for X-Chromosome Inactivation Determination

Another example of the use of methylation-sensitive enzyme-coupled PCR for diagnostic purposes is the application of the novel human androgen receptor assay (HUMARA) for analyzing X inactivation ***(16,35)***. This assay is a sensitive means for detecting nonrandom X inactivation. The HUMARA gene includes a CAG repeat which is highly polymorphic. In the assay, the CAG repeat is amplified together with its flanking DNA which contains two HpaII restriction sites. The close proximity of the restriction enzyme sites to the short tandem CAG repeat allows the PCR assay to distinguish between the maternal and paternal alleles according to the size of the CAG repeat of each allele and identify their methylation status. Genomic DNA is digested with HpaII before PCR amplification. In an active X chromosome the HpaII site is unmethylated and PCR amplification fails to yield a product. On the other hand, the inactive X chromosome is methylated and resistant to HpaII digestion, thus yielding a PCR product. In females with random X inactivation, the maternal and the paternal inactive X chromosomes will be amplified equally. However, in females who are nonsymptomatic X-linked-disease carriers, one of the X chromosomes is selectively inactivated and methylated, and in the HUMARA assay it will be selectively amplified.

1.2.2. Arbitrarily Primed PCR

A further development of the methylation-sensitive restriction enzyme-coupled PCR was the application to arbitrarily primed PCR (AP-PCR) in order to identify specific methylation changes at multiple sites in genomic DNA with unknown sequences ***(36)***. This method uses a single pair of primers without any prior information of the sequence studied. It involves restricting the DNA with a methylation-sensitive restriction enzyme before two cycles of low-stringency amplification followed by 30 or 40 cycles of high-stringency PCR. Reproducible fingerprints are generated . The random association of primers with genomic DNA at low annealing temperatures generates multiple PCR fragments. When the two primers flank a methylated HpaII restriction site in a single fragment, amplification takes place. However, no PCR product would be expected if the site is unmethylated. Methylation-sensitive AP-PCR has been applied to identify CpG islands that became hypermethylated in human lung cancer cells ***(37)*** and in bladder and colon tumor DNAs ***(38)***. Methylation-sensitive AP-PCR is rapid and simple. It can be used to screen methylation changes in different tissues and to isolate DNA fragments associated with such changes. This method can detect such changes in as little as 200 ng of genomic DNA.

2. Materials

2.1. Digestion of the Genomic DNA

1. A restriction enzyme that does not cut in the target sequence and its appropriate buffer.
2. λDNA for carrier.

2.2. Dialysis of the Digested DNA

1. Millipore VSWP mixed esters filters: 0.025-μm 25 m*M*.
2. Petri dishes.
3. Double-distiiled water (ddH_2O).

2.3. Digestion of the Dialyzed DNA with a Methylation-Sensitive Restriction Enzyme

1. A methylation-sensitive restriction enzyme that has a restriction site in the sequence studied, and its appropriate buffer.
2. A 37°C incubator.

2.4. Dialysis of the Methylation-Sensitive Restricted DNA

1. Millipore filters as specified above.
2. Petri dishes.

2.5. PCR Amplification

1. Taq DNA polymerase and 10× PCR buffer (usually supplied together, e.g., Promega Corporation).
2. A concentrated $MgCl_2$ (typically 15–25 m*M*) solution (*see* **Note 1**).
3. A mix of the four dNTPs (2 m*M* each).
4. Primers 300–500 μ*M* each in separate solutions (*see* **Note 2** and **Fig. 1**).
5. A thermal cycler.
6. Mineral oil (not needed if the thermal cycler has a heated lid).

2.6. Gel Electrophoresis

1. Agarose.
2. 10× TAE buffer: 2 m*M* Tris-acetate, 0.05 *M* ethylenediaminetetraacetic acid (EDTA).
3. 10-mg/mL ethidium bromide.
4. Gel electrophoresis apparatus.

2.7. Determination of Allele-Specific Methylation

1. A restriction enzyme that cleaves a polymorphic site unique to one allele.

3. Methods

3.1. Digestion of the Genomic DNA

Digest genomic DNA with a restriction enzyme that does not cut in the sequence studied. The amount of genomic DNA can range from 200 pg to 1 μg (*see* **Note 3**). For amounts less than 1 μg, λDNA (1 μg per sample) should be added as carrier. The purpose of the digestion is to reduce the size of the genomic DNA and thus increase accessibility of the methylation-sensitive enzyme to its site and to assure complete digestion.

3.2. Dialysis of the Digested DNA

Dialyze 100 ng–200 pg of the digested DNA in a final volume of 50 μL through a Millipore VSWP mixed esters filter as follows:

1. Place the disk of the Millipore filter on top of 12 mL of ddH_2O in a Petri dish.
2. Aliquot the digested DNA on top of the filter disk. Cover the Petri dish with its top to avoid evaporation.
3. Dialyze for 2 h at room temperature.
4. Collect the sample into an Eppendorf tube.

The purpose of the dialysis is to eliminate all the electrolytes of the DNA sample and thus produce optimal conditions in the buffer for the next restriction enzyme digest (*see* **Note 4**).

3.3. Digestion of the Dialyzed DNA with a Methylation-Sensitive Restriction Enzyme

Divide the dialyzed DNA into two portions: one for an uncut control and the second for the digestion of the next step.

1. Digest the dialyzed DNA with HpaII or another methylation-sensitive restriction enzyme in the appropriate buffer and according to the manufacturer's instructions.
2. Incubate for 16 h at 37°C.

3.4. Dialysis of the Methylation-Sensitive Restricted DNA

Dialyze through Millipore filters by repeating the details of **Subheading 3.2.** The purpose of the dialysis at this stage is to eliminate all the electrolytes from the digested DNA and provide optimal conditions for the PCR reaction (*see* **Note 4**).

3.5. PCR Amplification

1. Perform the PCR amplification in a 100-μL mixture as follows: 1× PCR buffer, 1.5 m*M* $MgCl_2$ (but *see* **Note 1**), 200 μmol of each of the four dNTPs, 6 μmol of each of the two primers, 1–2.5 U Taq polymerase.
2. Mix all the components well.
3. Add 25 μL of the DNA collected after the dialysis of **Subheading 3.4.** (*see* **Note 5**).
4. Make up the volume of the reaction to 100 μL with ddH_2O. A control with unmethylated DNA to confirm full cleavage with the methylation-sensitive restriction enzyme (*see* **Note 6**) and an additional control without DNA should be included. Each assay should also include a PCR amplification with the control of the nested primers.
5. Add 4–5 drops of mineral oil to each sample (this is not necessary if using a thermal cycler with a heated lid). The mineral oil should completely cover the surface of the reaction mixture, to prevent evaporation.
6. PCR amplify in a thermal cycler with the following steps:
   a. 4 min at 95°C (denaturation)
   b. 1 min at 95°C (denaturation)
   c. 2 min at 55°C (annealing to primers)
   d. 3 min at 72°C (elongation)
   Perform 35–40 cycles (each cycle is from step b to step d). These are general conditions; however, optimization of conditions (particularly MgCl2 concentration and annealling temperature) is recommended for each pair of primers (*see* **Note 7**).

3.6. Gel Electrophoresis

After amplification, load 20 μL of each sample onto 2% agarose gel containing ethidium bromide. Electrophorese and identify the band of the product under UV illumination (*see* **Note 8**).

3.7. Determination of Allele-Specific Methylation

For identifying allele-specific methylation, restrict the PCR product with an enzyme that cleaves a polymorphic site unique to one allele. Electrophorese on an agarose gel to distinguish between the two alleles and determine whether methylated or nonmethylated.

4. Notes

1. $MgCl_2$ concentration will need to be optimized for each specific PCR reaction.
2. The design of the primers is one of the most critical steps in the assay. The primers should be 20–24 bp in length and flank the methylation-sensitive restriction site being examined. The sequence of the primers should not include the same methylation-sensitive restriction site. Nested primers should be constructed to serve as an internal standard and to ensure the specificity of the PCR reaction. One primer can be shared with one of the flanking primers. The second primer should avoid the methylation-sensitive site so that the PCR amplification product will not include the site (*see* **Fig. 1**).
3. The first digestion step can be performed on relatively large amounts of DNA (1 µg) if available. The digested DNA should be aliquoted and stored at –20°C. The desired number of aliquots can be removed for each experiment.
4. It is advisable to measure the sample volume after each of the dialysis steps, since dialysis may cause an increase in sample volume. When this occurs, a correction of DNA concentration should be made.
5. Because the amount of genomic DNA in each sample is very small, contamination can mask the true results. Consequently, it is advisable to keep the genomic DNA separate from any other DNA samples.
6. To ensure full cleavage with the methylation-sensitive enzyme prior to the PCR amplification, it is suggested to amplify another sequence of the genomic DNA, preferably a CpG island that includes the same methylation-sensitive restriction site as that assayed in the experiment.
7. The initial denaturation step (95°C for 4 min) of the PCR amplification is of critical importance for PCR yield, especially with the very small amounts of DNA for which the PCR method is the method of choice.
8. Optional: Southern blotting can be performed in cases where you cannot observe the expected product as a band(s) on the ethidium bromide-stained gel.

References

1. Cedar, H. (1988) DNA methylation and gene activity. *Cell* **53,** 3–4.
2. Doerfler, W. (1983) DNA methylation and gene activity. *Annu. Rev. Biochem.* **52,** 93–124.
3. Siegfried, Z. and Cedar, H. (1997) DNA methylation: a molecular lock. *Curr. Biol.* **7,** 305–307.
4. Schorderer, D. F. and Gartler, S. M. (1992) Analysis of CpG suppression in methylated and nonmethylated species. *Proc. Natl. Acad. Sci. USA* **89,** 957–961.

5. Bird, A. P. (1986) CpG-rich islands and the function of DNA methylation. *Nature* **321,** 209–213.
6. Eden, S. and Cedar, H. (1994) Role of DNA methylation in the regulation of transcription. *Curr. Opin. Genet. Dev.* **4,** 255–259.
7. Li, E., Beard, C., and Jaenisch, R. (1993) Role for DNA methylation in genomic imprinting. *Nature* **366,** 362–365.
8. Baylin, S. B., Herman, J. G., Graff, J. R., Vertino, P. M., and Issa, J. P. (1998) Alterations in DNA methylation: a fundamental aspect of neoplasia, in *Advances In Cancer Research* (Vandewoude, G. and Klein, G., eds.), Academic, San Diego, CA, pp. 141–196.
9. Brandeis, M., Kafri, T., Ariel, M., Chaillet, J. R., McCarrey, J., Razin, A., and Cedar, H. (1993) The ontogeny of allele-specific methylation associated with imprinted genes in the mouse. *EMBO J.* **12,** 3669–3677.
10. Razin, A. and Cedar, H. (1994) DNA methylation and genomic imprinting. *Cell* **77,** 473–476.
11. Ariel, M., Robinson, E., McCarrey, J. R., and Cedar, H. (1995) Gamete-specific methylation imprints on the *Xist* gene. *Nature Genet.* **9,** 312–315.
12. Monk, M., Boubelik, M., and Lehnert, S. (1987) Temporal and regional changes in DNA methylation in the embryonic, extraembryonic and germ cell lineages during mouse embryo development. *Development* **99,** 371–382.
13. Kafri, T., Ariel, M., Brandeis, M., Shemer, R., Urven, L., McCarrey, J., Cedar, H., and Razin, A. (1992) Developmental pattern of gene-specific DNA methylation in the mouse embryo and germline. *Genes Dev.* **6,** 705–714.
14. Brandeis, M., Ariel, M., and Cedar, H. (1993) Dynamics of DNA methylation during development. *BioEssays* **15,** 1–5.
15. Shemer, R., Kafri, T., O'Connell, A., Eisenberg, S., Breslow, J. L., and A., R. (1991) Methylation changes in the *apo AI* gene during embryonic development of the mouse. *Proc. Natl. Acad. Sci. USA* **88,** 10,300–10,304.
16. Kubota, T., Nonoyama, S., Tonoki, H., Masuno, M., Imaizumi, K., Kojima, M., Wakui, K., Shimadzu, M., and Fukushima, Y. (1999) A new assay for the analysis of X-chromosome inactivation based on methylation-specific PCR. *Hum. Genet.* **104,** 49–55.
17. Willard, H. F. (1995) The sex chromosomes and X chromosome inactivation, in *The Metabolic and Molecular Bases of Inherited Diseases* (Scriver, C. R., Beaudet, A. L., Sly, W. S., and Valle, D., eds.), McGraw-Hill, New York, pp. 718–737.
18. Southern, E. M. (1975) Detection of specific sequences among DNA fragments separated by gel electrophoresis. *J. Mol. Biol.* **98,** 503–517.
19. Bird, A. P. (1978) Use of restriction enzyme to study eukaryotic DNA methylation. II. The symmetry of methylation sites supports semiconservative copying of the methylation pattern. *J. Mol. Biol.* **118,** 49–60.
20. Maxam, A. M. and Gilbert, W. (1980) Sequencing and labeling DNA with base-specific chemical cleavages. *Meth. Enzymol.* **65,** 499.

21. Saluz, H. P. and Jost, J. P. (1993) Major techniques to study DNA methylation, in *DNA Methylation: Molecular Biology and Biological Signficance* (Jost, J. P. and Saluz, H. P., eds.), Burkhausen-Verlag, Basil, pp. 11–26.
22. Frommer, M., McDonald, L. E., Millar, D. S., Collis, C. M., Watt, F., Grigg, G. W., Molloy, P. L., and Paul, C. L. (1992) A genomic sequencing protocol that yields a positive display of 5-methylcytosine residues in individual DNA strands. *Proc. Natl. Acad. Sci. USA* **89,** 1827–1831.
23. Clark, S. J., Harrison, J., Paul, C. L., and Frommer, M. (1994) High sensitivity mapping of methylated cytosines. *Nucleic Acids Res.* **22,** 2990–2997.
24. Singer-Sam, J., LeBon, J. M., Tanguay, R. L., and Riggs, A. D. (1990) A quantitative HpaII-PCR assay to measure methylation of DNA from a small number of cells. *Nucl. Acids Res.* **18,** 687–692.
25. Singer-Sam, J., Yang, T. P., Mori, N., Tanguay, R. L., Le Bon, J. M., Flores, J. C., and Riggs, A. D. (1990) DNA methylation in the 5′ region of the mouse PGK-1 gene and a quantitative PCR assay for methylation, in *Nucleic Acid Methylation* (Clawson, G., Willis, D., Weissbach, A., and Jones, P., eds.), Liss, New York, pp. 285–298.
26. Herman, J. G., Graff, J. R., Myohanen, S., Nelkin, B. D., and Baylin, S. B. (1996) Methylation specific PCR: a novel PCR assay for methylation status of CpG islands. *Proc. Natl. Acad. Sci. USA* **93,** 9821–9826.
27. Herman, J. G., Umar, A., Polyak, K., Graff, J. R., Ahuja, N., Issa, J.-P., J., Markowitz, S., Willson, J. K. V., Hamilton, S. R., Kinzler, K. W., Kane, M. F., Kolodner, R. D., Vogelstein, B., Kunkel, T. A., and Baylin, S. T. (1998) Incidence and functional consequences of hMLH1 promoter hypermethylation in colorectral carcinoma. *Proc. Natl. Acad. Sci. USA* **95,** 6870–6875.
28. Kubota, T., Das, S., Christian, S. L., Bayling, S. B., Herman, J. G., and Ledbetter, D. H. (1997) Methylation-specific PCR simplifies imprinting analysis. *Nature Genet.* **16,** 16–17.
29. Shibata, H., Ueda, T., Kamiya, M., Yoshiki, A., Kusakabe, M., Plass, C., Held, W. A., Sunahara, S., Katsuki, M., Muramatsu, M., and Hayashizaki, Y. (1997) An oocyte-specific methylation imprint center in the mouse *U2afbp-rs/U2af1-rs1* gene marks the establishment of allele-specific methylation during preimplantation development. *Genomics* **44,** 171–178.
30. Stoger, R., Kubicka, P., Liu, C.-G., Kafri, T., Razin, A., Cedar, H., and Barlow, D. P. (1993) Maternal-specific methylation of the imprinted mouse *Igf2r* locus identifies the expressed locus as carrying the imprinting signal. *Cell* **73,** 61–71.
31. Glenn, C. C., Porter, K. A., Jong, M. T. C., Nicholls, R. D., and Driscoll, D. J. (1993) Functional imprinting and epigenetic modification of the human *SNRPN* gene. *Hum. Mol. Genet.* **2,** 2001–2002.
32. Reis, A., Dittrich, B., Greger, V., Buiting, K., Lalande, M., Gillessen-Kaesbach, G., Anvret, M., and Horsthemke, B. (1994) Imprinting mutations suggested by abnormal DNA methylation patterns in familial Angelman and Prader-Willi syndromes. *Am. J. Hum. Genet.* **54,** 741–747.

33. Chotai, K. A. and Payne, S. J. (1998) A rapid, PCR based test for differential molecular diagnosis of Prader-Willi Angelman syndromes. *J. Med. Genet.* **35,** 472–475.
34. Yu, Y., Xu, F., Peng, H., Fang, X., Zhao, S., Li, Y., Cuevas, B., Kuo, W.-L., Gray, J. W., Siciliano, M., Mills, G. B., and Bast, R. C., Jr. (1999) *NOEY2 (ARhi)*, an imprinted putative tumor suppressor gene in ovarian and breast carcinomas. *Proc. Natl. Acad. Sci. USA* **96,** 214–219.
35. Allen, R. C., Zogbi, H. Y., Moseley, A. B., Rosenblatt, H. M., and Belmont, J. W. (1992) Methylation of *HpaII* and *HhaI* sites near the polymorphic CAG repeat in the human androgen-receptor gene correlates with X chromosome inactivation. *Am. J. Hum. Genet.* **51,** 1229–1239.
36. Welsh, J. and McClelland, M. (1990) Fingerprinting genomes using PCR with arbitrary primers. *Nucleic Acids Res.* **18,** 7213–7218.
37. Kohno, T., Kawanishi, M., Inazawa, J., and Yokota, J. (1998) Identification of CpG islands hypermethylated in human lung cancer by the arbitrarily primed-PCR method. *Hum. Genet.* **102,** 258–264.
38. Gonzalgo, M. L., Liang, G., Spruck III, C. H., Zingg, J.-M., Rideout, W. M. III, and Jones, P. A. (1997) Identification and characterization of differentially methylated regions of genomic DNA by methylation-sensitive arbitrarily primed PCR. *Cancer Res.* **57,** 594–599.

16

Bisulfite-Based Methylation Analysis of Imprinted Genes

Sabine Engemann, Osman El-Maarri, Petra Hajkova, Joachim Oswald, and Joern Walter

1. Introduction

1.1. Genomic Imprinting and DNA Methylation

Genomic imprinting is an epigenetically controlled form of gene regulation leading to the preferential expression of one parental gene copy. To date, approximately 40 imprinted genes have been described that are exclusively or predominantly expressed from either the paternal or the maternal allele *(www.mgu.har.mrc.ac.uk/imprinting/implink.html)*. Changes in the imprinted expression of such genes result in developmental abnormalities; in the human they are associated with several diseases and various types of cancer ***(1–3)***.

Allele-specific expression of imprinted genes requires an epigenetic marking process that allows the transcription machinery to distinguish between paternal and maternal alleles. Although the molecular links between the mechanisms of marking and expression control remains unclear, it is increasingly evident that allele-specific DNA methylation is a key player in this process.

Almost every imprinted gene contains one or more regions displaying an allele-specific methylation pattern ***(4)***. The most convincing result demonstrating a link between methylation and allele-specific gene expression comes from studies in which the methyltransferase gene *Dnmt1* was inactivated. In *Dnmt1–/–* mice, inappropriate expression of several imprinted genes, including *H19*, *Igf2*, and *Igf2r*, was observed ***(5)***. Furthermore, diseases linked to abnormal expression of imprinted genes, such as Beckwith-Wiedemann syndrome, or Prader-Willi and Angelman syndromes, are frequently connected with alterations of the parent-of-origin-specific methylation patterns ***(1–4)***.

From: *Methods in Molecular Biology, vol. 181: Genomic Imprinting: Methods and Protocols*
Edited by: A. Ward © Humana Press Inc., Totowa, NJ

The methylation analysis of imprinted genes is therefore a valuable tool to localize regions important for allele-specific marking and to detect epigenetic mutations resulting in abnormal development.

In comparison to a variety of other methods used to analyze the methylation status of imprinted genes, the bisulfite-based genomic sequencing has several clear advantages. First, unlike methods that employ methylation-sensitive restriction enzymes, it allows sequence-independent analysis of CpGs within the analyzed region. Second, it provides detailed information about methylation patterns in imprinted genes or regions and allows the determination of qualitative and quantitative changes in methylation patterns during development. Third, almost any DNA polymorphism can be used to distinguish the parental alleles. The method does not depend on the availability of polymorphic allele-specific restriction sites. Fourth, even the smallest amount of DNA or cells can be analyzed.

The original bisulfite genomic sequencing method developed by Frommer et al. *(6)* has revolutionarized the field of methylation analysis by providing a qualitatively and quantitatively novel experimental approach. The method allows direct and positive determination of methylation sites in genomic DNA, as only methylated cytosines are detected as cytosines (see below). Products of PCR-amplified bisulfite-treated DNA can be sequenced directly (to detect average methylation levels; **refs.** *7,8*) or cloned in *Escherichia coli* and sequenced subsequently to obtain information about the methylation pattern of individual chromosomes. Furthermore, information about the methylation pattern of individual DNA strands can be obtained, as the strands are no longer complementary after bisulfite treatment and hence are amplified and sequenced separately.

Several modifications of the original bisulfite sequencing protocol have been published, improving the sensitivity and quality of the results *(9–13)*. The following detailed protocol is routinely used in our laboratory to determine methylation patterns in imprinted genes.

1.2. Principles of the Method

The protocol is a modified version of the original bisulfite-based methylation analysis technique described by Frommer et al. *(6)*. The main difference is that the material under investigation, either isolated DNA or intact cells (e.g., isolated primordial germ cells, oocytes, or early embryos) is routinely embedded into low-melting-point (LMP) agarose. All subsequent modification steps can be performed in the agarose, in which the DNA is physically trapped. The embedding greatly facilitates the handling of the probes (no DNA isolation is required) and reduces the loss of template during the procedure. This

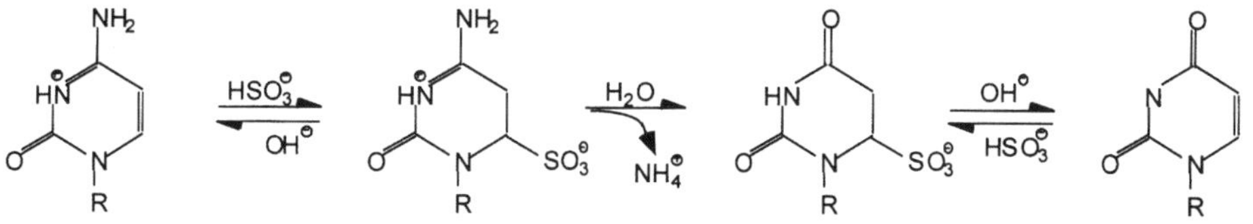

I II III

Fig. 1. Chemistry of (I) sulfonation at position C6 of cytosine, (II) irreversible hydrolytic deamination at position C4, generating 5-sulfonate-uracil, and (III) subsequent desulfonation under alkaline conditions. Note that methylation at position C5 impedes sulfonation at position C6 (step I).

facilitates manipulation of the minute quantities of DNA or cells/tissue, which is desirable when investigating methylation patterns during early embryogenesis or tissue-specific imprinting. Moreover, embedding reduces the reannealing of denatured DNA strands, therefore ensuring the highest quality and reproducibility in the bisulfite reaction. Different versions of the procedure are used when working with intact cells (**Subheading 3.1.1.**) and purified genomic DNA (**Subheading 3.1.2.**).

When working with intact cells, the cells are embedded into agarose, lysed, and treated with proteinase K to make the genomic DNA accessible for the subsequent bisulfite treatment. The inclusion of an endonuclease restriction (with a methylation-insensitive enzyme) of the embedded DNA is recommended, to obtain smaller DNA fragments (of about 3–6 kb), which enhances the spatial separation of complementary DNA strands after the subsequent denaturation step. Agarose-embedded (and digested) DNA is denatured by alkaline treatment and boiling. Nonmethylated cytosine residues in single strands are modified in the presence of 2.5 *M* sodium bisulfite, then converted to uracil residues by subsequent alkaline treatment (**Fig. 1**). Then the DNA is washed extensively and stored in minimal volumes of buffer. The sequence of interest is finally amplified by (nested) PCR. Following the bisulfite treatment, DNA strands are no longer complementary, therefore they are amplified and analyzed separately (**Fig. 2**). The PCR products can be used directly for sequencing, which allows the quantification of average values of cytosine methylation at distinct positions. Alternatively, the PCR product may be cloned and individual clones sequenced, in order to reveal information about individual chromosomes.

The following protocol is based on a previously published procedure *(12)*. We have included several minor modifications in order to make the sample handling easier and less time consuming.

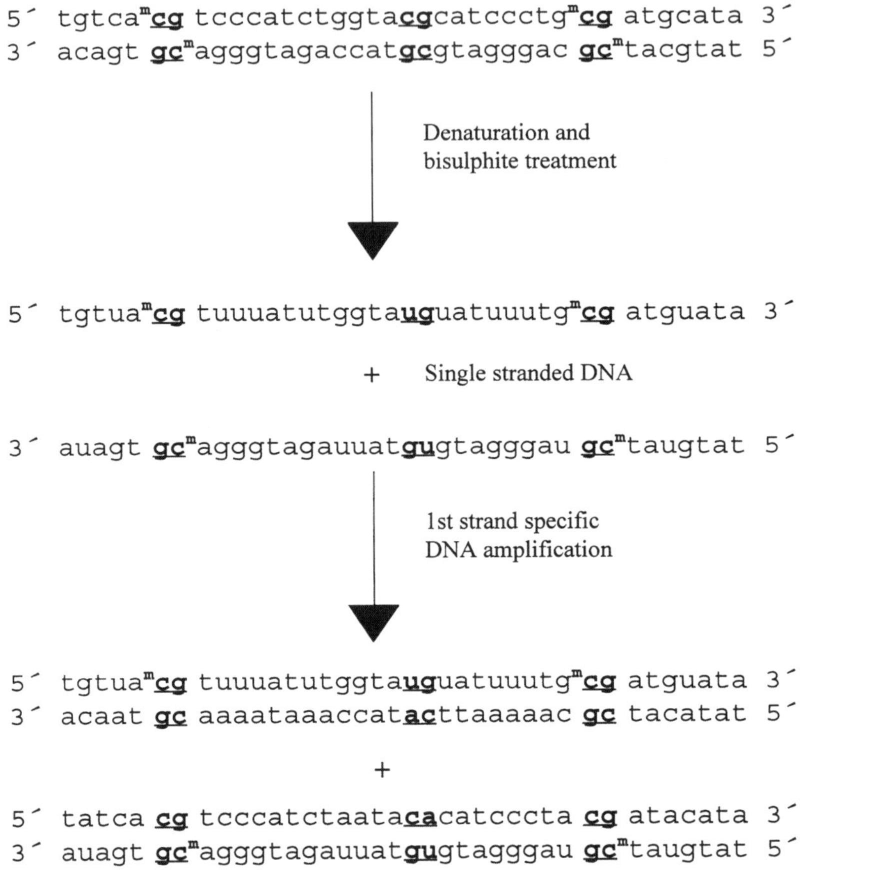

Fig. 2. Schematic diagram of the bisulfite conversion of a DNA sequence; note that the upper and lower strands are no longer complementary after the bisulfite treatment (a = adenine, c = cytosine, ^{m}c = 5-methyl-cytosine, g = guanine, t = thymine, u = uracil) The amplification products are shown for the upper and as well for the lower strand after the complete bisulfite treatment.

1.3. Chemistry of the Bisulfite Reaction

The reaction of cytosine residues with sodium bisulfite leading to selective conversion to uracils follows three major steps (**Fig. 1**): (1) sulfonation, (2) deamination, and (3) desulfonation ***(14)***.

1. *Reversible sulfonation of cytosine residues to cytosine-6-sulfonate.* The sulfonation is favored at low pH and low temperature; at 0°C the equilibrium state is reached within 20 min.

2. *Irreversible hydrolytic deamination of cytosine-6-sulfonate to uracil-6-sulfonate.* This reaction is favored at higher concentrations of sodium bisulfite and at higher temperatures; the pH optimum is between pH 5 and 6.
3. *Reversible desulfonation of uracil-6-sulfonate to uracil.* The elimination reaction is favored at high pH.

Only non-base-paired cytosines, i.e., in single-stranded DNA, can be efficiently modified by sodium bisulfite. Cytosines in nondenatured, double-stranded DNA are almost refractory to the reaction ***(15)***. Furthermore, under the conditions described, the reaction is highly selective and almost complete for nonmethylated cytosine residues, whereas nearly 100% of 5-methylcytosines remain unconverted ***(15)***.

2. Materials

2.1. Embedding of Material into Agarose and Bisulfite Reaction

1. Trypsin (Biochrom), 0.25% (w/v) in phosphate-buffered saline (PBS).
2. Mineral oil (heavy white mineral oil, Sigma).
3. LMP agarose (SeaPlaque agarose, FMC) 2% in PBS and in water.
4. Tris-EDTA (TE) buffers, pH 7.0 and pH 8.0: 10 m*M* Tris-HCl, 1 m*M* ethylenediaminetetraacetic acid (EDTA).
5. Proteinase K (Boehringer Mannheim).
6. Hydroquinone (Sigma).
7. 40 µg/mL phenylmethylsulfonyl fluoride (PMSF, purchased from Sigma) in TE buffer.
8. 2.5 *M* sodium bisulfite solution (pH 5). Prepare as follows: dissolve 1.9 g of sodium metabisulfite (Merck) in a mix of 2.5 mL H_2O and 750 µL of 2 *M* NaOH (freshly prepared), dissolve 55 mg of hydroquinone in 500 µL of H_2O at 50°C, and mix the two solutions (*see* **Notes 1–3**).
9. Phosphate-buffered saline (PBS).
10. Lysis solution: 10 m*M* Tris-HCl, 10 m*M* EDTA, 1% sodium dodecyl sulfate (SDS), 20 µg/mL proteinase K (*see* **Notes 4** and **5**).
11. Restriction enzymes and buffers.
12. NaOH (0.2 *M*, 0.4 *M*, 2.0 *M*).

2.2. Purification and Cloning of PCR Products

1. Taq polymerase (Boehringer Mannheim).
2. Geneclean II (Bio 101) or comparable kit for purification of PCR fragments from agarose gels.
3. TA cloning kit (Invitrogen) with *INV F′* ultracompetent *E. coli* cells (*see* **Note 6**).

3. Methods

3.1. Bisulfite-Based Cytosine Methylation Analysis of Agarose-Embedded Material

3.1.1. Preparation of Cells for Bisulfite Treatment

The following procedure should be used when working with limited amounts of tissue or with only a few cells for which DNA isolation is difficult. When larger quantities of cellular material are available—from biopsies, paraffin-embedded tissues, sperm samples, or other sources—we recommend isolating the genomic DNA using standard procedures and then following the protocol of **Subheading 3.1.2.**

1. When starting with tissue samples, this material should be trypsinized to obtain a suspension of single cells. In the case of individually collected cells (oocytes, zygotes, etc.), proceed directly to **step 2**.
2. Wash and recover the cells in a 1× PBS solution at a maximum density of 60 cells/µL.
3. Mix 3 µL of the cell suspension with 6 µL of hot (80°C) 2% (w/v) LMP agarose, prepared in 1× PBS. Use a 2-mL Eppendorf tube.
4. Add 500 µL of heavy mineral oil, incubate in a boiling water bath for 20 min, and transfer to ice (additional 30 min) to solidify the agarose/cell mixture.
5. Incubate the agarose bead in 500 µL of the lysis solution, overlaid with the mineral oil, at 37°C overnight.
6. Remove the lysis solution and the oil and inactivate proteinase K by adding 500 µL of 1× TE pH 7.0 containing 40 µg/mL PMSF (2× 45 min) at room temperature. This step is optional.
7. Remove the solution and wash with 1× TE (pH 8) for 2× 15 min.
8. Equilibrate against 100 µL of restriction buffer for 2× 15 min.
9. Remove the solution and add 100 µL of 1× restriction buffer containing 20 units of restriction endonuclease and incubate overnight. (Alternatively, add 50 units for 1 h digestion).
10. Remove the restriction buffer and incubate with 500 µL of 0.4 *M* NaOH for 2× 15 min.
11. Wash with 1 mL of 0.1 *M* NaOH for 5 min.
12. Remove all the solution and overlay with 500 µL of mineral oil.
13. Boil the bead in a water bath for 20–30 min to separate the individual DNA strands.
14. Chill on ice for 30 min to resolidify the agarose bead.
15. Add 1 mL of the 2.5 *M* sodium bisulfite solution (ice-cold). The agarose bead should enter the aqueous (lower) phase (*see* **Note 3**).
16. Proceed with the bisulfite treatment (**Subheading 3.1.2., step 8**).

3.1.2. Bisulfite Treatment of Isolated DNA

1. Digest the genomic DNA with a suitable restriction enzyme (which does not cut within the region to be amplified) in a volume of 21 µL. In order to achieve a complete bisulfite conversion we recommend not using more than 700 ng DNA for the restriction, so that the DNA content of each agarose-DNA bead formed later (*see* **step 8**) does not exceed 100 ng.
2. Boil for 5–10 min in a water bath.
3. Chill on ice and quickly spin down.
4. Add 4 µL of 2 *M* NaOH (final concentration 0.3 *M* NaOH) and incubate for 15 min at 50°C.
5. Mix with 2 vol (50 µL) of hot (liquid) 2% (w/v) LMP agarose (prepared in water).
6. Pipet 1 mL of the 2.5 *M* sodium bisulfite solution into a 2-mL Eppendorf tube and overlay with 750 µL of heavy mineral oil (tubes should be kept for 30 min on ice before proceeding) (*see* **Note 3**).
7. Pipet up to seven 10-µL aliquots of the DNA–agarose mixture into ice-cold mineral oil to form beads (each bead containing up to 100 ng of DNA). Make sure that all beads have entered the aqueous phase; beads can be pushed into the bisulfite solution using a pipet tip (*see* **Notes 7–10**).
8. Leave on ice for 30 min.
9. Incubate at 50°C for 3.5 h.
10. Remove all solution; wash with 1 mL of 1× TE (pH 8) for 2 × 15 min.
11. Incubate in 500 µL of 0.2 *M* NaOH, 2 × 15 min.
12. Remove NaOH solution and wash with 1 mL of 1× TE (pH 8), 3 × 10 min. Store in a small volume of TE (pH 8) at 4°C (beads are stable for at least several weeks).
13. Before PCR amplification, wash the beads with H_2O for 2 × 15 min.

3.2. PCR Amplification of Bisulfite-Treated DNA

3.2.1. Primer Design

The following guidelines for primer design for the amplification of bisulfite-treated DNA should be considered when performing a bisulfite-based methylation analysis of imprinted genes.

1. The region to be analyzed should contain sequence polymorphisms (e.g., G-to-A transitions), which allows the identification of the parental alleles after bisulfite treatment. This may limit the analysis to either the lower or upper strand.
2. A "bisulfite-converted" DNA sequence should be generated using computational support by substituting all C residues for T residues except at CpG sites. Such a sequence file can be used by any primer-designing software to test the selected

primers to avoid hairpin structures, false priming sites, and possible primer dimers.

3. Overlapping of the primers with CpG dinucleotides and hence inclusion of wobble positions (T/C) should be strictly avoided especially at the 3′ end of the oligos. This should ensure that PCR products of the region of interest are amplified regardless of whether they contain a high or low degree of nonconverted CpG residues after the bisulfite treatment.
4. The length of the oligos should be between 25–30 nucleotides to achieve a high PCR specificity.
5. The primers should be located in an originally cytosine-rich region so that they selectively amplify converted DNA (*see also* **Notes 11–12**).
6. Extensive T and A stretches in both primers, which are common to bisulfite-treated DNA, should be avoided to minimize the formation of primer dimers.
7. If a T/A cloning vector is used for cloning of amplified products, it is recommended that the 5′ end of the oligomers be T or A. (This will favor the nonspecific addition of flanking adenine residues by the Taq polymerase at the end of the synthesized DNA strands.)

3.2.2. Optimizing PCR Conditions

1. The PCR conditions for amplifying bisulfite-treated material should be carefully optimized. The bisulfite treatment reduces the sequence specificity (by changing all non-CpG cytosines to uracils) and thus the selectivity for primer annealing.
2. It is recommended that the length of the product does not exceed 600–700 bp, as longer fragments may be more difficult to amplify from bisulfite-treated DNA (due to depurination and fragmentation of DNA as a result of low pH during the bisulfite treatment).
3. A nested or at least a seminested approach for amplifying the target region is recommended to increase the sensitivity when working with limited numbers of cells and to ensure the specificity of the product.
4. To avoid any contamination with previous PCR products, the bisulfite treatment and handling of the DNA or cells should be carried out in a separate room and using separate pipets.
5. We advise that a gradient PCR cycler be used to optimize the annealing temperature for the PCR reaction (with the aim of obtaining as high an annealing temperature as possible in the final reaction).

3.2.3. Cloning and Sequencing

1. To increase the efficiency of cloning, the specific PCR product should be purified from any nonspecific band(s) or primer dimers by agarose gel elution.
2. Cloning of the PCR product can be improved by additional incubation of the purified product in the presence of dATP and Taq polymerase for 5 min at 95°C, followed by 60 min at 72°C. This will increase the percentage of DNA molecules with flanking A residues at the 3′ end (this step is optional).

3. To verify positive clones, we routinely apply a colony PCR protocol. Products of the correct size can be subsequently sequenced using internal primers.
4. According to our experience, blue/white screening of colonies is not always reliable (especially when short fragments are cloned). In such situations it is recommended to analyze all colonies, as the blue ones may also contain an insert (*see also* **Note 5**).

3.2.4. Drawbacks of the Bisulfite-Based Methylation Analysis

Although the bisulfite-based methylation analysis is a powerful tool to obtain detailed genomic methylation data, it is connected with specific experimental or technical problems which are briefly discussed below.

1. In order to perform a bisulfite-based methylation analysis, detailed sequence information about the genomic region of interest is required.
2. The upper and lower strands of the bisulfite-treated DNA samples are analyzed separately. Therefore it is impossible (except in the case of single-cell analysis) to obtain data about the original double-stranded DNA.
3. Amplifications (or cloning of PCR products) from the upper and lower strands may not work equally well in all cases.
4. In cases of analyses of non-CpG methylation on nonsymmetrical methylation patterns, as, for example, in plants and fungi, it may be difficult to design primers for the PCR amplification of the bisulfite-treated DNA. In this case primers can be designed that contain either C or T at the respective positions. However, the use of such primers with "wobble" positions greatly reduces the specificity of the PCR reaction and may cause a bias in the amplification of certain products (mostly those products that were not fully converted).
5. A systematic analysis by Warnecke et al. *(17)* nicely demonstrated that the choice of primers might cause a bias in the PCR reaction, such that either a low or highly methylated template DNA is predominantly amplified. The problem of biased amplification or cloning has to be tested individually, and several control experiments should be carried out. First, different templates with a known content of methylated cytosine residues should be mixed in different ratios and the bisulfite treatment and amplification steps carried out as usual. The distribution of nonconverted and converted cytosine residues in the analyzed products will then allow determination of whether, and to what magnitude, a bias has occurred. One strategy to avoid such problems is to perform independent experiments (including different techniques) to analyze the methylation profile of a given template, for example, by conventional Southern blot hybridization or Ms-SNuPE *(8)*. Both techniques can be very helpful in obtaining an independent estimation of the real ratio of modified and unmodified cytosines within the sequence of interest.
6. During the cloning procedure, a selection against a specific subset of PCR products may occur. This problem might be overcome by the use of different cloning vectors or a different strain of cells for the cloning.

7. Quite frequently, especially if only a few cells are used for the analysis, the observed methylation patterns after the sequencing of single clones show a clonal distribution. To rule out that this is due to the preferential amplification and cloning of only a single or a few converted chromosomal fragments, the bisulfite treatment and in consequence also the following PCR and analysis steps have to be repeated at least once.

4. Notes

1. The sodium bisulfite solution and NaOH solutions should always be prepared fresh and stored for not longer than 24 h before use.
2. Batches of commercially available sodium bisulfite are mixtures of sodium bisulfite and sodium metabisulfite. The ratio between the substances may vary among different batches. We recommend using pure sodium metabisulfite, which facilitates accurate preparation of solutions with the desired molarity.
3. Bisulfite and hydroquinone solutions are light sensitive, thus should be protected from light in all steps.
4. To favor the dissolving of the chemicals during the preparation of the solutions, these may be heated up to 50°C.
5. Common laboratory solutions and buffers for molecular biology, such as SDS, EDTA, PBS, Tris-HCl (pH 8.0), NaOH, and TE, were prepared according to **ref. *16***.
6. For some PCR fragments amplified from bisulfite-treated DNA, we observed a clonal selection against fully converted templates. In those cases we were able to overcome the problem using a different cloning vector system (e.g., *pGEM-T*, Promega) in combination with different *E. coli* cells.
7. If ice crystals appear during incubation of the bisulfite–hydroquinone solution on ice, proceed normally; this will not affect the results.
8. During pipetting of the agarose–DNA mixture into the mineral oil, some or all of the mixture might remain inside the pipet tip. This is usually because the agarose–DNA mixture has become too cold: the mixture should be kept at 50°C–65°C until formation of the beads. Also, as the agarose–DNA mixture is discharged into the cold mineral oil, the pipet tip should be only slightly inserted in the cold oil layer and the content should be discharged rapidly.
9. If the agarose beads dissolve after entering the bisulfite solution, the layer of mineral oil is not cold enough. To prevent that, the tubes containing mineral oil should be preincubated on ice for at least 20 min or, alternatively, kept at –20°C for 10 min (in that case the bisulfite solution should be added separately after the formation of the beads). Moreover, only heavy mineral oil of pure quality (e.g., from Sigma) should be used. If the problem nevertheless persists, we recommend that the concentration of LMP agarose be increased.
10. During bead formation, two or more beads can collide and fuse to form one big bead. To avoid this, the beads should be pipetted into opposite sides of the tube and the number of beads added should not exceed four per tube.

11. In cases when unconverted sequences are observed frequently, the following should be considered: (a) Primers are not selective enough for converted DNA. The primers should be located in a C-rich region to increase the selectivity of amplification toward fully converted sequences. (b) Incomplete bisulfite conversion may be caused by an excess of DNA in the reaction. The maximum recommended amount of DNA is 100 ng per bead. (c) The DNA was not properly denatured. Make sure that denaturation steps and desulfonation steps are carried out using fresh NaOH solution and sodium bisulfite solution.
12. Failure of PCR amplification may be caused by (a) inefficient bisulfite conversion (*see* **Subheading 3.1.2.**); (b) insufficient amount of template DNA; (c) size of a desired PCR product—try to amplify a smaller fragment; (d) low sensitivity of the amplification—a nested PCR approach is recommended, or, alternatively, the use of a different set of primers.

Acknowledgments

This work was supported by the Deutsche Forschungsgemeinschaft Wa1029/1–2 and the European Union BMH4-CT96-0050.

References

1. Constancia, M., Pickard, B., Kelsey, G., and Reik, W. (1998) Imprinting mechanisms. *Genome Res.* **8,** 881–900.
2. Nicholls, R. D., Saitoh, S., and Horsthemke, B. (1998) Imprinting in Prader-Willi and Angelman syndromes. *TIGS* **14,** 194–200.
3. Tycko, B. (1997) DNA methylation in genomic imprinting. *Mutat. Res.—Rev. in Mutat. Res.* **386,** 131–140.
4. Reik, W. and Walter, J. (2001) Genomic imprinting; parental influence on the genome. *Nature Rev. Genet.* **2,** 21–32.
5. Li, E. B. C. and Jaenisch, R. (1992) Targeted mutation of the DNA methyltransferase gene results in embryonic lethality. *Cell* **69,** 915–926.
6. Frommer, M., McDonald, L. E., Millar, D. S., Collis, C. M., Watt, F., Grigg, G. W., Molloy, P. L., and Paul, C. L. (1992) A genomic sequencing protocol that yields a positive display of 5-methylcytosine residues in individual DNA strands. *Proc. Natl. Acad. Sci. USA* **89,** 1827–1831.
7. Paul, C. L. and Clark, S. J. (1996) Cytosine methylation: Quantitation by automated genomic sequencing and GENESCAN analysis. *BioTechniques* **21,** 126–133.
8. Gonzalgo, M. and Jones, P. A. (1997) Rapid quantitation of methylation differences at specific sites using methylation-sensitive single nucleotide primer extension (Ms-SNuPE) *Nucleic Acids Res.* **25,** 2529–2531.
9. Clark, S. J., Harrison, J., Paul, C. L., and Frommer, M. (1994) High sensitivity mapping of methylated cytosines. *Nucl. Acids Res.* **22,** 2990–2997.
10. Feil, R., Walter, J., Allen, N. D., and Kelsey, G. (1994) Developmental control of allelic methylation in the imprinted mouse *Igf2* and *H19* genes. *Development* **120,** 2933–2943.

11. Raizis, A. M., Schmitt, F., and Jost, J. P. (1994) A bisulfite method of 5-methylcytosine mapping that minimises template degradation. *Anal. Biochem.* **226,** 161–166.
12. Olek, A. Oswald, J., and Walter, J. (1996) A modified and improved method for bisulfite based cytosine methylation analysis. *Nucleic Acids Res.* **24,** 5064–5066.
13. Paulin, R., Grigg, G. W., Davey, M. W., and Piper, A. A. (1998) Urea improves efficiency of bisulfite-mediated sequencing of 5-methylcytosine in genomic DNA. *Nucl. Acids Res.* **26,** 5009–5010.
14. Hayatsu, H., Wataya, Y., Kai, K., and Iida, S. (1970) Reaction of sodium bisulfite with uracil, cytosine, and their derivatives. *Biochemistry* **9,** 2858–2864.
15. Wang, R. Y.-H., Gehrke, C. W., and Ehrlich, M. (1980) Comparison of bisulfite modification of 5-methyldeoxycytidine and deoxycytidine residues. *Nucleic Acids Res.* **8,** 4777–4790.
16. Sambrook, G., Fritsch, E. F., and Maniatis, T. (1988) *Molecular Cloning and Laboratory Manual*, 2nd ed., Cold Spring Harbor Laboratory Press, Cold Spring Harbor, NY.
17. Warnecke, P. M., Stirzaker, C., Melki, J. R., Millar, D. S., Paul, C. L., and Clark, S. J. (1997) Detection and measurement of PCR bias in quantitative methylation analysis of bisulfite-treated DNA. *Nucl. Acids Res.* **25,** 4422–4426.

17

Direct Analysis of Chromosome Methylation

D borah Bourc'his and Evani Viegas-Péquignot

1. Introduction

DNA methylation is a possible candidate for a genomic imprinting marker in mammals. This epigenetic modification of DNA satisfies several essential criteria for the identification of the parental origin of individual alleles and larger portions of the genome: DNA methylation is stably propagated in somatic cells during cell division, it is reversible, it may inactivate the target sequence, and male and female gametes have different methylation patterns (reviewed in **ref. *1***).

Various methods for identifying methylated or unmethylated cytosines on DNA have been proposed. Their specificity, sensitivity, resolution, and potential artefacts determine their field of application and affect the interpretation of data. Despite their inherent limitations, these methods each provide a different level of information about the distribution of 5-MeC in mammalian genomes.

Chromosome methylation analysis is a method for the large-scale screening of genome methylation that has several advantages over global methylation analysis using extracted genomic DNA. Overall, methylation is placed in the context of chromosome architecture and can therefore be related to chromosome components, such as centromeres, telomeres, heterochromatin, and euchromatin, for which information concerning sequence mapping, replication timing, or the content of a particular protein may be available. This topological assessment of chromosome methylation makes it possible to identify relationships between functional or structural parameters on the largest scale of genome organization. In addition, chromosome methylation profiles can be evaluated in individual cells and in parental sets of chromosomes, provided they carry adequate parental chromosome markers and, as chromosomes can

From: *Methods in Molecular Biology, vol. 181: Genomic Imprinting: Methods and Protocols*
Edited by: A. Ward © Humana Press Inc., Totowa, NJ

be fixed just after each S phase, temporal changes in methylation related to DNA replication can be easily followed.

1.1. Detection of Chromosome Methylation

Two major alternatives exist for mapping the distribution of 5-MeC in mammalian chromosomes. Methylated sites can be identified using antisera *(2,3)* or monoclonal *(4)* antibodies against 5-MeC, followed by immunofluorescence or immunoperoxidase detection. Unmethylated sites can be evaluated by *in situ* digestion of chromosomes with methyl-sensitive restriction enzymes such as *HhaI* or *HpaII*, followed by labeling the cut site with nonradioactive (or radioactive) markers and immunocytochemical detection *(5–7)*.

5-MeC antibody binding is a convenient way of rapidly labeling hemimethylated and methylated DNA. After enzyme digestion, only unmethylated sites are detected and the specific sequence recognition of the enzymes may severely bias extrapolations of overall DNA methylation profiles. This brings us back to the problem of data interpretation in the context of a given approach. Apparently conflicting results may, in fact, reflect only misinterpretation or overestimation.

1.2. Specificity and Sensitivity of Chromosome Methylation Analysis

Chromosome methylation profiles have been established for normal and methylation-defective cells from somatic, germinal, and embryonic tissues. Specific patterns have been observed on identification of methylated (5-MeC antibody labeling) or unmethylated sites (restriction digestion).

The major difficulty in chromosome analysis is the choice of appropriate experimental conditions that will preserve the target and chromosome morphology while ensuring optimal accessibility. Conditions that prove to be very efficient for the detection of targets on isolated DNA may fail to give satisfactory results with chromosomes, in which DNA is tightly packaged and surrounded by proteins. Thus, to identify methylated or unmethylated sites, chromosomes are prepared under harsh fixation conditions (methanol or ethanol/acetic acid fixative), known to extract most of the chromatin proteins *(8)*. The accessibility of the DNA may then be increased by additional UV *(4)*, enzymatic, or alkaline treatments *(9)*, but, in any case, experimental protocols must be carefully chosen to prevent DNA loss or degradation and damage to chromosome architecture.

Convenient control experiments can be performed to ensure the specificity of chromosome labeling. They may include: (1) methylase-specific treatments (or a universal methylase) prior to chromosome digestion with a methyl-sensitive enzyme or antibody binding; (2) *in situ* digestion with a methyl-insensitive restriction endonuclease, and (3) the use of checking procedures (e.g., omission, competition, blocking, or adsorption of reagents) currently used in classical immunocytochemical methods.

The resolution and sensitivity of chromosome analyses are determined by chromosome structure and are limited principally by the commercial availability of efficient stains (e.g., stable fluorochromes) and by current microscopy techniques. Comparative analysis of the chromosome methylation patterns generated by 5-MeC monoclonal antibody and sequence-specific methylation evaluated by classical molecular approaches has shown that chromosome analysis provides direct and reliable information about the methylation status of highly and moderately repeated sequences located in constitutive heterochromatin and in euchromatin, respectively *(4)*. Repeated sequences include most of the potentially methylated CpG sites in the genomes of humans and other mammals *(10)*. Current chromosome methylation analysis cannot detect individual CpGs and single-copy genes because only clustered methylated and unmethylated CpGs can be identified at the chromosome level.

Antisera directed against 5-MeC and indirect immunofluorescence have been used with mouse and human chromosomes and have demonstrated preferential binding to constitutive heterochromatin, composed of highly repeated DNA and located mainly in centromeric or pericentromeric regions *(2,3)*. More recently, it has been shown using a monoclonal antibody that somatic euchromatin (chromosome arms) can also display specific binding *(4)*. A weak but reproducible R-like banding pattern, with some T bands stronger than others, is generated on human chromosomes (**Fig. 1A**). This labeling was correlated with the methylation of interspersed repeated Alu sequences that preferentially map to R bands and are heavily methylated in somatic cells *(11,12)*. Both euchromatin and heterochromatin have been found to be labeled in species as diverse as mice (**Fig. 1B**) and fish *(13)*, in mouse Dnmt1 (methyltransferase 1) mutants (**ref.** ***14*** and unpublished results) and in ICF syndrome (**Fig. 1A**, *inset*), the only genetic disorder known to involve a defective genomic methylation pattern *(4)*. *In situ* enzyme digestion of mammalian somatic chromosomes (**Fig. 1C**, **D**, and **E**) has provided specific information about differential *HhaI* methylation patterns on human X chromosomes *(5)* and about differences in the methyl content of parental and hybrid species *(7)*.

In early development, a differential chromosome methylation pattern related to the parental origin of chromosomes is observed just after fertilization (**Fig. 2**). This chromosome imprinting is erased during sucessive cleavages by a passive mechanism revealed by the progressive decrease in the numbers of asymmetrically labeled chromosomes, composed of hemimethylated and demethylated DNA strands *(15)*. *HpaII* and *HhaI* digestions *in situ* of mouse embryonic chromosome have confirmed these results.

Fig. 1. Distribution of methylated (5-MeC antibody binding) and unmethylated (*HhaI* digestion) sites on somatic human and mouse chromosomes (*see facing page*). **(A)** Chromosomes from lymphocyte cultures of normal individuals. Euchromatin is heterogeneously labeled (e.g., chromosome 7 displays an R-like banding pattern, long arrow) and constitutive heterochromatin is intensely labeled (juxtacentromeric regions of chromosomes 1 and 16, small arrows). *Inset* shows chromosome from patients carrying a constitutional methylation defect (ICF syndrome). Constitutive heterochromatin (chromosomes 1 and 16) and facultative heterochromatin (inactive X chromosome, Xi) are faintly labeled, indicating hypomethylation. **(B)** Chromosomes from mouse fibroblast cultures. As in human cells, euchromatin shows a banding pattern (long arrow) and constitutive heterochromatin is strongly methylated (small arrow). **(C, D, E)** Chromosomes from lymphocyte culture of normal human females after *HhaI* digestion and biotin 14-dATP end-labeling. (C) streptavidin–Texas red staining. Euchromatin displays a banding pattern, similar to R banding; short arm of acrocentic chromosomes (long arrows) are in general labeled, and the two X chromosomes are differently stained. The Xi (below) is more stained than the Xa (active X). (D, E) *HhaI* digestion and immunoperoxidase labeling. The Xi is shown before (D) and after (E) DAB intensification. (*see* refs. ***5***,***6***).

Fig. 2. Chromosome methylation patterns of embryos (*see facing page*). Methylated sites were revealed by indirect immunofluorescence labeling with 5-MeC monoclonal antibody. *HpaII* unmethylated sites were revealed by immunoperoxidase staining after biotin 14-dATP end-labeling. **(A, B)** Metaphases from mouse (A) and human (B) embryos at the one-cell stage. Two distinct sets of chromosomes are observed. Chromosomes of paternal origin (small arrow) are faintly labeled, and those of maternal origin are intensely labeled (long arrow). Maternally inherited chromosomes display an R-like banding pattern in euchromatin. Centromeric heterochromatin is intensely labeled in most chromosomes of the paternal set, whereas it is brighly stained in only a few chromosomes of the maternal set. **(C)** Chromosomes from a two-cell mouse embryo. An asymmetrical labeling (arrows) of the two sister chromatids is clearly observed in chromosomes from maternal origin. *Insets:* Asymmetrical chromosomes of two-cell embryos obtained after *HpaII* digestion and immunoperoxidase labeling. **(D)** Chromosomes from an eight-cell embryo. Most of the chromosomes are symmetrically and weakly labeled, chromosomal asymmetry is rarely observed at this stage (arrows). From the one-cell stage to the eight-cell stage the decreasing in fluorescence labeling is associated with DNA replication indicating the existence of a passive demethylation (*see* **ref. *15***).

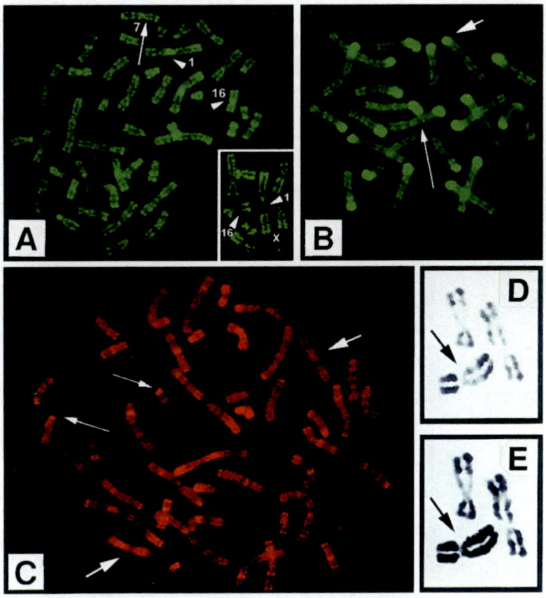

Fig. 1. Distribution of methylated (5-MeC antibody binding) and unmethylated (*HhaI* digestion) sites on somatic human and mouse chromosomes (*see facing page*).

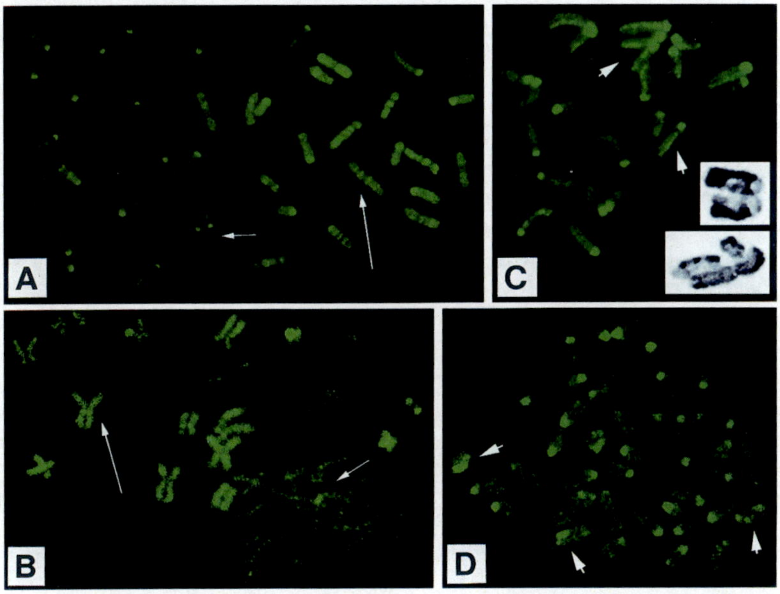

Fig. 2. Chromosome methylation patterns of embryos (*see facing page*).

2. Materials

2.1. Chromosome Preparations

2.1.1. Human Lymphocyte Cultures

1. Culture medium: TC199 (Seromed) supplemented with streptomycin and penicillin (BioMérieux) (final concentrations of 100 IU/mL and 50 μg/mL, respectively), phytohemagglutinin (PHA, Gibco BRL), and 20% human serum.
2. Thymidine (Sigma) at a final concentration of 0.3 mg/mL.
3. Colchicine: 10 mg/L stock solution (Eurobio).
4. Hypotonic buffer: Human serum/sterile distilled water (1/6, v/v).
5. Ethylenediaminetetraacetic acid (EDTA): 40-mg/mL stock solution.
6. Fixative: Ethanol/acetic acid (3/1, v/v).

2.1.2. Fibroblast Cultures

1. Culture medium: RPMI 1640 containing L-glutamine (Gibco BRL), kanamycin (final concentration of 80 UG/mL), and supplemented with 10% fetal calf serum.
2. Colchicine: 10-mg/L stock solution (Eurobio).
3. Hypotonic buffer: Human or fetal calf serum/sterile distilled water (1/6, v/v).
4. EDTA: 40 mg/mL stock solution.
5. 1× Trypsine-EDTA (Gibco BRL).
6. PBS: 0.2 *M* phosphate buffer, 0.15 *M* NaCl, pH 7.3–7.4, filtered before use.
7. Fixative: Ethanol/acetic acid (3/1, v/v).

2.1.3. Embryonic Cells

1. PMSG and hCG (Chorulon) hormones (Intervet).
2. Recovery medium: M2 (Sigma).
3. Hyaluronidase (Sigma): Final concentration of 300 μg/mL in M2 medium.
4. Culture medium: M16 (Sigma).
5. Light paraffin oil (Merck).
6. Colchicine: 10-mg/L stock solution (Eurobio).
7. Hypotonic buffer: Fetal calf serum/sterile distilled water (1/6, v/v).
8. EDTA: 40-mg/mL stock solution.
9. Fixative: Methanol/acetic acid (3/1, v/v).
10. Stereomicroscope with understage illumination.

2.2. 5-MeC Antibody Binding to Chromosomes

1. PBS solution (*see* **Subheading 2.1.2.**).
2. PBT: 0.1% Tween-20, 0.15% bovine serum albumun (BSA), in phosphate-buffered saline (PBS), filtered before use.
3. Germicidal 30-W lamp (UV light, 254 nm).

4. Antibodies: anti-5MeC monoclonal antibody ***(16)***, RAM-FITC (anti-mouse conjugate, Nordic Immunology).
5. Antifading *p*-phenylenediamine (PPD) free base (Sigma): 0.1% solution in PBS/glycerol (1/9, v/v), adjusted to pH 8 with 0.1 *M* NaOH, stored at –20°C.

2.3. Direct Restriction Enzyme Digestion of Chromosomes

1. Methyl-sensitive restriction enzymes: *HhaI* and *HpaII* (New England Biolabs).
2. Terminal deoxynucleotidyltransferase (TdT) (Boehringer) and T4 polymerase (New England Biolabs).
3. Methyl-sensitive enzyme digestion and T4 polymerase buffers prepared as recommended by the manufacturers.
4. TdT buffer: 100 m*M* sodium cacodylate, pH 7, 1 m*M* $CoCl_2$/0.2 m*M* dithiothreitol, 0.1% BSA. Prepare just before use and protect against light.
4. 2× SSC (0.3 *M* NaCl/30 m*M* sodium citrate) and 0.1× SSC (15 m*M* NaCl/1.5 m*M* sodium citrate).
5. PBS and PBT as described under **Subheadings 2.1.2.** and **2.2.**
6. Deoxynucleotide dGTP, dCTP, dTTP: 1 m*M* stock solution.
7. Biotin 14-dATP (Gibco BRL): 1 m*M* stock solution.
8. Antibodies: anti-biotin polyclonal antibody (Vector laboratories), RAG (rabbit anti-goat) conjugated to a fluorochrome (fluorescein or rhodamine), RAG-IgG(H+L) peroxidase-conjugate (Nordic Immunology), and streptavidin-TR (Texas red) (Gibco BRL).

2.4. Microscopy

1. Conventional microscope equipped for epifluorescence and transmitted light illumination (ordinary and phase-contrast equipment).
2. Filter sets specific for the fluorochromes (fluorescein and rhodamine or Texas red).
3. Color slide film, Kodak 400 ASA.
4. High-resolution CCD (charged coupled device)-cooled camera (Hamamatsu), Explorer (Alcatel-Samba system), and Adobe Photoshop software.

3. Methods

3.1. Chromosome Preparations

3.1.1. Short-Term Lymphocyte Cultures

1. Add 0.5 mL of adult whole blood (collected in a Venoject tube containing lithium heparin) to 7.5 mL of culture medium (TC 199, antibiotics, PHA, and human serum) and incubate for 72 h at 37°C.
2. Add 30 µL of colchicine (final concentration of 0.04 µg/mL) to each 8 mL of culture. Incubate at 37°C for 2 h to obtain metaphase chromosomes.
3. Centrifuge for 10 min at 400*g* at room temperature.

4. Remove the supernatant and add gradually to the pellet 10 mL of hypotonic buffer prewarmed at 37°C. The hypotonic solution should be added in three or four steps and mixed gently with a Pasteur pipet.
5. Add 100 µL of EDTA (400-µg/µL final concentration), mix, and incubate for 10 min at 37°C.
6. Add 1 mL of fixative, and resuspend gently and centrifuge for 10 min at 400*g* at room temperature.
7. Remove the supernatant, add 1 mL of fixative, and gently resuspend the pellet. Repeat this action several times to increase the amount of fixative gradually. In total, 8 mL of fixative/tube should be used. Centrifuge for 10 min at 400*g* at room temperature.
8. Remove the supernatant and add 6 mL of fixative to the pellet. Gently resuspend and store the cell suspension at 4°C overnight.
9. Centrifuge for 10 min at 400*g* at room temperature.
10. Replace the fixative (final volume of 4 mL) at least twice.
11. Centrifuge for 10 min at 400*g* at room temperature.
12. Discard the supernatant and add about 0.5 mL of fixative to obtain a concentrated cell suspension.
13. Put 1 drop of this preparation on a precleaned, soaked slide kept in distilled water at 4°C until use. If chromosome spread is not dense enough, increase the number of drops on subsequent slides.
14. Air-dry the slides for at least 24 h and store slides at –20°C. Slides may be stored for several months.

3.1.2. Fibroblast Cultures

1. Remove the culture medium from a culture (25-cm^2 flask) of confluent fibroblasts.
2. Rinse with 2–3 mL PBS.
3. Add 2 mL of trypsin-EDTA to the attached cells, incubate at 37°C until they detach, and then divide the cell suspension into two new flasks and add 6 mL of culture medium to each.
4. Incubate for 24 or 48 h at 37°C, then add 30 µL of colchicine and incubate at 37°C for 2 h (*see* **Note 1**).
5. Rinse the culture with 2–3 mL PBS.
6. Add 2 mL of trypsin-EDTA to the attached cells.
7. Centrifuge for 10 min at 400*g*.
8. The remaining steps are identical to those described for lymphocyte preparation.

3.1.3. Preimplantation Embryo Cultures

Chromosomes are prepared ***(15)*** according to a modified version of the air-drying method of Tarkowski ***(17)***. Embryos at the pronuclear stage are fixed 28–32 h after injecting hCG, when the pronuclei have disappeared completely, with or without mitotic arrest (colchicine treatment), in the night of d 0. For all other cell stages (two-cell to blastocyst stages), colchicine treatment is recommended. Hypotonic and fixative treatments are performed under a stereomicroscope with understage illumination, and each embryo is treated individually.

1. Make mature female mice superovulate by intraperitoneal injection of 5 IU PMSF, followed by an injection of 5 IU hCG 46–48 h later. Allow the mice to mate overnight, kill the females with a vaginal plug, and remove the fertilized eggs by oviduct puncture.
2. Release the cumulus cells by hyaluronidase treatment (300 µg/mL in M2 medium).
3. Rinse embryos with M2 medium and transfer them to microdrops (100 µL) of M16 medium under paraffin oil. Culture embryos in 35-mm Petri dishes. Incubate at 37°C in an atmosphere containing 5% CO_2.
4. Incubate with 10 µL of colchicine (final concentration of 1 µg/mL) for 9–12 h to obtain chromosomes from embryos at 1-, 2-, 4-, and 8-cell stages and for 7 h and 4 h to obtain chromosomes from morula and blastocyst embryos, respectively.
5. Place the embryos singly in 500 µL of hypotonic solution at room temperature. Transfer the embryo to a new bath of 500 µL of hypotonic solution until no cellular membrane limits are visible (*see* **Note 2**). The time required to achieve this depends on the stage of development and may be from 1 to 5 min (blastocyst). Hypotonic treatment is performed in 4-well multidishes (Nunc). EDTA is added during hypotonic treatment, at a final concentration of 400 µg/mL for 4-cell to 16-cell embryos and of 800 µg/mL for morula and blastocyst embryos.
6. After hypotonic treatment, transfer each embryo onto a clean microscope slide in a droplet of hypotonic solution.
7. Place 1 drop of methanol/acetic acid fixative (3/1) on the embryo using a fine Pasteur pipet and blow to quickly evaporate the fixative. Place several embryos separately on each slide and use a diamond pen to mark the position of each embryo.
8. Air-dry for 24 h and then inspect the quality of the preparation with a phase-contrast micoscope and store the slides at –20°C until use.

3.2. 5-MeC Antibody Binding to Chromosomes

Several procedures for increasing the accessibility of methylated chromosome regions to 5-MeC antibody have been proposed, including alkaline, pepsin/HCl, and other enzyme treatments *(9)*. In our hands, reproducible results have been obtained only by controlled UV light treatment *(4)* as recommended in the early work of Miller et al. *(2)*. Optimal results were obtained using slides stored for at least 1 wk at –20°C and defrosted at room temperature before use.

1. Immerse slides in PBS (*see* **Note 3**) in a Petri dish and place them under a germicidal lamp for 15 h (chromosomes from lymphocytes and fibroblasts) or

8 h (embryo chromosomes). Place the slides at 30–35 cm from the germicidal lamp (*see* **Note 4**).
2. Immerse the slides briefly in cold PBT.
3. Add 100 µL of 5-MeC antibody diluted 1/10 in PBT to the slides and cover with a precleaned 24 × 32 mm glass coverslip. Incubate for 45 min at room temperature in a dark, humid chamber.
4. Wash the slides in PBT for 5 min at room temperature.
5. Incubate the slides (covered with a 24 × 32 mm glass coverslip) with 100 µL of RAM-FITC antibody diluted 1/40 in PBT for 45 min at room temperature.
6. Wash the slides with PBS (about 1 mL, using a Pasteur pipet).
7. Put a few drops of PBS onto the slides and cover with a 24 × 60 mm glass coverslip. Store the slides at 4°C and add 10 µL of the antifading PPD solution, pH 8, before microscope observation. Photomicrographs may be obtained directly using a fluorescein filter or a dual-color bandpass filter. However, as the fluorescent signals are not very intense, the use of a high-resolution CCD-cooled camera is recommended.

3.3. In Situ Chromosome Digestion

Direct methyl-sensitive enzyme digestion of metaphase chromosomes provides information about the chromosomal organization of unmethylated sites. The method is based on the ability of modifying enzymes such as terminal deoxynucleotidyltransferase (TdT) and T4 polymerase to use the terminus generated by restriction endonucleases as a primer for the polymerization of labeled nucleotides (biotinylated or digoxygenin labeling; *see* **Notes 5** and **6**) *(5,6)*. The labeled tails are immunocytochemically detected by a first specific antibody (anti-biotin or anti-digoxygenin) and then by an IgG-peroxidase or fluorescein-conjugate or directly with streptavidin-TR (biotin tails) (*see* **Note 7**).

3.3.1. HhaI Digestion

HhaI cleaves the GCGC sequence provided that the internal C is unmethylated generating a 3′ hydroxyl terminus. Terminal deoxynucleotidyltransferase (TdT) then has the ability to use this terminus as primer for polymerization in the presence of a labeled nucleotide (e.g., biotin 14-dATP).

1. Equilibrate the slides for 10 min at room temperature with the appropriate restriction enzyme buffer, as recommended by the manufacturer.
2. Apply to each slide 50 µL of 0.1 U/µL HhaI enzyme diluted in the appropriate buffer. Cover with a precleaned glass coverslip (24 × 32 mm) and incubate for 30 min at 37°C in a humid chamber.
3. Rinse the slides in *HhaI* buffer for 5 min and then equilibrate the slides with freshly prepared TdT buffer.

4. Add to each slide 50 µL of a mixture containing 0.25 U/µL TdT diluted in the reaction buffer, 40 µmol biotin 14-dATP, and 30 µmol of each unlabeled deoxynucleotide (dGTP, dCTP, dTTP). Cover with a glass coverslip and incubate for 30 min at 37°C in a humid chamber.
5. Wash the slides successively in 2× SSC for 30 min at room temperature, 0.1× SSC for 30 min at 43°C, and finally in 2× SSC for 20 min at room temperature.
6. Incubate the slides in PBT for 10 min at room temperature.
7. Add 100 µL of anti-biotin IgG antibody diluted 1/500 in PBT. Cover with a glass coverslip and incubate for 45 min at 37°C.
8. Rinse the slides twice for 5 min each in PBT at room temperature.
9. Apply 100 µL of the second antibody, anti-goat IgG peroxidase- or FITC-labeled, diluted 1/40 in PBT. Cover with a glass coverslip. For a TR detection , after **step 8**, incubate and apply 100 µL of streptavidin-TR, diluted in PBT.
10. Incubate for 45 min at 37°C.
11. Wash the slides in PBS for 8 min for peroxidase detection. For FITC detection, simply rinse the slides with a few drops of PBS and store covered with a 24 × 60 mm coverslip, at 4°C. Just before microscope observation, add antifading PPD solution, pH8. Microscopy is performed as described for 5-MeC antibody binding (**Subheading 3.2.**, **step 7**).
12. For peroxidase detection, develop the slides in the dark by incubating for about 5 min in a 0.5-mg/mL DAB solution in PBS and adding 100 µL of 30% hydrogen peroxide immediately before use. The time of DAB incubation may be adjusted; it varies from 4 to 7 min.
13. Rinse the slides thoroughly in distilled water and air dry. Before the standard light or phase-contrast microscopy, apply several drops of PBS to chromosomes and cover with a 24 × 60 mm coverslip.
14. DAB intensification using gold reaction and silver precipitation may be performed to enhance labeling (*see* **Note 8**).

3.3.2. HpaII Digestion

HpaII cleaves the CCGG sequence if the internal C is unmethylated. T4 polymerase is used after *HpaII* digestion, its 3′–5′ exonuclease activity being used to expose the 5′ extremity of DNA. This extremity is then used as a template for the 5′–3′ polymerase activity of T4 polymerase. The labeled nucleotide incorporated during the polymerization is then detected by immunocytochemistry.

1. Equilibrate the slides for 10 min at room temperature with the restriction enzyme buffer recommended by the manufacturer.
2. Apply to each slide 50 µL of 0.2 U/µL *HpaII* enzyme diluted in the appropriate buffer. Cover with a glass coverslip and incubate for 30 min at 37°C in a humid chamber.
3. Rinse the slides in *HpaII* buffer for 5 min.
4. Equilibrate the slides with T4 polymerase buffer for 10 min at room temperature.

5. Add 50 µL of 0.2-U/µL T4 polymerase diluted in the buffer supplied by the manufacturer, and allow the exonuclease reaction to proceed for 30 min at 37°C on the slide cover with a glass coverslip (24 × 32 mm), in a humid chamber.
6. Rinse the slides in T4 buffer for 5 min at room temperature.
7. For polymerization, add 50 µL of a mixture of 0.2 U/µL of T4 polymerase, 40 µmol biotin 14-dATP, and 30 µmol of each each unlabeled deoxynucleotide (dGTP, dCTP, dTTP) to the slides. Cover with a glass coverslip and incubate for 30 min at 37°C in a humid chamber.
8. Wash the slides and carry out immunodetection as described for *HhaI* (**steps 5–14**).

4. Notes

1. Cultures may be synchronized using thymidine ***(18)*** to increase the number of metaphase or prometaphase cells per slide: add thymidine at a final concentration of 0.3 mg/mL after 48 or 72 h of culture. Rinse the cells after 15–17 h, and incubate them in culture medium without thymidine for 7 h. Add colchicine during 2 h to obtain metaphase chromosomes and for 30 min to obtain prometaphase chromosomes.
2. The presence of oil during the hypotonic treatment of embryo cells may disturb the quality of chromosome spreading. The transfer of embryos to a second hypotonic bath is strongly recommended.
3. PBS, PBT, and other buffers are freshly prepared and filtered before use to avoid background.
4. UV treatment before 5-MeC antibody binding must be performed in an open environment to avoid slide and PBS warming.
5. In some cases, nick-translation, rather than end-labeling, may be used to incorporate labeled nucleotides into chromosome cut-sites ***(7)***. However, resolution and specificity seem to be higher with end-labeling methods.
6. Radioisotopes and autoradiography have also been used after enzyme digestion. However, nonradioactive methods have several advantages, including safety, high stability, rapidity, and high resolution.
7. Other procedures for chromosome digestion have been described, involving the use of a high concentration of enzyme followed by simple Giemsa staining ***(19,20)***. However, the resulting labeling is compromised by the disruption of chromosome morphology and DNA degradation or loss due to the very high enzyme concentration and by the nonspecificity of Giemsa staining, in which only the remaining and possibily slightly digested chromosomal DNA is labeled.
8. The DAB (3,3′-diaminobenzidine tetrahydrochloride) peroxidase substrate signal could be intensified by a gold reaction and silver precipitation, resulting in a very sensitive and high-contrast signal ***(5)***.

References

1. Tighlman, S. M. (1999) The sins of the fathers and mothers: genomic imprinting in mammalian development. *Cell* **96,** 185–193.

2. Miller, O. J., Schnedl, W., Allen, J., and Erlanger, B. F. (1974) 5-methylcytosine localized in mammalian constitutive heterochromatin. *Nature* **251,** 636–637.
3. Okamoto, A., Miller, D.A., Erlanger, B.F., and Miller, O.J. (1981) Polymorphism of 5-methylcytosine-rich DNA in human acrocentric chromosomes. *Hum. Genet.* **58,** 255–259.
4. Miniou, P., Jeanpierre, M., Blanquet, V., Sibella, V., Bonneau, D., Herbelin, C., Fischer, A., Niveleau, A., and Viegas-Péquignot, E. (1994) Abnormal methylation pattern in constitutive and facultative (X inactive chromosome) heterochromatin of ICF patients. *Hum. Mol. Genet.* **3,** 2093–2102.
5. Viegas-Pequignot, E., Dutrillaux, B., and Thomas, G. (1988) Inactive X has the highest concentration of unmethylated *HhaI* sites. *Proc. Natl. Acad. Sci. USA* **85,** 7657–7660.
6. Miniou, P., Bourc'his, D., Molina Gomes, D., Jeanpierre, M., and Viegas-Péquignot, E. (1997) Undermethylation of Alu sequences in ICF syndrom: molecular and *in situ* analysis. *Cytogenet. Cell Genet.* **77,** 308–313.
7. O'Neill, R. J. W., O'Neill, M. J. O., and Graves, J. A. M. (1998) Undermethylation associated with retroelement activation and chromosome remodelling in an interspecific mammalian hybrid. *Nature* **393,** 68–72.
8. Jeppeson, P., Mitchell, A., Turner, B., and Perry, P. (1992) Antibodies to defined histone epitopes reveal variations in chromatin conformation and underacetylation of centromeric heterochromatin in human metaphase chromosomes. *Chromosoma* **101,** 322–332.
9. Barbin, A., Montpellier, C., Kokalj-Vocak, N., Gibaud, A., Niveleau, A., Malfoy, B., Dutrillaux, B., and Bourgeois, C. (1994) New sites of methylcytosin-rich DNA detected on metaphase chromosomes. *Hum. Genet.* **94,** 684–692.
10. Yoder, J. A., Walsh, C. P., and Bestor, T. H. (1997) Cytosine methylation and the ecology of intragenomic parasites. *Trends Genet.* **13,** 335–340.
11. Hellmann-Blumberg, U., Hintz, M. F., Gatewood, J. M., and Schmid, C. W. (1993) Developmental differences in methylation of human Alu repeats. *Mol. Cell Biol.* **13,** 4523–4530.
12. Kochanek, S., Renz, D., and Doerfler, W. (1993) DNA methylation in the Alu sequences of diploid and haploid primary human cells. *EMBO J.* **12,** 1141–1151.
13. de Almeida-Toledo, L. F., Viegas-Péquignot, E., Coutinho-Barbosa, A. C., Foresti, F., Niveleau, A., and de Almeida Toledo-Finlo, S. (1998) Localization of 5-methylcytosine in metaphase chromosomes of diploid and triploid pacu fish, *Piaractus mesopotamicus* (Pisces, Characiformes). *Cytogenet. Cell Genet.* **83,** 21–24.
14. Nan, X., Tate, P., Li, E., and Bird, A. (1996) DNA methylation specifies chromosomal localization of MeCP2. *Mol. Cell Biol.* **16,** 414–421.
15. Rougier, N., Bourc'his, D., Molina Gomes, D., Niveleau, A., Plachot, M., Pàldi, A., and Viegas-Péquignot, E. (1998) Chromosome methylation patterns during mammalian preimplantation development. *Genes Deve.* **12,** 2108–2113.
16. Reynaud, C., Bruno, C., Boullanger, P., Grange, J., Barbesti, S., and Niveleau, A. (1991) Monitoring of urinary excretion of modified nucleosides in cancer patients using a set of six monoclonal antibodies. *Cancer Lett.* **61,** 255–261.

17. Tarkowski, A. K. (1966) An air-drying method for chromosome preparation from mouse eggs. *Cytogenics* **5,** 394–400.
18. Viegas-Péquignot, E. and Dutrillaux, B. (1978) Une méthode simple pour obtenir des prophases et des prométaphases. *Ann. Genet.* **21,** 122–128.
19. Bianchi, N. O., Vidal-Rioja, L., and Cleaver, J. E. (1986) Direct visualization of the sites of DNA methylation in human and mosquito chromosomes. *Chromosoma* **94,** 362–366.
20. Ferruci, L., Mezzanotte, R., Vanni, R., Stuppia, R., Guanciali-Franchi, P., Calabrese, G., Palka, G., Bianchi, U., and Summer, T. (1988) Effect of *HpaII* and *MspI* restriction endonucleases on chronic myelogenous leukemia chromosomes. *Cancer Genet. Cytogenet.* **34,** 251–256.

18

In Vitro Methylation of Predetermined Regions in Recombinant DNA Constructs

Ilse Van den Wyngaert, Roger L. P. Adams, and Stefan U. Kass

1. Introduction

DNA methylation at position 5 in the cytosine ring in the sequence CpG can be detrimental to the transcription of a variety of genes in higher eukaryotes *(1,2)*. Although the significance of this transcriptional repression is currently under debate *(3,4)*, there is little disagreement that it plays an important role in genomic imprinting and X-chromosome inactivation *(5,6)*. To study the effects of DNA methylation on transcription in an experimental system, bacterial DNA methyltransferases have been used widely in order to mimic the DNA methylation pattern of eukaryotic genes. However, usually every target site in a given recombinant DNA molecule will be subject to DNA methylation by making use of those enzymes. This might result in an exaggeration of the effects of DNA methylation, as most recombinant DNA molecules contain a high degree of prokaryotic DNA, which is rich in CpGs. This methylated CpG-rich DNA can contribute to the effects of DNA methylation by formation of a repressive chromatin structure *(7,8)*. In addition, selective DNA methylation is required to distinguish the effects of DNA methylation on transcription initiation and transcript elongation *(8,9)*. Thus, there is a requirement for a method to generate recombinant DNA molecules that are methylated in a predetermined region. The chapter following this one will describe a method that makes use of ligation of methylated DNA fragments into unmethylated vector DNA. This method relies on the availability of suitable restriction sites, which allow directional cloning of the fragment and, in addition, requires a highly efficient ligation reaction. The method described in this chapter generates CpG-methylated regions in recombinant plasmids by making use

From: *Methods in Molecular Biology, vol. 181: Genomic Imprinting: Methods and Protocols*
Edited by: A. Ward © Humana Press Inc., Totowa, NJ

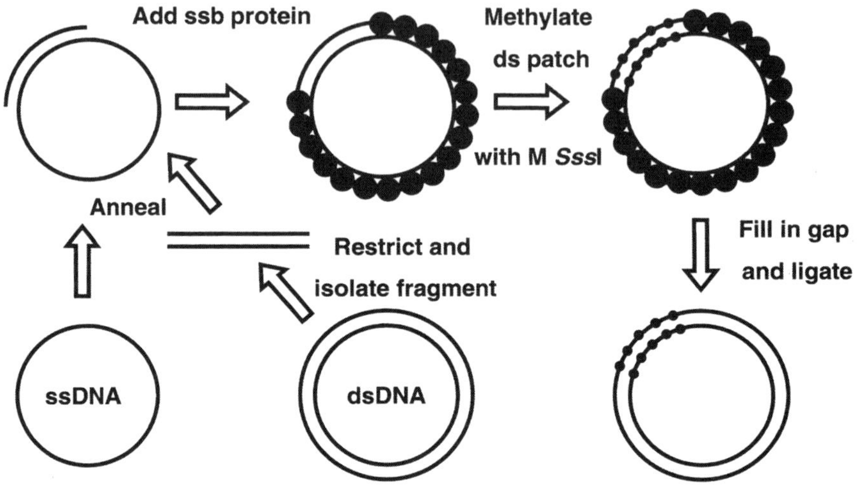

Fig. 1. Regional methylation of plasmid DNA. Single-stranded DNA (ssDNA) and a restriction fragment are isolated from the same plasmid and annealed. A single-stranded DNA-binding (ssb) protein is added and the double-stranded DNA (dsDNA) patch methylated using MSssI DNA methyltransferase. The gap is then filled in ligated.

of targeted methylation on a double-stranded patch in a single-stranded DNA molecule.

1.1. Principle of the Method

Any double-stranded DNA (dsDNA) fragment originating from the plasmid can be annealed to a single-stranded DNA (ssDNA) plasmid molecule to generate a single-stranded molecule with a dsDNA patch. A single-stranded DNA-binding protein is added in order to prevent the CpG methyltransferase M.*Sss*I ***(10)*** from methylating or binding to the ssDNA. Thereafter DNA methyltransferase M.*Sss*I is used to methylate all CpGs in the dsDNA patch. Subsequently the ssDNA gap is filled in and a dsDNA plasmid is generated that contains a region of methylated DNA (*see* **Fig. 1** for a schematic representation of the method). The amounts and quantities specified in this method are sufficient to generate approximately 6–10 µg of double-stranded plasmid DNA. This is sufficient to carry out at least duplicate transfections into tissue culture cells. We recommend carrying out mock methylation reactions which will serve as controls in the verification of DNA methylation as well as in transcription assays.

2. Materials

2.1. DNA

1. 5 µg (approximately 3 pmol) of single-stranded (ss) plasmid DNA (*see* **Note 1**).
2. Approximately 9 pmol fragment DNA is required (*see* **Note 2**).

2.2. Annealing of dsDNA Fragment to ssDNA Plasmid

1. 99% ethanol.
2. 75% ethanol.
3. 3 *M* Na-acetate, pH 6.9.
4. Annealing buffer: 20 m*M* Tris-HCl, pH 7.5, 10 m*M* $MgCl_2$, 50 m*M* NaCl, 1 m*M* DTT.
5. 95°C water bath/heating block.
6. Annealing bath/incubator (from 70°C to 30°C in 1 h).

2.3. DNA Methylation Reaction

1. T4 gene 32 protein (10 µg/µL; ssDNA-binding protein, Roche Molecular Biochemicals).
2. DNA methyltransferase *Sss*I (2 units/µL; New England Biolabs).
3. 10× methylation buffer: 100 m*M* Tris-HCl, pH 7.9, 100 m*M* $MgCl_2$, 500 m*M* NaCl, 10 m*M* DTT).
4. S-adenosyl-L-methionine (32 m*M*; New England Biolabs).
5. Proteinase K (25 mg/mL; Roche Molecular Biochemicals).
6. Phenol, TE-equilibrated.
7. Chloroform.
8. Ethanol (99% and 75%).
9. 10% sodium dodecyl sulfate (SDS).
10. 0.5 *M* ethylenediaminetetraacetic acid.

2.4. Filling in and Ligation Reaction

1. Fill-in buffer: 100 µmol each of dATP, dCTP, dGTP, dTTP, 1 m*M* DTT, 50 m*M* Tris-HCl, pH 7.5, 10 m*M* $MgCl_2$).
2. Klenow fragment, 3′-5′exo- (5 units/µL; New England Biolabs).
3. T4 DNA ligase (400 units/µL, New England Biolabs).
4. ATP, 100 m*M*.
5. 37°C incubator/oven.
6. Nucleotide Removal Kit (Qiagen).

2.5. Agarose Gel Electrophoresis and Southern Blot Analysis

It is advised to verify the DNA methylation of the constructs using methylation-sensitive restriction enzymes with a subsequent agarose gel electrophoresis and Southern blot analysis. Standard protocols were used for these procedures *(11)*.

3. Methods

3.1. Annealing of dsDNA Fragment to ssDNA Plasmid

1. Mix 3 pmol ssDNA plasmid with approximately 9 pmol of a restriction fragment derived from the same plasmid.
2. Add 1/10 volume of 3 *M* Na-acetate (pH 6.9), add 2.5× volume of 99% ethanol.
3. Mix, store for 10 min on ice, then centrifuge for 10 min at 17,900 relative centrifugal force (rcf) (note that all centrifugations are carried out in a microcentrifuge and 17,900 rcf corresponds to 13,000 rpm), room temperature.
4. Discard supernatant, add 500 µL ethanol (75%), vortex, store for 10 min on ice, then centrifuge as before.
5. Repeat **step 4**, then air-dry the DNA pellet for 10 min.
6. Dissolve the DNA pellet in 50 µL annealing buffer.
7. Heat for 5 min at 95°C.
8. Allow to anneal for 1 h from 70°C to 30°C.
9. Add 1/10 volume of 3 *M* Na-acetate (pH 6.9), and 2.5× volumes of 99% ethanol.
10. Mix, store for 10 min on ice, then centrifuge for 10 min, 15,000*g*, room temperature.
11. Discard the supernatant, add 500 µL ethanol (75%), vortex, store for 10 min on ice, then centrifuge as before.
12. Repeat **step 11**, then air-dry DNA pellet for 10 min.

3.2. DNA Methylation Reaction

1. Dissolve the DNA pellet in 119 µL H_2O.
2. Add 15 µL 10× methylation buffer
3. Add 5 µL T4 gene 32 protein (10 µg/µL).
4. Incubate 15 min, 37°C.
5. Add 10 µL DNA methyltransferase *Sss*I.
6. Add 0.75 µL S-adenosyl-methionine (32 m*M*; final concentration: 160 µ*M*).
7. Incubate for 16 h at 37°C (preferably in an oven, to avoid condensation).
8. Add 7.5 µL SDS 10%, 1.5 µL EDTA 0.5 *M*, 1.5 µL proteinase K (25 mg/mL); incubate for 30 min at 55°C.
9. Add 160 µL phenol and 160 µL chloroform, vortex, centrifuge 10 min at 17,900 rcf, room temperature.
10. Transfer aqueous phase to fresh tube, add 320 µL chloroform, vortex, and centrifuge as before.
11. Transfer aqueous phase to fresh tube, add 1/10 volume of 3 *M* Na-acetate (pH 6.9), and 2.5× volumes of 99% ethanol.

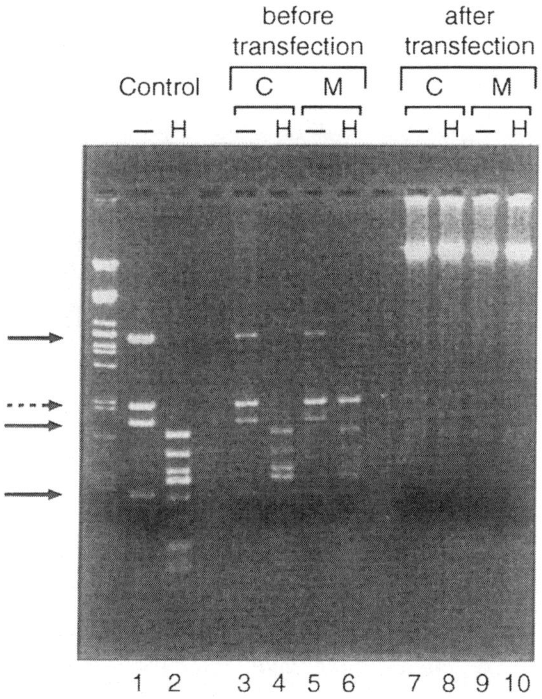

Fig. 2. Aragose gel analysis of regionally methylated plasmid DNA. A 1.1-kb restriction fragment was used to generate a methylated patch. The DNA was restricted with the appropriate restriction enzymes as indicated and resolved on a 1.2% gel. *Lanes 1 and 2*, control plasmid; *lanes 3–6*, mock-methylated (C) and methylated (M) constructs before transfection into cells; *lanes 7–10*, total DNA isolated from mammalian cells after transfection with constructs. All samples were digested to release the fragment used to generate the methylated patch (dashed arrow). H indicates digestion with *Hpa*II.

12. Mix, store for 10 min on ice, then centrifuge for 10 min, 15,000*g*, room temperature.
13. Discard supernatant, add 500 µL ethanol (75%), vortex, store for 10 min on ice, then centrifuge as before.
14. Repeat **step 13**, then air-dry the DNA pellet for 10 min.

3.3. Filling in and Ligation Reaction

1. Dissolve the DNA pellet in 92 µL fill-in buffer.
2. Add 5 µL Klenow enzyme.

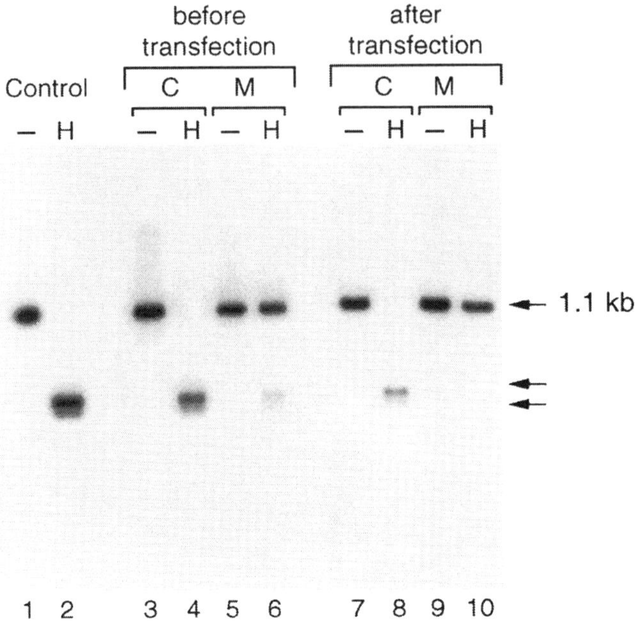

Fig. 3. Southern blot analysis of regionally methylated plasmid DNA. The gel from **Fig. 2** was subjected to Southern blot analysis using the 1.1-kb fragment as a probe. Labeling is as for **Fig. 2**. Exposure of the X-ray film was longer for lanes 7–10 in order to obtain comparable band intensities.

3. Incubate for 2 h at 37°C.
4. Add 2 µL T4 DNA ligase and incubate for 5 h at room temperature.
5. Purify the mixture using the Nucleotide Removal Kit (Qiagen).

3.4. Agarose Gel Electrophoresis and Southern Blot Analysis of Regionally Methylated Fragments

To test whether DNA methylation went to completion in the predetermined region, the DNA was digested prior or subsequent to transfection into mammalian tissue culture cells with the methylation-sensitive restriction enzyme *Hpa*II. **Figure 2** shows an agarose gel of a restriction digest of a regionally methylated construct. A 1.1-kb DNA fragment was used in the methylation reaction, and it can be seen that this is resistant to digestion by *Hpa*II (**Fig. 2**, lane 6) when methylated, but not in the control (mock-methylated) sample (**Fig. 2**, lane 4). The gel from **Fig. 2** was subjected to Southern blot analysis in order to reveal methylation patterns of DNA after transfection into mammalian cells.

Figure 3 shows that upon *Hpa*II digestion, the methylated DNA remains largely undigested (**Fig. 3**, lane 6), whereas control plasmid (**Fig. 3**, lane 2) and mock-methylated control (**Fig. 3**, lane 4) reveal the expected digestion pattern.

Furthermore, after transfecting the plasmid DNA into cells (**Fig. 3**, lanes 7–10), the same digestion pattern is observed, indicating that the preimposed DNA methylation on the 1.1-kb fragment was stable for at least 2 d after transfection.

Before transfection, small amounts of *Hpa*II digestion products are seen (**Fig. 3**, lane 6) which are due to incomplete methylation of the threefold excess of fragment DNA. During transfection this fragment DNA is degraded and thus not detectable anymore (**Fig. 3**, lane 10).

4. Notes

1. Most plasmid vectors contain an f1 origin of replication, which allows the isolation of single-stranded DNA (ssDNA) from plasmid vectors. Blondel and Thillet *(12)* describe an efficient method that we have used routinely.
2. Restriction fragments are annealed at a molar ratio of 3:1 (fragment DNA:ssDNA). Therefore, for each fragment to be methylated, approximately 9 pmol fragment DNA is required. High-quality fragment DNA can be isolated from agarose gel slices using the QIAquick gel extraction kit (Qiagen).

References

1. Kass, S. U., Pruss, D., and Wolffe, A. P. (1997) How does DNA methylation inhibit transcription? *Trends Genet.* **13,** 444–449.
2. Ng, H. H. and Bird, A. (1999) DNA methylation and chromatin modification. *Curr. Opin. Genet. Develop.* **9,** 158–163.
3. Walsh, C. P. and Bestor, T. H. (1999) Cytosine methylation and mammalian development. *Genes Dev.* **13,** 26–34.
4. Simmen, M. W., Leitgeb, S. Charlton, J., Jones, S. J. M., Harris, B. R., Clark, V. H., and Bird, A. (1999) Nonmethylated transposable elements and methylated genes in a chordate genome. *Science* **283,** 1164–1167.
5. Brannan, C. I. and Bartolomei, M. S. (1999) Mechanisms of genomic imprinting. *Curr. Opp. Genet. Dev.* **9,** 164–170.
6. Lyon, M. F. (1999) X-chromosome inactivation. *Curr. Biol.* **9,** R235–R237.
7. Kass, S. U., Goddard, J. P., and Adams, R. L. P. (1993). Inactive chromatin spreads from a focus of methylation. *Mol. Cell Biol.* **13,** 7372–7379.
8. Hsieh, C. L. (1997) Stability of patch methylation and its impact in regions of transcriptional initiation and elongation. *Mol. Cell Biol.* **17,** 5897–5904.
9. Kass, S. U., Landsberger, N., and Wolffe, A. P. (1997). DNA methylation directs a time-dependent repression of transcription initiation. *Curr. Biol.* **7,** 157–165.
10. Renbaum, P., Abrahamove, D., Fainsod, A., Wilson, G. G., Rottem, S., and Razin, A. (1990) Cloning, characterization, and expression in *Escherichia coli* of the gene

coding for the CpG DNA methylase from *Spiroplasma sp.* Strain MQ1 (M.*Sss*I). *Nucl. Acids Res.* **18,** 1145–1152.

11. Sambrook, J., Fritsch, E. F., and Maniatis, T., eds. (1989) *Molecular Cloning: A Laboratory Manual.* Cold Spring Harbor Laboratory Press, Cold Spring Harbor, NY.
12. Blondel, A. and Thillet, J. (1991) A fast and convenient way to produce single-stranded DNA from a phagemid. *Nucleic Acids Res.* **19,** 181.

19

In Vitro Methylation of Specific Regions in Recombinant DNA Constructs by Excision and Religation

Ghislaine Dell, Marika Charalambous, and Andrew Ward

1. Introduction

The first imprinted genes were identified in the early 1990s (e.g., **refs. *1,2***) and there are now over 40 mammalian genes known to be regulated by genomic imprinting (for an up-to-date list, *see* **ref. *3***). The details of the mechanism that discriminates between the active and silent alleles of these genes, based on their parent of origin, may differ from one imprinted gene to the next, but must include some form of epigenetic mark that distinguishes alleles that have passed through the male or female germline ***(4–7)***. The addition of methyl groups to cytosine residues of CpG dinucleotides might provide such a mark, since regions of differential methylation have been identified in the vicinity of many of the known imprinted genes ***(8,9)***. Moreover, analysis of imprinted gene expression in a methyltransferase knockout ($Dnmt1^{-/-}$) mouse has shown that the imprint is lost in a number of cases, resulting in either two silent alleles (*Igf2*, *Igf2r* and *Kvlqt*) or two expressed alleles (*H19*, $p57^{kip2}$, *Snrpn*, and *Xist*) ***(4,10–12)***. Although there may be exceptions (for instance, imprinted expression of *Mash2* is maintained in $Dmnt1^{-/-}$ embryos; **ref. *13***), differential methylation is likely to be an important aspect of the imprinting mechanism that is relevant to most of the imprinted genes in mammals. There is accumulating evidence that methylation is also important for the imprinting of plant genes (reviewed in **ref. *14***).

In this chapter we describe a method that can be used to change the methylation status of defined regions in recombinant DNA so that effects on gene expression can be assessed following transfection of plasmid constructs into

From: *Methods in Molecular Biology, vol. 181: Genomic Imprinting: Methods and Protocols*
Edited by: A. Ward © Humana Press Inc., Totowa, NJ

cultured cells. Using conventional molecular techniques *(15)*, the region of DNA to be methylated is excised from the parent construct, purified, and then methylated in vitro using the CpG methylase *Sss*I. The methylated region is then ligated back into the construct backbone, yielding a plasmid that is methylated only in the region of interest. Using this method we have been able to modify a variety of recombinant DNA constructs for testing in transfection assays.

2. Materials

2.1. Plasmid to Be Methylated

It is very important that the region to be excised and methylated is bounded by two different restriction sites that are each unique within the parent construct. This is to ensure that once the insert has been methylated, it can be ligated back into the construct backbone in the correct orientation. It also ensures that the vector cannot recircularise during ligation without taking up insert (although it does not exclude the possibility that two vectors might ligate to each other). A large amount of plasmid DNA starting material is needed (at least 20 μg is recommended), especially if the region to be methylated is small in relation to the remainder of the construct (*see* **Note 1**), and should be prepared using a method that yields plasmid of a quality suitable for mammalian cell transfection.

2.2. Enzymes

1. Restriction enzymes to excise the region of interest (used with buffers supplied by the manufacturer).
2. *Sss*I methylase (2000 units/mL; New England Biolabs), supplied with 10× buffer (500 m*M* NaCl, 100 m*M* $MgCl_2$, 10m *M* DTT, 100 m*M* Tris-HCl, pH 7.9) and 32 m*M* S-adenosyl methionine (SAM) methyl donor cofactor (*see* **Note 2**).
3. Enzymes to check extent of methylation reaction. Widely available methyl-sensitive restriction enzymes are *Hha*I and *Hpa*II (*see* **Note 3**), recognizing the sequences GCGC and CCGG, respectively (used with buffers supplied by the manufacturer).
4. T4 DNA ligase. Use high-concentration preparations to maximize ligation efficiency (e.g., 20,000–30,000 Weiss units/mL enyme supplied by Promega or New England Biolabs). Aliquot the 10× reaction buffer (100 m*M* $MgCl_2$, 10 m*M* DTT, 10 m*M* ATP, 500 m*M* Tris-HCl, pH 7.5, 250 μg/mL bovine serum albumin) supplied with enzyme into single-use amounts, as it deteriorates with freeze-thawing.

2.3. DNA Purification

1. QIAquick gel extraction kit (QIAGEN).
2. QIAquick nucleotide removal kit (QIAGEN).

2.4. Electrophoresis

1. Agarose (ultrapure, electrophoresis grade; Gibco-BRL).
2. 1× TAE buffer: 0.04 *M* Tris-acetate, 0.001 *M* ethylenediaminetetraacetic acid (EDTA), pH 8.0. Note that a 50× TAE stock can be stored at room temperature and diluted as required.
3. Horizontal gel apparatus and power supply. Two sizes of gel combs are needed for molding of either small wells, with approximately 20 to 30 μL volume capacity (for analysis of restriction reactions), or large wells, with approximately 200-μL volume capacity (for purification of relevant restriction fragments).
4. DNA markers of known size and concentration (e.g., phage λ DNA digested with *Hind*III is widely available).
5. 10× gel loading buffer: 0.4% (w/v) bromophenol blue, 0.4% (w/v) xylene cyanol FF, 25% (v/v) ficoll (type 400, Pharmacia).
6. 10-mg/mL ethidium bromide stock solution for gel staining, and UV light source for visualization of stained DNA.

2.5. Miscellaneous

1. Absolute ethanol, for diluting DNA purification buffer solutions and for DNA precipitation steps.
2. 3 *M* sodium acetate, pH 5.5, for DNA precipitations.
3. Scalpel for removing bands from gels.

3. Method

1. Digest 20 μg each of the insert-containing and vector-containing plasmid DNAs (*see* **Note 4**) overnight with the chosen restriction enzymes. Use a reaction volume of 100 μL, containing 50 U of each restriction enzyme. If the enzymes require sequential digestion (i.e., because they have incompatible reaction conditions), carry out each reaction overnight, precipitating the DNA in between by adding 1/10 volume 3 *M* sodium acetate, pH 5.5, and 3 vol absolute ethanol. Mix gently, by inverting the tube several times, and place at –20°C for at least 1 h. Collect precipitated DNA by microcentrifugation at 10,000–14,000 rpm (9,000–17,000*g*) for 10 min, wash the pellet with 70% ethanol, spin again, dry, and resuspend in sterile distilled water.
2. After each phase of the digestion, remove 2–3 μL (around 500 ng) of each sample, add 1 μL 10× loading buffer, and make up the volume to 10 μL with water. Run this out on a gel to ensure that the digests are complete. Pour a 1% (w/v) agarose gel, first dissolving the agarose in 1× TAE buffer and then adding 0.5 μg/μL ethidium bromide, using a comb with small wells (in adjacent wells run one sample of 500 ng undigested plasmid for comparison and one sample of DNA size markers). Run the gel, submerged in 1× TAE buffer, at 100–150 V for 90–120 min.
3. Having confirmed that the restriction digests have gone to completion (*see* **Note 5**), add 1/10th volume 10× gel loading buffer to the remainder of the digest

and load this onto another 1% agarose (w/v) gel using a comb with large wells so that the entire sample can be separated (*see* **Note 6**).

4. Run the digest at 100 V until the required DNA fragment is well separated from any other fragments (progress can be monitored at intervals by viewing the gel with a UV light source, then returning it to the electrophoresis tank and continuing the run if necessary).
5. Once good separation has been achieved, excise the required DNA fragments (i.e., vector, or construct backbone, as well as the insert, or region to be methylated) from the gel using a scalpel. Trim the gel slice on all sides to remove as much as possible of the excess, unstained agarose. DNA can then be recovered from the gel slice using a QIAquick gel extraction kit (Qiagen) as detailed by the manufacturer (*see* **Note 7**), but note the following details: not more than 300 mg of agarose should be loaded per QIAquick spin column, and the final elution step is performed with 50 μL elution buffer (EB, 10 m*M* Tris HCl, pH 8.5) per column.
6. Run 1 μL of the vector and insert DNA on a 1% agarose gel to check its integrity. On the same gel, run DNA markers containing bands of known size and concentration to allow an estimation of the yield of the purified DNA fragment.
7. At this point, the vector (construct backbone) can be stored at –20°C until it is needed for the ligation (**step 13**).
8. Divide the insert (region to be methylated) DNA into two tubes. One will be methylated and the other mock-methylated, to serve as a control.
9. Set up the methylation reactions. Given here is a 50-μL specimen reaction (*see* **Notes 8** and **9**): 25 μL DNA, 5 μL 10× reaction buffer, 5 μL (2000 U/μL) *Sss*I methylase (replace this with water or enzyme storage buffer for the mock-methylated tube), 15 μL sterile distilled water, 1 μL SAM.
10. Incubate the methylation reactions for 4 h at 37°C. It is then vital that the SAM be replenished, as it is heat-labile and deteriorates quite quickly at this temperature. Add another 1 μL of the SAM and return the tubes to 37°C overnight (*see* **Note 10**).
11. Digest approximately 100 ng of the mock-methylated and methylated reaction with a methylation-sensitive restriction enzyme that will allow determination of whether the methylation reaction was successful and complete. Common enzymes to use are *Hha*I and *Hpa*II, as they each recognize a four-base motif that is frequently present in DNA sequences (*see* **Note 3**). Digest for 1–2 h at 37°C using 5–10 U enzyme. Run the DNA on a 1% agarose gel; the mock-methylated DNA should digest while the methylated DNA remains intact. If the methylation is not complete, the overnight incubation can be repeated. Add 10,000 U more enzyme, 1 μL fresh SAM, and sufficient buffer and sterile distilled water to make the volume up to 100 μL.
12. Once the methylation is complete, the DNA can be purified away from enzymes and salts using a QIAquick nucleotide removal kit (Qiagen), following the manufacturer's instructions but with the final elution step performed using 50 μL elution buffer (EB, 10 m*M* Tris HCl, pH 8.5). Between 1 and 5 μL of the sample

is then checked on a 1% agarose gel to confirm integrity and to estimate the yield by comparison with DNA markers of known concentration.

13. Ligate the insert (both methylated and mock-methylated fragments) with vector at a molar ratio of 3–5 : 1 insert : vector. Use all of the available insert (*see* **Note 11**) and adjust the amount of vector accordingly. Ligase buffer should be added to provide 1× final concentration, with 1 μL ligase (20–30 U/μL), with any extra reaction volume made up with sterile distilled water and the reactions incubated overnight at 4°C.
14. DNA can again be purified away from enzymes and salts using a QIAquick nucleotide removal kit (Qiagen), as in **step 12**.
15. It is necessary to check the progress of the ligation to see how much of the desired product is present. This is not always straightforward, as there are usually multiple possible ligation products. Digest the ligation mix with one of the enzymes used in the original insert/vector preparation. Cut an amount approximating 250 ng (base this calculation on the amount of DNA that went into the ligation reaction) for 2 h at 37°C, then run it on a 1% agarose gel at around 50 V until the ligation products have separated as fully as possible. In addition to single-insert ligation products, you may also see on the gel bands that represent (a) single unligated insert fragments, (b) two insert fragments ligated together, or (c) two vector fragments ligated together. Typically, the yield of vector containing a single insert is around 30–40% of the ligation product (this is one reason that it is important to have a mock-methylated control sample). The described protocol has been successful for inserts of 0.8–2.8 kb, being ligated back into vector constructs of 6–14 kb.
16. The DNA constructs are now in a form that allows the effects of methylation of specific regions to be examined following their introduction into cells (*see* **Note 12**).

4. Notes

1. It is recommended that a large quantity of DNA be used from the outset, because there are several purification steps at which some loss of material is unavoidable. Beginning with 20–50 μg DNA will usually result in around 2–5 μg each of the methylated and mock-methylated construct samples. It is possible to prepare insert in batches that are then pooled in the ligation reaction (*see* **Note 11**).
2. S-adenosyl methionine (SAM) is unstable at 37°C and when subjected to repeated freeze–thaw cycles. Consequently, it is advisable to prepare several aliquots from the tube supplied with the enzyme on the first occasion that it is thawed and use a fresh aliquot each time these reactions are carried out.
3. A comprehensive database of restriction enzymes and their properties, including details of recognition sequences and methylation sensitivity, is maintained at the following location of the World Wide Web: *http://rebase.neb.com/rebase/*.
4. Sufficient DNA must be cut (a) to allow generation of a mock-methylated as well as a methylated plasmid (the mock-methylated plasmid will serve as a control, given that the ligations will not be 100% efficient) and (b) to allow for the loss of

some DNA during the purification steps. In some instances the insert-containing and vector-containing constructs will be the same plasmid, and it may be possible to purify both fragments from the same digest. Exceptions to this are cases in which the two fragments are similar in size and cannot be adequately resolved on an agarose gel. This problem can be circumvented by carrying out two separate reactions, including in each an enzyme that cuts one of the fragments into two smaller pieces, allowing the other fragment to be cleanly excised from the gel (at **step 5**).

5. For restriction digests that fail to reach completion, more enzyme can be added and the reaction continued for several more hours. However, the volume of enzyme in the reaction should never exceed 10% of the total reaction volume, because the glycerol used in enzyme storage buffers can interfere with enzyme activity at higher concentrations. Where the addition of extra enzyme would exceed this 10% limit, or for digests that fail completely, protein can be removed by phenol/chloroform extraction and the DNA recovered following precipitation (as in **step 1**). Repeat the digest, using a new batch of restriction enzyme if necessary.
6. Some care is needed to ensure that an optimal amount of digested DNA is loaded into the available wells of the agarose gel. A high concentration of DNA is desirable, as this will ensure a good yield of the purified fragment. However, if the concentration is too high, it will be impossible to achieve good separation of the required fragment from other products of the restriction digest. Aim for about 1 μg of digested DNA for every 1 mm of well (measured across the gel), in the first instance, but note that quantities may need to be adjusted empirically for each particular digest (for example, more DNA can be loaded in cases where the size of the required fragment differs greatly from that of any additional fragments).
7. A number of kits are available for the purpose of DNA purification from agarose; the one recommended here works well. An effective alternative to commercial kits is the following, adapted from the method described by Heery ***(16)***: First, a spin column is made from one 1.5-mL and one 0.5-mL microfuge tube. Remove the lids from both tubes and pierce the bottom of the smaller tube using a narrow-gage syringe needle. Plug the small tube with a ball (approximately 5 mm diameter) of polyallomer wool (this is used as the filter material in tropical fish tanks and can be obtained from pet shops; it can be sterilized by autoclaving) and place the gel slice in the small tube, on top of this plug. Place the small tube inside the large tube and microcentrifuge at 10,000 rpm (9,000*g*) for 10 min. Buffer containing the DNA fragment will collect in the 1.5-mL microfuge tube, while the gel matrix will remain above the polyallomer wool plug (the eluted DNA can be visualized in solution under UV light). Next, the eluted DNA can be purified away from any contaminating small molecules by passing it through a column made from a 1-mL syringe barrel (again plugged with a ball of polyallomer wool), containing Sephadex-G50 (hydrate 10 g in 150 mL 1× TE buffer: 0.01 *M* Tris-HCl pH 7.6, 0.001 *M* EDTA pH 8.0, and sterilize by autoclaving). Place the filled syringe barrel in a 15-mL plastic disposable centrifuge tube and spin in a bench-top centrifuge at 250*g* (around 1200 rpm) for 2 min to remove

excess buffer. Transfer the column to a fresh 15-mL tube, add the solution eluted from the gel slice to the top of the column, and recentrifuge at 250*g* for 2 min. A solution containing most of the purified DNA fragment should elute at this step (this can again be monitored under UV light), and more of the DNA fragment can usually be recovered by adding 100 µL 1× TE buffer and centrifuging once more.

8. The methylation reaction can be scaled up successfully to a volume of at least 150 µL.
9. The *Sss*I enzyme mimics mammalian methyltransferase, in that it methylates cytosine residues in CpG dinucleotide pairs, but differs in that it will use both fully methylated and hemimethylated DNA as substrate ***(17)***. In the absence of Mg^{2+}, *Sss*I is highly processive, tending to methylate cytosines at all CpGs along a DNA template molecule. It should be possible to produce random, distributed patterns of methylation by increasing the Mg^{2+} content of the reaction buffer ***(18)***, but we have not investigated this in our own experiments.
10. The methylase reaction shown will usually result in complete methylation of the DNA fragment. If this is not the case, then a fresh aliquot of SAM, and/or a new batch of enzyme, may be required (*see* **Note 2**).
11. Ligation reactions containing 20 µg of either methylated or mock-methylated insert, together with 10 µg vector work well. Ligations on this scale may require pooling of two or more batches of methylated DNA fragments, but allow efficient preparation of large quantities of the final product.
12. In our experiments the methylated constructs and controls have been assayed for gene expression following their introduction into various mammalian cell lines. A high transfection efficiency is desirable, since the yield of each construct preparation is typically very low (2–5 µg of the correct ligation product). To achieve high transfection efficiencies with small amounts of plasmid DNA (0.25–1 µg), we favor the use of FuGENE 6 transfection reagent (Roche). Gene expression from the transfected constructs can be measured in a transient assay or following stable maintenance of the exogenous DNA. In the latter case, the pattern of DNA methylation established in vitro can be maintained through DNA replication in vivo in at least some cell types ***(19)***.

Acknowledgments

We are grateful for support from the BBSRC, Cancer Research Campaign, and International Association for Cancer Research.

References

1. Barlow, D. P., Stoger, R., Herrmann, B. G., Saito, K., and Schweifer, N. (1991) The mouse insulin-like growth factor type-2 receptor is imprinted and closely linked to the *Tme* locus. *Nature* **349,** 84–87.
2. DeChiara, T. M., Efstratiadis, A., and Robertson, E. J. (1991) Parental imprinting of the mouse insulin-like growth factor II gene. *Cell* **64,** 849–859.

3. Beechey, C. V., Cattanach, B. M., and Selley, R. L. (2000) MRC Mammalian Genetics Unit, Harwell, Oxfordshire. World Wide Web Site: Mouse Imprinting Data and References (URL: *http://www.mgu.har.mrc.ac.uk/imprinting/implink.html*).
4. Caspary, T., Cleary, M. A., Baker, C. C., Guan, X.-J., and Tilghman, S. M. (1998) Multiple mechanisms regulate imprinting of the mouse distal chromosome 7 gene cluster. *Mol. Cell Biol.* **18,** 3466–3474.
5. Brannan, C. I. and Bartolomei, M. S. (1999) Mechanisms of genomic imprinting. *Curr. Opin. Genet. Dev.* **9,** 164–170.
6. Tilghman, S. M. (1999) The sins of the fathers and mothers: genomic imprinting in mammalian development. *Cell* **96,** 185–193.
7. Sleutels, F., Barlow, D. P., and Lyle, R. (2000) The uniqueness of the imprinting mechanism. *Curr. Opin. Genet. Dev.* **10,** 229–233.
8. Razin, A. and Cedar, H. (1994) DNA methylation and genomic imprinting. *Cell* **77,** 473–476.
9. Razin, A. and Shemer, R. (1995) DNA methylation in early development. *Hum. Mol. Genet.* **4,** 1751–1755.
10. Li, E., Beard, C., and Jaenisch, R. (1993) Role for DNA methylation in genomic imprinting. *Nature* **366,** 362–365.
11. Beard, C., Li, E., and Jaenisch, R. (1995) Loss of methylation activates Xist in somatic but not in embryonic cells. *Genes Dev.* **9,** 2325–2334.
12. Shemer, R., Birger, Y., Riggs, A. D., and Razin, A. (1997) Structure of the imprinted mouse *Snrpn* gene and establishment of its parental-specific methylation pattern. *Proc. Natl. Acad. Sci. USA* **94,** 10,267–10,272.
13. Tanaka, M., Puchyr, M., Gertsenstein, M., Harpal, K., Jaenisch, R., Rossant, J., and Nagy, A. (1999) Parental origin-specific expression of *Mash2* is established at the time of implantation with its imprinting mechanism highly resistant to genome-wide demethylation. *Mech. Dev.* **87,** 129–142.
14. Vinkenoog, R., Spielman, M., Adams, S., Dickinson, H. G., and Scott, R. J. (2000) Genomic imprinting in plants, in *Methods in Molecular Biology—Genomic Imprinting* (Ward, A., ed.), Humana Press, Totowa, NJ, chapter 25, pp. 327–370.
15. Sambrook, J., Fritsch, E. F., and Maniatis, T. (1989) *Molecular Cloning—A Laboratory Manual*, Cold Spring Harbor Laboratory, Cold Spring Harbor, NY.
16. Heery, D., Gannon, F., and Powell, R. (1990) A simple method for subcloning DNA fragments from gel slices. *Trends Genet.* **6,** 173.
17. Renbaum, P., Abrahamove, D., Fainsod, A., G. G., W., Rottem, S., and Razin, A. (1990) Cloning, characterisation, and expression in *Escherichia coli* of the gene coding for the CpG DNA methylase from *Spiroplasma sp.* strain MQ1 (M.*Sss*I) *Nucleic Acids Res.* **18,** 1145–1152.
18. Matsuo, K., Silke, J., Gramatikoff, K., and Schaffner, W. (1991) The CpG-specific methylase *Sss*I has topoisomerase activity in the presence of Mg^{2+}. *Nucleic Acids Res.* **22,** 5354–5359.
19. Holmgren, C., Kanduri, C., Dell, G. C., Ward, A., Mukhopadhya, R., Lobanenkov, V., and Ohlsson, R. (2001) CpG methylation regulates the *H19* insulator. *Curr. Biol.* **11,** in press.

20

Detection of Methyl-Sensitive DNA-Binding Proteins with Possible Involvement in the Imprinting Phenomenon

Kerstin Otte and Björn Rozell

1. Introduction

1.1. Cytosine Methylation in Gene Regulation and Imprinting

The molecular basis of parental imprinting is still unknown, but strong evidence points to DNA methylation as one of the mechanisms involved. Expression of imprinted genes was found to be altered in transgenic mice deficient in DNA methyltransferase 1 (Dnmt1) *(1)*. Recently two new methyltransferases have been identified, Dnmt2 and Dnmt3 *(2,3)*, but it remains unclear which of these enzymes has *de novo* activity and differentially methylates imprinted genes. Parent-specific methylation patterns have been detected in regions of the human and mouse imprinted genes (4–6). These regions may present cis-acting elements recognized by methylation-sensitive protein factors that control the allele-specific transcription of imprinted genes. Recently a family of methyl-CpG binding proteins was described *(7)*. The five members of this family share a similar methyl-CpG binding domain. One of these proteins, MeCP2, is known to be necessary for embryonic development, although the effects on imprinted genes are not as drastic as those seen in Dnmt1 null mutants. MeCP2 binds to methyl-CpGs without apparent sequence specificity and mediates transcriptional repression *(8)*. This silencing of transcription is mediated by recruitment of a histone deacetylase complex to chromatin-bound MeCP2, causing a local deacetylation of histones and remodeling of chromatin structure *(9,10)*. However, no involvement of MeCP2 in the recognition of differentially methylated regions of imprinted genes has yet been shown.

From: *Methods in Molecular Biology, vol. 181: Genomic Imprinting: Methods and Protocols*
Edited by: A. Ward © Humana Press Inc., Totowa, NJ

To date, only a few potential methylation-sensitive proteins binding to differential methylated regions have been described. (1) A methylation-dependent factor was shown to bind to a GC-rich sequence in the promoter region of the imprinted mouse *Xist* gene ***(11)***, which is required for transcription. (2) In a previous study we investigated one of the differential methylated regions (DMRs) of the imprinted *Igf2* gene ***(12)***. The transcribed paternal allele of *Igf2* is methylated in this region, while the silenced maternal allele is unmethylated on specific CpG residues ***(5,6)***. Since this methylation is tissue-specific and correlates directly with expression, the presence of a silencer element under epigenetic control has been suggested ***(6)***. We revealed a stem-loop structure in this DMR that was shown to bind a specific methylation-sensitive protein in a developmental fashion (**Fig. 1**). One possible role of the stem-loop structure could be to act as a structural silencer element that is recognized by specific protein factors depending on its methylation status. In the light of recent evidence showing that inverted repeats directly influence gene silencing by methylation, the presence of a conserved inverted repeat in the *Igf2* gene suggests a potential involvement in the methylation of the DMR ***(13,14)***. (3) Several groups have recently shown that the CTCF silencer protein binds in a methylation-sensitive manner to an imprinting control region located between the oppositely imprinted *Igf2* and *H19* genes ***(15–17)***.

1.2. Gel Mobility-Shift Assay to Identify Methylation-Sensitive Proteins

This chapter presents the gel mobility-shift assay as a method to identify methylation-sensitive DNA binding proteins. This assay can be a very effective point of first entry into a more detailed investigation of the proteins involved in recognition and binding to methylated sequences.

In the gel mobility-shift assay, the binding of a protein to a radiolabeled DNA fragment reduces the mobility of the DNA in a nondenaturing polyacrylamide gel and thus results in a complex that can be distinguished electrophoretically from the unbound probe. If the protein component is sensitive to methylation of cytosine residues of the target DNA, this complex should not form. Comparison of methylated and unmethylated DNA fragments in this assay can therefore identify methylation-sensitive proteins (**Fig. 1**). This method can be performed with whole-cell or nuclear extracts to initially identify methylation-dependent binding proteins or with partially or completely purified proteins to assess whether they are sensitive to cytosine methylation. Here we describe in detail the extraction of proteins, preparation and labeling of the DNA to be analyzed, and the DNA-protein binding reaction itself.

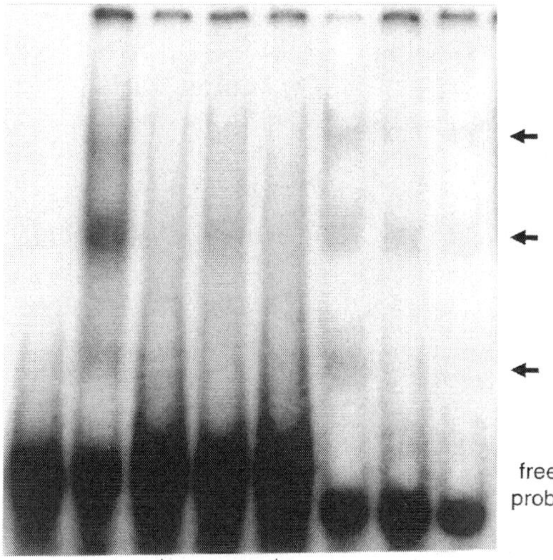

Fig. 1. Methylation-sensitive protein binding to a murine stem-loop structure examined by gel mobility-shift assay. Three distinct bands (indicated by arrows) represent binding of proteins to the native oligonucleotide (lane 2). This binding was reduced upon methylation of all CpGs (lane 4). Furthermore, protein binding was crucially dependent on the presence of the loop as the stem area (S) or the half-sites (HS) were only weakly bound (lanes 6–8) ***(12)***. Reactions contain nuclear protein extract from mouse neonatal liver (mouse: lanes 2–8). Native oligonucleotide (native) was added in lanes 1–3. The methylated counterpart (methyl) was added in lanes 4 and 5. No protein extract (ne) was added to the sample in lane 1. Competition analysis was performed by addition of 100× molar excess of unlabeled oligonucleotide (lanes 3 and 5). Lane 6 contains oligonucleotide corresponding to the stem area only (S), and lanes 7 and 8 one of the half-sites each (HS).

2. Materials

2.1. Protein Preparation

1. Phosphate buffered saline (PBS): Dissolve 8 g NaCl, 0.2 g KCl, 1.44 g Na_2HPO4, 0.24 g KH_2PO_4 in 800 mL dH_2O. Adjust pH to 7.4 with HCl. Add dH_2O to 1 L.
2. Buffer A: 15 m*M* NaCl, 5 m*M* $MgCl_2$, 60 m*M* KCl, 0.1 m*M* ethylene glycol-bis-N,N,N′,N′-tetraacetic acid (EGTA), 15 m*M* Tris-HCl (pH 7.4), 15 m*M* dithiothreitol (DTT), 0.1 m*M* PMSF (*see* **Note 1**).

3. Sucrose cushion: 1.2 *M* sucrose, 5% glycerol, buffer A, in a microfuge tube.
4. Buffer B: 10 m*M* HEPES pH 7.9, 1.5 m*M* $MgCl_2$, 10 m*M* KCl, 0.5 m*M* DTT (*see* **Note 2**), 0.5 m*M* phenylmethylsulfonyl fluoride (PMSF) (*see* **Note 1**).
5. Buffer C: 20 m*M* HEPES (pH 7.5), 1 m*M* PMSF (*see* **Note 1**), 2 m*M* DTT (*see* **Note 2**), 400 m*M* KCl, and 20% glycerol.

2.2. Preparation of DNA Probe

1. 1× methylase buffer: 50 m*M* NaCl, 10 m*M* Tris-HCl, 10 m*M* $MgCl_2$, 1 m*M* DTT (*see* **Note 2**). Adjust pH to 7.9.
2. *Sss*I methylase and 5-adenosylmethionine (SAM) (both from New England Biolabs).
3. 10× CIP buffer: 0.5 m*M* Tris-HCl (pH 9.0), 10 m*M* $MgCl_2$, 1 m*M* $ZnCl_2$, 10 m*M* spermidine.
4. 1× kinase buffer: 50 m*M* Tris-HCl (pH 7.6), 10 m*M* $MgCl_2$, 5 m*M* DTT, 0.1 m*M* EDTA.

2.3. Binding Reaction

1. 5× binding buffer: 50 m*M* HEPES (pH 7.5), 50% glycerol, 250 m*M* KCl, 0.5 m*M* EDTA, 5 m*M* DTT (*see* **Note 2**), 12.5 m*M* $MgCl_2$.
2. 5× TBE: 14 g boric acid, 10 mL 0.5 *M* EDTA, 27 g Tris-base, make up to 500 mL with dH_2O (*see* **Note 3**).
3. Polyacrylamide gel: To prepare a 4% mini gel (8 × 10 cm), mix 0.8 mL 30% acrylamide/bisacrylamide (37.5/1), 1.2 mL 5 × TBE (*see* **Note 3**), 4 mL dH_2O. Add 40 µL 10% ammonium persulfate (*see* **Note 4**) and 16 µL N,N,N′,N′-tetramethylenediamine (TEMED) and allow the gel to polymerize for approximately 1 h.

3. Methods

3.1. Protein Preparation

Nuclear and whole-cell extracts may be prepared from cell cultures and tissues. The preparation of extracts from whole cells enables the entire DNA-binding protein content of the cell to be examined. Whole-cell extracts are easier to prepare, requiring less steps in their preparation, which makes their use favorable when a tissue sample is limiting. It is also necessary to prepare whole-cell extracts when the tissue has been frozen. On the other hand, preparation of nuclei gives between 10- and 100-fold enrichment of a nuclear-localized protein, depending on the size of cells.

3.1.1. Preparation of Nuclei from Tissues

Samples should be kept on ice at all times, and centrifugation should be carried out at 4°C.

1. Dissect tissue of interest into PBS on ice, and rinse in PBS several times to remove blood cells.

2. Homogenize tissue samples (around 10 mg) in 200 μL ice-cold buffer A containing 0.3 *M* sucrose, 10% glycerol.
3. Overlay the homogenate on a 1 mL sucrose cushion.
4. Centrifuge the samples for 20 min in a microcentrifuge at 1500*g* (4000 rpm) at 4°C.
5. Resuspend the pelleted nuclei in buffer A/0.3 *M* sucrose/10% glycerol, and store at –70°C.
6. To prepare protein extracts follow **steps 2–4** under **Subheading 3.1.3.**

3.1.2. Preparation of Nuclei from Cell Lines

Samples should be kept on ice at all times, and centrifugation should be carried out at 4°C.

1. Harvest 5×10^7 to 1×10^8 cells and centrifuge at 250*g* for 10 min.
2. Wash with PBS.
3. Centrifuge at 250*g* for 10 min.
4. Resuspend in 5 vol buffer B.
5. Incubate for 10 min.
6. Centrifuge at 250*g* for 10 min.
7. Resuspend in 3 vol buffer B. Add Nonidet P-40 to 0.05% and homogenize to release the nuclei. Check the successful release of nuclei by microscopy.
8. Centrifuge at 250*g* for 10 min to pellet the nuclei.
9. To prepare protein extracts, follow **steps 2–4** under **Subheading 3.1.3.**

3.1.3. Preparation of Whole-Cell Extracts

Samples should be kept on ice at all times, and centrifugation should be carried out at 4°C.

1. If a cell line is to be used, harvest cells and centrifuge at 250*g* for 10 min. If tissue is to be used, dissect and wash in ice-cold PBS.
2. Homogenize sample in 3 vol buffer C with a pessle.
3. Centrifuge at 20,000*g* for 1 h.
4. Determine protein concentration, for example, by Bradford assay ***(18)***.
5. Aliquot and store at –70°C.

3.2. Preparation of DNA Probe

To examine protein binding to a DNA sequence, both restriction fragment or oligonucleotide probes may be used. If the protein-binding site is known or a previously identified factor is to be studied, a short oligonucleotide probe representing the binding site should be used. Oligonucleotides can be synthesized with internal methyl-cytosine residues and are dephosphorylated at their 5′ ends. They therefore can be readily end-labeled. Restriction fragments have to be methylated and dephosphorylated in vitro before labeling.

3.2.1. In Vitro Methylation and Dephosphorylation of Restriction Fragments

1. Digest the DNA of interest and purify the DNA fragment by gel electrophoresis.
2. Mix 1 µg of restriction fragment, 1 U *Sss*I methylase, and 5 mmol S-adenosylmethionine (SAM) in methylase buffer (*see* **Note 5**). *Sss*I methylates all cytosine residues within the dinucleotide recognition sequence 5′-CG-3′.
3. Incubate for 1 h or longer at 37°C (*see* **Note 5**).
4. Extract DNA with and equal volume of phenol/chloroform/isoamyl alcohol (25/24/1).
5. Precipitate the sample with 2 vol 100% ethanol and 0.1 vol 3 *M* sodium acetate for 30 min at –20°C.
6. Centrifuge at 15,000 rpm for 30 min, wash the pellet in 70% ethanol, and resuspend in dH_2O.
7. Treat an aliquot of the methylated DNA with a methyl-sensitive restriction enzyme (e.g., *Hpa*II does not cut methylated DNA within the 4-base recognition sequence, 5′-CCGG-3′) to check the degree of methylation.
8. By agarose gel electrophoresis, a high degree of uncut fragment should be observed.
9. To dephosphorylate the restriction fragments, add 5 µL 10× CIP buffer and 0.01 U of calf intestinal phosphatase per picomole DNA ends and make up to 50 µL.
10. Incubate at 37°C for 30 min.
11. Extract DNA with an equal volume of phenol/chloroform/isoamyl alcohol (25/24/1).
12. Take the upper aqueous layer and precipitate the DNA with 2 vol of ethanol and 0.1 vol 3 *M* sodium acetate for 30 min at –20°C.
13. Centrifuge at 20,000*g* (15,000 rpm) for 30 min, wash the pellet in 70% ethanol, and resuspend in dH_2O. Store at –20°C.

3.2.2. Radioactive Labeling of DNA

The easiest way of labeling oligonucleotides is to add a ^{32}P-labeled phosphate to the 5′ end using T4 kinase. Single-stranded oligonucleotides have to be annealed with the complementary oligonucleotide before they are labeled. Restriction fragments can also be labeled using this method once the terminal phosphate groups have been removed.

3.2.3. Annealing of Oligonucleotides

1. Mix 2 pmol of each oligonucleotide in 10 m*M* HEPES.
2. Denature at 95°C for 5 min.
3. Cool down to room temperature over a few hours.
4. Store at –20°C.

3.2.4. End-Labeling of DNA

1. Mix 2 pmol of annealed oligonucleotide or dephosphorylated restriction fragment with 2 µL γ^{32}P-dATP in 1× kinase buffer and add 5 U of T4 DNA kinase.
2. Incubate 30 min at 37°C.
3. Precipitate DNA with 4 vol 100% ethanol and 0.1 vol 4 *M* LiCl at –70°C for 30 min.
4. Centrifuge at 15,000 rpm for 30 min and wash the pellet in 70% ethanol.
5. Resuspend DNA in 100 µL dH_20 and store at 4°C.

3.3. Binding Reaction

Conditions for the binding of proteins to DNA depend on the stability of the formed protein/DNA complex. The method described here uses incubation on ice for 30 min, but longer incubation periods or reactions at room temperature may be advisable for certain complexes and should therefore be determined experimentally. Poly dIdC is added as nonspecific competitor for general DNA-binding proteins. The amount to be used in the binding reaction may vary depending on the concentration of proteins and abundance and affinity of the factor studied. It is advisable to determine the amount giving the best results experimentally.

An approach to determine the specificity of the binding activity is to test whether an oligonucleotide containing the consensus sequence effectively abolishes the gel mobility-shift complex by competition with the radiolabeled fragment. Ideally, these studies should be conducted with the wild-type sequence, a variety of point-mutated derivatives, and an unrelated sequence. Usually a range of up to a 1000-fold excess of competitor DNA is added to the binding reaction (**Fig. 1**).

1. Prepare a 4% polyacrylamide gel (*see* **Subheading 2.3.3.**). Prerun the gel in freshly prepared 0.5× TBE at 150 V for 1 h before electrophoresis. Exchange buffers before loading the probes.
2. For a 15 µL binding reaction, mix 5 µg of protein, 30 fmol radiolabeled DNA, and 500 ng poly dIdC (Pharmacia), in 1× binding buffer.
3. Incubate on ice for 30 min.
4. Run 10 µL of each reaction on a 4% polyacrylamide gel. Bromphenol blue in glycerol may be added to a spare track as marker. The samples are run at 150 V. The length of the run depends on the length of the oligonucleotide or restriction fragment used.
5. Dry the gel on a filter paper (80°C for 1 h under vacuum) and autoradiograph overnight.

4. Notes

1. PMSF functions as a protease inhibitor. It should be prepared as a stock solution of 10 m*M* in isopropanol and aliquots are kept at –20°C. Since it is rapidly inactivated in aqueous solutions, it should be added to the buffer immediately before use.
2. DTT should be prepared as a 1 *M* stock solution in water and stored at –20°C. DTT solutions should not be autoclaved.
3. TBE older than 1 wk may inhibit the binding reaction of proteins to DNA and should therefore be prepared freshly.
4. The 10% ammonium persulfate (APS) solution is not stable for longer than a week and should be stored at 4°C.
5. SAM is unstable at 37°C and should be replenished in reactions incubated for longer than 4 h.

References

1. Li, E., Beard, C., and Jaenisch, R. (1993) Role for DNA methylation in genomic imprinting. *Nature* **366,** 362–365.
2. Okano, M., Xie, S., and Li, E. (1998) Cloning and characterization of a family of novel mammalian DNA (ctosine-5) methyltransferases. *Nature Genet.* **19,** 219–220.
3. Okano, M., Xie, S., and Li, E. (1998) Dnmt2 is not required for de novo and maintenance methylation of viral DNA in embryonic stem cells. *Nucleic Acids Res.* **26,** 2536–2540.
4. Brandeis, M., Kafri, T., Ariel, M., Chaillet, J. R., McCarrey, J., Razin, A., and Cedar, J. (1993) The ontogeny of allele-specific methylation associated with imprinted genes in the mouse. *EMBO J.* **12,** 3669–3677.
5. Schneid, H., Seurin, D., Vasquez, M. P., Gourmelen, G., Cabrol, S. and Le Bouc, Y. (1993) Parental allele specific methylation of the human insulin-like growth factor-II gene and Beckwith-Wiedemann syndrome. *J. Med. Genet.* **30,** 353–362.
6. Feil, R., Walter, J., Allen, N. D., and Reik, W. (1994) Developmental control of allelic methylation in the imprinted mouse *IGFII* and *H19* genes. *Development* **120,** 2933–2943.
7. Hendrich, B. and Bird, A. (1998) Identification and characterization of a family of mammalian methyl-CpG binding proteins. *Mol. Cell Biol.* **18,** 6538–6547.
8. Nan, X., Campoy, J., and Bird, A. (1997) MeCP2 is a transcriptional repressor with abundant binding sites in genoic chromatin. *Cell* **88,** 471–481.
9. Nan, X., Ng, H. H., Johnson, C. A., Laherty, C. D., Turner, B. M., Eisenman, R. N., and Bird, A. (1998) Transcriptional repression by the methyls-CpG-binding protein MeCP2 involves a histone deacetylase complex. *Nature* **393,** 386–389.
10. Jones, P. L,, Veenstra, G. J. C., Wade, P. A., Vermaak, D., Kass, S. U., Landsberger, N., Strouboulis, J., and Wolffe, A. P. (1998) Methylated DNA and MeCP2 recruit histone deacetylase to repress transcription. *Nature Genet.* **19,** 187–191.

11. Huntriss, J., Lorenzi, R., Purewal, A., and Monk, M. (1997) A methylation-dependent DNA-binding activity recognising the methylated promoter region of the mouse Xist gene. *Biochem. Biophys. Res. Commun.* **235,** 730–738.
12. Otte, K., Choudhury, D., Charalambous, M., Engstrom, W., and Rozell, B. (1998) A conserved structural element in horse and mouse IGF2 genes binds a methylation sensitive factor. *Nucleic Acids Res.* **26,** 1605–1612.
13. Luff, B., Pawlowski, L., and Bender, J. (1999) An inverted repeat triggers cytosine methylation of identical sequences in *Arabidopsis*. *Mol. Cell* **3,** 505–511.
14. Selker, E. U. (1999) Gene silencing: repeats that count. *Cell* **97,** 157–160.
15. Bell, A. C. and Felsenfeld, G. (2000) Methylation of a CTCF-dependent boundary controls imprinted expression of the *Igf2* gene. *Nature* **405,** 482–485.
16. Hark, T., Scheonherr, C. J., Katz, D. J., Ingram, R. S., Levorse, J. M., and Tilghman, S. M. (2000) CTCF mediates methylation-sensitive enhancer-blocking activity at the *H19/Igf2* locus. *Nature* **405,** 486–489.
17. Kanduri, C., Pant, V., Loukinov, D., Pugacheva, E., Qi, C.-F., Wolffe, A., Ohlsson, R., and Lobanenkov, V. (2000) CTCF functionally associates with the insulator upstream of the *H19* gene in a parent of origin-specific and methylation-sensitive manner. *Curr. Biol.* **10,** 853–856.
18. Bradford, M. M. (1976) A rapid and sensitive method for the quantitation of microgram quantities of protein utilizing the principle of protein-dye binding. *Anal. Biochem.* **72,** 248–254.

21

Probing Chromatin Structure with Nuclease Sensitivity Assays

Richard I. Gregory, Sanjeev Khosla, and Robert Feil

1. Introduction

To further our understanding of genomic imprinting it will be essential to identify key control elements, and to investigate their regulation by both epigenetic modifications (such as DNA methylation) and trans-acting factors. So far, sequence elements that regulate parental allele-specific gene expression have been identified in a number of imprinted loci, either because of their differential DNA methylation or through functional studies in transgenic mice *(1,2)*. A systematic search for allele-specific chromatin features constitutes an alternative strategy to identify elements that regulate imprinting. The validity of such an in vivo chromatin approach derives from the fact that in several known imprinting control-elements, a specialized organization of chromatin characterized by nuclease hypersensitivity is present on only one of the two parental chromosome *(3)*. For example, the differentially methylated 5′-portion of the human *SNRPN* gene—a sequence element that controls imprinting in the Prader-Willi and Angelman syndromes' domain on chromosome 15q11-q13—has strong DNase-I hypersensitive sites on the unmethylated paternal chromosome *(4)*. A differentially methylated region that regulates the imprinting of *H19* and that of the neighboring insulin-like growth factor-2 gene on mouse chromosome 7 was also found to have parental chromosome-specific hypersensitive sites *(5,6)*. The precise nature of the allelic nuclease hypersensitivity in these and other imprinted loci remains to be determined in more detail, for example, by applying complementary chromatin methodologies *(7,8)*. However, it is commonly observed that a nuclease hypersensitive site corresponds to a small region where nucleosomes are absent or partially disrupted.

From: *Methods in Molecular Biology, vol. 181: Genomic Imprinting: Methods and Protocols*
Edited by: A. Ward © Humana Press Inc., Totowa, NJ

This lack of canonical nucleosomes is caused by binding of nonhistone proteins. These are most often a variety of different nuclear factors, some of which are responsible for the observed hypersensitivity ***(9–11)***. In addition to the identification and the mapping of nuclease hypersensitive sites, in vivo nuclease-sensitivity assays allow one to investigate how chromatin is organized throughout larger domains. The differences detected in "generalized nuclease sensitivity" between the parental chromosomes may be indicative of specific intranucleosomal modifications (such as core-histone acetylation) ***(12–14)***. This chapter presents nuclease-sensitivity assays that allow one to analyze chromatin in a parental chromosome-specific manner. They all determine the disappearance of DNA fragments, or the appearance of specific digestion products, as a consequence of nuclease digestion in nuclei. The protocols presented are based on commonly used methodologies ***(15,16)*** that involve purification of nuclei from tissues or cultured cells ***(17–19)***, followed by incubation of the nuclei with different amounts of the nuclease of choice, typically at stepwise increasing enzyme concentrations in a series of 5–10 tubes. After termination of the nuclease digestions by the addition of a "stop solution," genomic DNA is extracted from the nuclei and redigested with a restriction enzyme. The restriction enzyme-digested DNA samples are analyzed by Southern hybridization subsequently, in order to deduce to what extent the chromosomal region of interest has become digested by the nuclease (in the nuclei).

1.1. Appropriate Cells and Tissues

In chromatin studies on imprinted genes it is crucial to be able to distinguish the parental chromosomes in a chromosomal region of interest. One way of achieving this is by comparing cells that have only either maternal copies or paternal copies of the region of interest. For example, parthenogenetic cells or embryos can be compared with androgenetic cells or embryos. Alternatively, animals may be available for comparative studies that are either maternally or paternally disomic for the chromosomal region of interest. There are, however, a number of factors that may complicate comparisons between different cell lines and tissues. First, such comparisons necessarily imply performing different experiments. Given the many variables involved in nuclease sensitivity assays, in practice it is difficult to reproduce exactly the same conditions for each experiment. In addition, when different cell lines are being compared, one would need to ensure that they are of the same cell type and have been grown under the same conditions for a comparable period of time. This may be particularly important for undifferentiated early-embryonic cells, for which it has been found that extended in-vitro culture can affect chromatin organization and DNA methylation at imprinted loci ***(6,20)***. Given these limitations, it is possibly best to compare parental chromosomes within a single experiment,

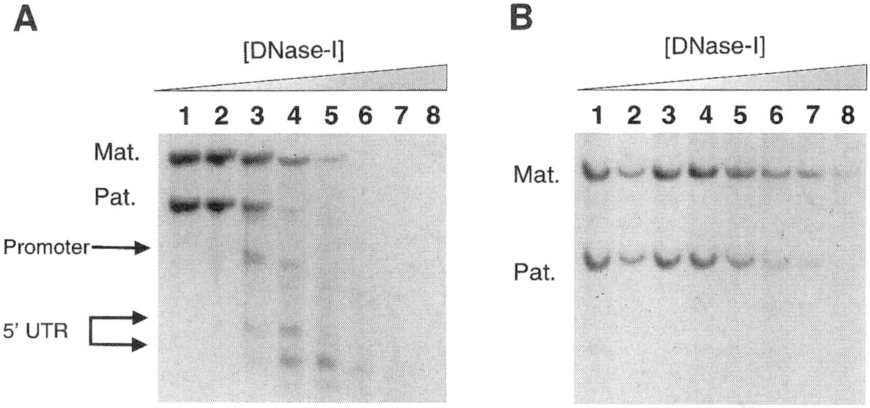

Fig. 1. Parental allele-specific DNase-I cleavage in the mouse *U2afl-rs1* gene. Nuclei were purified from (*M. musculus* × *M. spretus*) F1 embryonic stem cells ***(13)*** as described under Subheading 3.1.2., and incubated with increasing concentrations of DNase-I according to Protocol A (0, 100, 200, 300, 400, 600, 750, 1000 U/mL for lanes 1–8, respectively). DNA was then purified, and 20 μg of each DNA sample was redigested with *Bgl*II (1A) or *Bgl*II+*Sac*I (1B). After electrophoresis on a 1% agarose gel the DNA was transferred onto nylon membrane and Southern hybridization was performed with a probe from the *U2afl-rs1* gene. (A) An autoradiogram that shows the use of a *Bgl*II RFLP to assay allele-specific nuclease sensitivity in *U2afl-rs1* and its upstream sequences ***(13)***. The maternal (*M. musculus*-specific) fragment (Mat.) is less sensitive to DNase-I than the paternal (*M. spretus*-specific) fragment (Pat.). This results mostly from the presence of paternal hypersensitive sites (indicated with arrows) in the promoter and the 5′-UTR of the *U2afl-rs1* gene. (B) In this *Bgl*II+*Sac*I digestion, the 3′-part of the *U2afl-rs1* gene is assayed. Although no DNase-I hypersensitive sites are detected upon hybridization with the *U2afl-rs1* probe, a two- to fourfold difference in "generalized sensitivity" is apparent between the maternal (Mat.) and paternal (Pat.) alleles.

that is, within the same batch of purified nuclei. This can be achieved by studying animals or cells that have maternal and paternal complements that are genetically different for the chromosomal region of interest. Genetic nonequivalence facilitates identification of restriction fragment length polymorphisms (RFLPs) and other sequence polymorphisms (see below) that allow one to tell the parental chromosomes apart. In several recent imprinting studies, interspecific hybrid mice and cells have been generated for this reason, for instance by crossing *Mus m. domesticus* with *Mus m. castaneus* or ***(21)*** *Mus spretus* mice ***(6,18)***. **Figure 1** presents a typical example of a nuclease sensitivity assay on an interspecific hybrid mouse, an assay in which an RFLP is used to distinguish the parental alleles in the Southern hybridization step.

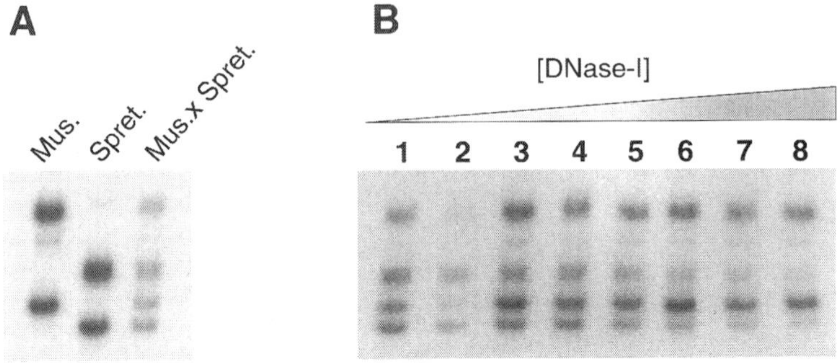

Fig. 2. SSCP-based analysis of DNase-I sensitivity on purified nuclei. **(A)** Primers from the 5′-portion of *U2af1-rs1* were used to PCR-amplify a 293-bp fragment from different genomic DNA samples. PCR products were denatured. Upon SSCP gel electrophoresis, denatured PCR fragments amplified from the *M. musculus* (Mus.) allele migrate differently than the *M. spretus* (Spret.)-specific PCR fragments. Note that in the PCR product amplified from (*M. musculus* × *M. spretus*) F1 hybrid DNA (Mus. × Spret.) the *M. musculus*- and the *M. spretus*-specific single-stranded bands are of the same intensities. **(B)** Allele-specific DNase-I sensitivity in kidney cells. Nuclei were isolated from (*M. musculus* × *M. spretus*) F1 adult kidney, and DNase-I digestions were performed according to protocol A (at 0, 100, 200, 300, 400, 600, 750, 1000 U/mL for lanes 1–8, respectively). PCR amplifications and SSCP analysis were performed as in **Fig. 2A**. Note that in the region analyzed (the 5′ portion of *U2af1-rs1*), the paternal chromosome (*M. spretus*) is more sensitive to DNase-I than the maternal chromosome (*M. musculus*)

1.2. Southern Blotting or PCR-Based Assays

In spite of the fact that the parental genomes are genetically different in interspecific and other hybrid mice, suitable RFLPs are not always detectable in chromosomal regions of interest. It is therefore important to use alternative means to discriminate the parental alleles—for example, by applying the polymerase chain reaction (PCR) *(22–23)*. Under **Subheading 3.4.**, we describe a PCR-based assay that allows one to distinguish the parental alleles *(24)*. Briefly, it consists of PCR amplification from a region in which there is a single- or a multinucleotide polymorphism between the maternal and the paternal genome. After amplification, maternal and paternal chromosome-specific PCR products are distinguished by single-strand conformation polymorphism (SSCP), a method widely used in human genetics for mutation detection *(25–27)*. An example of how PCR-SSCP can be applied to analyze parental allele-specific DNase-I sensitivity is presented in **Fig. 2**.

Both Southern hybridization and PCR-SSCP can be used to analyze DNA samples that have been extracted from nuclease-treated nuclei (*see* Protocols

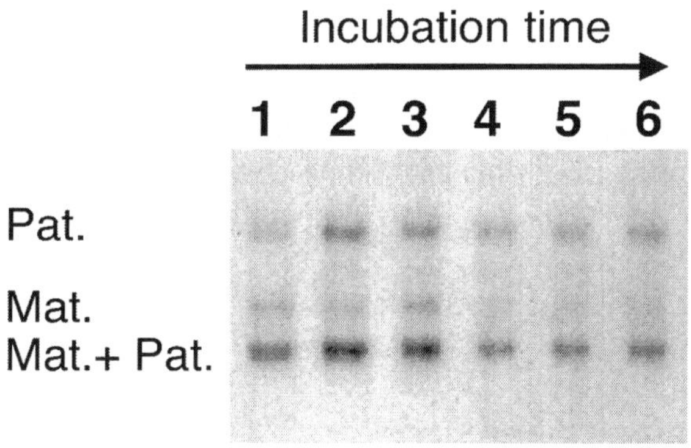

Fig. 3. SSCP-based analysis of nuclease sensitivity on limited numbers of cells. Nuclei from congenic mouse fibroblasts, F1 (C57BL/6 × *M. spretus*) for chromosome 11, were incubated with DNase-I as described for Protocol B. Samples 1–8 correspond to 0, 0.5, 1, 2, 3, 5, 8, and 10 min of incubation at 25°C, respectively. After DNase-I digestion, a fragment from the *U2af1-rs1* gene was amplified by PCR. Amplification products were denatured followed by SSCP gel electrophoresis. This differentiates the maternal (Mat.) and paternal (Pat.) alleles for one of the two single-stranded molecules.

A, C, and D). However the two detection methods have their own advantages and limitations. Southern blot analysis, using RFLPs, assays larger (1–15 kb) chromosomal regions, whereas PCR-SSCP considers allelic nuclease sensitivity in localized small (150–300 bp) regions. For the visualization of nuclease hypersensitive sites, Southern blotting should be the method of choice (*see* **Fig. 1A**). For the analysis of relative levels of generalized nuclease sensitivity, the two methodologies are equally valid and the choice will depend on the availability of RFLPs and other sequence polymorphisms between the parental chromosomes. Finally, PCR-SSCP, being based on the amplification from template genomic DNA, uniquely allows analysis of small numbers of cells. For certain cell types and tissues, and in case one sets out to study chromatin structure in early embryos, the availability of cells is limited, and in these instances nuclease digestions cannot be performed on purified nuclei. To circumvent the problem of limited cell availability, we and others ***(19,24)*** developed assays in which DNase-I digestions are performed on cells that have been permeabilized with a nonionic detergent (*see* Protocol B). After termination of the nuclease digestion, the permeabilized cells are treated with proteinase K, and PCR amplifications are performed directly thereafter using the cell lysate as a template. PCR products are analyzed by SSCP gel electrophoresis to determine the parental allele-specific levels of DNase-I sensitivity. An example of this PCR-SSCP-based approach to study allelic nuclease sensitivity in limited numbers of cells is presented in **Fig. 3**.

1.3. DNase-I, MNase, or Restriction Endonucleases

Many exogenous nucleases and chemical treatments can be used to study chromatin structure *(15,16,28)*. In this chapter we focus on the use of DNase-I, restriction endonucleases, and micrococcal nuclease (MNase). These are the most commonly used DNA nucleases and each has its own specificity, thereby allowing analysis of different aspects of chromatin. DNase-I is a nuclease that introduces single-strand cuts into double-stranded DNA; it also cuts single-stranded DNA but much less efficiently *(16)*. The enzyme does not have any sequence (base) preferences but seems to digest more readily certain structural modifications of DNA *(29)*. For studies of DNase-I sensitivity in nuclei, it is therefore advisable to digest naked DNA in a control experiment (for example, using purified genomic DNA). However, since DNase-I normally digests all sequences at about the same efficiency, it constitutes an excellent general marker of chromatin structure. DNase-I is frequently used for the detection of hypersensitive sites in chromosomal regions of interest, by applying mild conditions of digestion. At higher concentrations one can analyze the overall organisation of larger chromatin domains by assaying its "generalized sensitivity" to DNase-I (*13,18*; **Fig. 1**). In contrast to DNase-I, the use of restriction endonucleases has the advantage that one knows the DNA sequences that are recognized by the enzyme. All restriction enzymes have their own recognition sequence (the restriction site) into which the enzyme introduces a double-strand cut. This allows for studies to focus on the nuclease sensitivity of chromatin at these specific sites *(5,21)*. However, not all restriction endonucleases are suitable for sensitivity assays on purified nuclei. For certain restriction endonucleases, optimal enzyme activity is achieved in digestion buffers that are not compatible with chromatin stability or allow for excessive digestion by endogenous nucleases (*see* **Subheading 2.2.3.**). The last DNA nuclease discussed here is MNase. This enzyme (it requires Ca^{2+} for its activity) introduces double-strand cuts into the DNA molecule and has both an endonuclease and a (much less important) exonuclease activity.

On naked DNA, MNase cuts pA and pT much more quickly than pC and pG; however, this nucleotide preference is influenced by the surrounding sequence context *(29)*. When studying its action on chromatin, it is therefore important to include a control experiment on naked DNA. Although in imprinting studies on purified nuclei, MNase can be used to analyze both hypersensitive sites and generalized sensitivity *(6,13)*, it specifically allows studies on nucleosomal positioning. This is because in chromatin the enzyme digests preferentially in the "linker DNA" between nucleosomes. In a typical experiment, chromatin is partially digested with MNase, after which genomic DNA is extracted and analyzed by Southern blot hybridization. Careful selection of the fragments that are used as probes allows one to determine the nucleosomal organization of a region of interest ***(6)***.

2. Materials

2.1. Purification of Nuclei from Tissues and Cells

1. Appropriate medium for culturing mammalian cells, for example, Dulbecco's modified Eagle's medium (DMEM) containing 10% (v/v) fetal calf serum.
2. Phosphate-buffered saline (PBS), pH 7.3.
3. 0.05% (w/v) trypsin solution (Sigma).
4. Homogenizers (we use a tissue grinder/homogenizer [from BDH] that has a glass mortar [tube] and a pestle with a hard plastic head. The clearance between pestle and mortar is 0.15–0.25 mm).
5. 14-mL polypropylene tubes (e.g., 17 × 100 mm Falcon® tubes).
6. Muslin cheese cloth.
7. Buffer I: 0.3 *M* sucrose, 60 m*M* KCl, 15 m*M* NaCl, 5 m*M* $MgCl_2$, 0.1 m*M* ethylene glycol-bis N,N,N′,N′-tetraacetic acid (EGTA), 15 m*M* Tris-HCl (pH 7.5), 0.5 m*M* dithiothreitol (DTT), 0.1 m*M* phenylmethylsulfonyl fluoride (PMSF), and 3.6-ng/mL aprotinin (Sigma).
8. Buffer II: 0.3 *M* sucrose, 60 m*M* KCl, 15 m*M* NaCl, 5 m*M* $MgCl_2$, 0.1 m*M* EGTA, 15 m*M* Tris-HCl (pH 7.5), 0.5 m*M* DTT, 0.1 m*M* PMSF, and 3.6-ng/mL aprotinin, 0.4% (v/v) Nonidet® P40 (NP40, Sigma).
9. Buffer III: 1.2 *M* Sucrose, 60 m*M* KCl, 15 m*M* NaCl, 5 m*M* $MgCl_2$, 0.1 m*M* EGTA, 15 m*M* Tris-HCl (pH 7.5), 0.5 m*M* DTT, 0.1 m*M* PMSF, and 3.6-ng/mL aprotinin.
10. Parafilm® (Sigma).
11. Equipment: A high-speed centrifuge with a swinging bucket rotor that takes 14-mL polypropylene tubes.

2.2. Nuclease Digestions

2.2.1. DNase-I Digestion on Purified Nuclei (Protocol A)

1. DNase-I digestion buffer: 0.3 *M* sucrose, 60 m*M* KCl, 15 m*M* NaCl, 5 m*M* $MgCl_2$, 0.1 m*M* EGTA, 15 m*M* Tris-HCl (pH 7.5), 0.5 m*M* DTT.
2. DNase-I (e.g., Boehringer [Mannheim, Germany] grade I enzyme) at 10 U/μL (unit definition as defined by supplier) in 100 m*M* NaCl, 10 m*M* Tris-HCl, pH 7.5, 100-μg/mL bovine serum albumin (BSA), 50% (v/v) glycerol. 10 to 20 μL aliquots are frozen, and each aliquot should be used only once after thawing.
3. Water baths set at 25°C and 50°C.
4. Stop solution: 20 m*M* ethylenediaminetetraacetic acid (EDTA), pH 8.0, 1% (w/v) sodium dodecyl sulfate (SDS).
5. Proteinase K at 20 mg/mL in TE buffer: 10 m*M* Tris-HCl pH 7.5, 1 m*M* EDTA. Proteinase K solution needs to be made nuclease-free: incubate at 37°C for 15 min before use.

6. Phenol-Sevag (phenol/chloroform/2-amyl alcohol, 25/24/1 v/v/v).
7. Sevag (chloroform/2-amyl alcohol, 24/1 v/v).
8. 5 *M* NaCl.
9. 2-Propanol.
10. 70% (v/v) Ethanol.
11. TE buffer: 10 m*M* Tris-HCl pH 7.5, 1 m*M* EDTA.

2.2.2. DNase-I Digestion on Permeabilized Cells (Protocol B)

1. Cells grown in appropriate medium.
2. PBS, pH 7.3.
3. 0.05% (w/v) trypsin solution.
4. DNase-I digestion buffer (*see* **Subheading 2.2.1.**).
5. DNase-I digestion buffer containing 0.4% (v/v) NP40.
6. DNase-I (e.g., grade I enzyme from Boehringer Mannheim) at 10 U/μL.
7. Water baths set at 25°C and 50°C.
8. Proteinase K at 20 mg/mL in TE buffer: 10 m*M* Tris-HCl, pH 7.5, 1 m*M* EDTA.

2.2.3. Restriction Endonuclease Digestion on Purified Nuclei (Protocol C)

1. A DTT-containing endonuclease restriction buffer that is suitable for most restriction enzymes but does not affect the nuclease assay: 50 m*M* Tris-HCl (pH 7.9), 10 m*M* $MgCl_2$, 100 m*M* NaCl, 1 m*M* DTT.
2. A restriction enzyme of choice (e.g., *Msp*I) at a high concentration (40 or 100 U/μL).
3. Water baths set at 37°C and 50°C.

4–11. *See* **Subheading 2.2.1.**

2.2.4. MNase Digestion on Purified Nuclei (Protocol D)

1. MNase digestion buffer: 15 m*M* Tris-HCl, pH 7.5, 60 m*M* KCl, 15 m*M* NaCl, 1 m*M* $CaCl_2$, 1 m*M* $MgCl_2$, 0.15 m*M* 2-mercapto-ethanol, 0.15 m*M* spermine, 0.5 m*M* spermidine, 0.1 m*M* PMSF, 0.34 *M* sucrose.
2. MNase (from Pharmacia, at 150 U/μL in 50% (v/v) glycerol).
3. Water baths set at 37°C and 50°C.

4–11. *See* **Subheading 2.2.1.**

2.3. DNA Electrophoresis and Southern Hybridization

1. Restriction endonuclease of choice and its corresponding digestion buffer.
2. Ficoll loading buffer: 0.25% (w/v) bromphenol blue, 30% (w/v) Ficoll® 400, 2% (w/v) SDS (in 5 × TBE buffer: 0.45 *M* Tris-borate, 10 m*M* EDTA).
3. Horizontal gel electrophoresis tank for agarose gels of 20–25 cm in length.
4. 1× TBE buffer: 0.09 *M* Tris-borate, 2 m*M* EDTA, pH 8.0.
5. Nylon membrane (we use Hybond N^+ membrane; Amersham).
6. 0.25 *N* HCl.
7. 0.4 *N* NaOH.

8. Neutralization buffer: 1.5 *M* NaCl, 0.5 *M* Tris-HCl, pH 7.2, 1 m*M* EDTA.
9. UV cross-linker (e.g., Stratalinker, Stratagene).

2.4. PCR-SSCP Analysis

2.4.1. PCR Amplification for SSCP Analysis

1. Template DNA (2 μL when continuing from Protocol B, 50–100 ng of DNA when continuing from Protocol A).
2. Forward and reverse primers (stock solution: 100 μ*M*).
3. dNTP mix (stock solution at 10 m*M* for each dNTP).
4. α^{32}P-dCTP (10 μCi/μL, specific activity 3000 Ci/mmol).
5. 10× PCR amplification buffer (supplied with the enzyme).
6. Taq polymerase (at 5 U/μL).
7. Thermal cycler.

2.4.2. SSCP Analysis of Nuclease Sensitivity

1. Acrylamide solution for SSCP gels: we use 2 × MDE® solution (FMC, from Flowgen Ltd).
2. 0.6× TBE buffer: 0.054 *M* Tris-borate, 1.2 m*M* EDTA.
3. N,N,N′,N′-Tetramethylethylenediamine (TEMED).
4. 10% (w/v) Ammonium persulfate (APS), freshly prepared.
5. Whatman #1 filter paper.
6. A standard DNA sequencing gel apparatus with 31 × 38.5-cm glass plates, 0.4-mm spacers, and shark tooth comb.
7. PCR product.
8. Loading dye: 95% (v/v) formamide, 10 m*M* NaOH, 0.25% (w/v) bromphenol blue, 0.25% (w/v) xylene cyanol.
9. Whatman 3-MM paper.
10. Thin transparent plastic wrap (e.g, Saran® Wrap).
11. Gel dryer (e.g., Bio-Rad® model 583).

2.5. Image Analysis

1. X-ray films.
2. Cassettes with scintillation screens for exposure of X-ray films.
3. Imaging equipment for densitometric measurements on exposed X-ray films (e.g., Bio-Rad Geldoc-1000 apparatus).
4. Phosphor imager (e.g., Molecular Imager FX system from Bio-Rad).

3. Methods

3.1. Purification of Nuclei

3.1.1. Purification of Nuclei from Tissues

1. Dissect fresh tissue (e.g., half an adult mouse liver) and rinse it in PBS (*see* **Notes 1** and **2**).

2. Homogenize tissue in a prechilled glass homogenizer with 5–10 mL of ice-cold buffer I, until no clumps of cells persist (about 10–20 strokes). Filter the cell suspension through four layers of muslin cheesecloth that has been moistened with 2 mL of buffer I.
3. Transfer suspension into a 14-mL tube, and spin down the cells in a swing-out rotor (at 6000*g* for 20 min, at 4°C).
4. Pour off the supernatant and resuspend the cells in 2 mL of ice-cold buffer I. Add 2 mL of ice-cold buffer II, mix gently, and place on ice for 5 min.
5. Prepare two 14-mL tubes containing 8 mL of ice-cold buffer III each. Carefully layer 2 mL of the cell suspension (from **step 4**) onto each of these 8-mL sucrose cushions.
6. Centrifuge in a swing-out rotor, at 10,000*g*, for 20 min at 4°C (*see* **Note 3**).
7. Very carefully take off the supernatant with a Pasteur pipet. This is a critical step, and one should take care that the top solution (which contains NP40) does not come into contact with the nuclear pellet at the bottom of the tube (*see* **Note 3**).
8. Resuspend the nuclei pellet into an appropriate volume of DNase-I digestion buffer or restriction endonuclease buffer (*see* **Subheading 3.2.**).

3.1.2. Purification of Nuclei from Cultured Cells

1. Culture 5×10^7 to 5×10^8 cells in appropriate culture medium. Ensure that cells are not grown beyond semiconfluency.
2. Rinse cells in PBS, add 2 mL of trypsin solution (for adhering cells only), and incubate at 37°C. When trypsinization is complete (after ~2 min), add 5 mL of culture medium to the cells.
3. Divide suspension into two 14-mL tubes, and spin down cells in a swing-out rotor (4000*g*, 5 min at 4°C).

4–8. Identical to **steps 4–8** under **Subheading 3.1.1.** (*see* **Note 4**).

3.2. Nuclease Digestion

3.2.1. DNase-I Digestion on Purified Nuclei (Protocol A)

1. Resuspend each nuclear pellet (**Subheading 3.1.1., step 8**) in 1 mL of ice-cold DNase-I digestion buffer. Combine the two samples in one tube, to obtain 2 mL of nuclei suspension.
2. Aliquot 200 μL of the suspension in ten 1.5-mL Eppendorf tubes that are numbered 1–10. Place the tubes 1–10 on ice, and add appropriate volumes of DNase-I (e.g., add 0, 1, 2, 4, 6, 8, 10, 12, 16, and 20 μL to tubes 1–10, respectively) (*see* **Note 5**).
3. Place tubes 1–10 into a 25°C water bath and incubate for 10 min.
4. Stop the DNase-I digestions by adding 200 μL of stop solution to each tube.
5. Add proteinase K to a final concentration of 200 μg/mL and incubate overnight at 50°C.

6. Extract the DNA twice with Phenol-Sevag (*see* **Note 1**). This is followed by an extraction with Sevag. In case the DNA in the first few tubes of the series is too viscous to allow extraction, DNA should be slightly sheared by passage through a fine syringe needle (2–3 times).
7. Add 20 μL of 5 *M* NaCl, mix well, and then add 400 μL of 2-propanol and precipitate the DNA for 5 min at room temperature. After centrifugation in a bench-top microcentrifuge (20 min at full speed), rinse the DNA pellets with 70% ethanol. Air-dry the pellets and dissolve them in 50 μL of TE buffer.

3.2.2. DNase-I Digestion on Permeabilized Cells (Protocol B)

1. Collect cells (>5000) by trypsinization followed by centrifugation (described under **Subheading 3.1.2.**). Resuspend the cells in 60 μL of ice-cold DNase-I buffer and transfer to a 0.5-mL Eppendorf tube. Add 60 μL of DNase-I buffer containing 0.4% NP40 to obtain a total volume of 120 μL, mix gently, and incubate on ice for 5 min (*see* **Note 6**).
2. Add 1 μL of DNase-I (10 units) and incubate for 20 min at 25°C.
3. At sequential time points after the initiation of the DNase-I digestion (e.g., 0, 0.5, 1, 2, 4, 10 min, respectively), remove a 20-μL aliquot from the reaction and heat-inactivate it at 95°C for 20 min (in a thermal cycler with heated lid, to prevent condensation).
4. After heat-inactivation of all six 20-μL aliquots, add proteinase K to a concentration of 200 μg/mL, and incubate at 50°C overnight.
5. Heat-inactivate the proteinase K by incubation of the samples at 95°C for 1 h.

3.2.3. Restriction Endonuclease Digestion on Purified Nuclei (Protocol C)

1. Nuclei (purified as described in **Subheadings 3.1.1.** and **3.1.2.**) are resuspended in 1.6 mL of ice-cold endonuclease restriction buffer. Add restriction enzyme to a final concentration of 4–10 units/μL (Note: This should dilute the enzyme at least tenfold), and distribute 200 μL-samples among eight 1.5-mL Eppendorf tubes that sit on ice.
2. Incubate the eight tubes (labeled 1–8) for increasing periods of time at 37°C (e.g., for 0, 0.5, 1, 2, 4, 8, 16, and 30 min, in tubes 1–8 respectively). At the sequential time points the enzyme digestions are stopped by addition of 200 μL of Stop Solution.

3–5. Identical to **steps 5–7** in **Subheading 3.2.1.**

3.2.4. MNase Digestion on Purified Nuclei (Protocol D)

1. Nuclei (purified as described in **Subheadings 3.1.1.** and **3.1.2.**) are resuspended in 1.6 mL of ice-cold MNase digestion buffer and placed on ice.
2. Add 1.5 μL of MNase to the tube and distribute 200 μL-samples among eight 1.5-mL Eppendorf tubes that are placed on ice.

3. Incubate the eight tubes (labeled 1–8) for increasing periods of time at 37°C (e.g, for 0, 10, 30, 60, 120, 240, 300, and 600 s, in tubes 1–8, respectively). At the sequential time points, the enzyme digestions are stopped by addition of 200 µL of stop solution.

4–6. Identical to **steps 5–7** under **Subheading 3.2.1.**

3.3. DNA Electrophoresis and Southern Hybridization

To determine the nuclease sensitivity in a genomic region of interest, DNA samples are extracted from the nuclease digests (performed using Protocol A, C, or D), and are then studied by Southern analysis (*see* **Note 7**). Digestion of genomic DNA samples with restriction enzymes, gel electrophoresis, and hybridization with radioactive probes are performed using standard methodologies (see Chapter 14). However, there are a number of specific points that are important for the analysis of nuclease sensitivity in nuclei:

1. Take 20–30 µg of each purified DNA sample (tubes 1–10 from Protocol A, tubes 1–8 from Protocols C and D) to digest with the restriction enzyme of choice. Given the relatively large amount of DNA, digestions should be performed in a volume of 100–150 µL and for 16–18 h.
2. Precipitate the digested DNA samples (with an equal volume of 2-propanol, then rinse the pellet with 70% ethanol and dry the pellet) and dissolve the DNA in minimal volume (20 µL) of TE buffer.
3. Add 5 µL of a 5× loading dye that gives a homogeneous migration of the DNA in an agarose gel (we use Ficoll loading buffer), to obtain a total volume of 25 µL.
4. For the (usually 1%) agarose gel it is important to use a thin comb, so that the volume of the wells is less than 40 µL. This considerably improves the resolution of bands upon Southern analysis.
5. To separate nuclease digestion products, it is advisable to migrate the DNA on a long (20–25 cm) horizontal gel (1% [w/v] agarose in 1× TBE buffer).
6. Preferably hybridize the filter with a radioactively labeled end-probe for the specific region of interest.

3.4. PCR-SSCP Analysis of Nuclease-Digested Chromatin

1. Prepare, on ice, 240 µL of "master mix" containing 4.8 µL of each primer, 4.8 µL dNTP mix, 1 µL of [^{32}P]-dCTP, and 2.5 µL of Taq polymerase in 1× PCR amplification buffer. For stock solutions refer to **Subheading 2.4.1.** (*see* **Notes 8** and **9**).
2. Aliquot 28 µL of the ice-cold master mix into seven 0.5-mL Eppendorf tubes that are kept on ice.
3. Take 2 µL of each DNA sample corresponding to time points 1–6 of Protocol B, and add to tubes 1–6. PCR amplify for 30–40 cycles in a thermal cycler (tube 7 is a negative control without template; for details on how to amplify from genomic DNA, see **ref. *30***).

4. Prepare the solution for the nondenaturing MDE® gel (a polyacrylamide-like matrix, specifically optimized for SSCP): mix 25 mL of 2× MDE® solution, 6 mL of 10× TBE buffer, and 69 mL of deionized water. Filter through Whatman #1 filter paper and degas the solution for 5 min. Add 40 µL TEMED and 400 µL freshly prepared 10% APS.
5. Pour the gel immediately. Insert the shark-tooth comb with teeth pointing upward to form a single well the width of the gel, and clamp on all sides. Lay the gel flat, and polymerize the matrix for at least 60 min.
6. After polymerization, remove the clamps, the tape, and the comb. Place the glass plates into the sequencing gel apparatus. Replace the comb with the teeth pointing downward and just in contact with the gel surface. Add 0.6× TBE buffer to the top and bottom tanks.
7. Take 1 µL of PCR product and add 10 µL of loading dye. Heat-denature the sample to 94°C for 3 min, then place on ice.
8. Load 1–3 µL of the sample into the gel. Run the gel at 4–8 W for 24 h (at room temperature). (Alternatively, gels can be run at 4°C, at about 30 W for 2–4 h).
9. After electrophoresis (when the Bromphenol blue has reached the bottom of the gel), transfer the gel onto a sheet of Whatman 3MM paper and cover with plastic wrap. Dry in a gel dryer for 45 min at 80°C.
10. The gel is exposed to an X-film for 4–16 h at –70°C. In addition, one can expose the gel to a phosphor imager to determine the relative intensities of the bands. Typical results are shown in **Fig. 2** (*see* **Notes 10** and **11**).

4. Notes

1. Wear gloves throughout all procedures and respect other usual safety precautions, in particular when handling phenol, chloroform, and acrylamide solutions.
2. Nuclei purification and subsequent nuclease sensitivity assays (Protocols A and C) work best on fresh tissues. However, tissues that have been frozen in liquid nitrogen can be used as well. They should be broken into small pieces (in liquid nitrogen) immediately followed by homogenization in ice-cold buffer I.
3. To prevent chromatin degradation by endogenous nucleases, all steps of the nuclei purification procedure (**Subheading 3.1.**) should be performed on ice or at 4°C (precool the centrifuge rotor). At **step 4**, it is critical not to extend the incubation in NP40-containing buffer for more than 5 min. At **step 6**, the nuclei pellet should be white (e.g., for liver nuclei preparation, all red color (hemoglobin) should remain in the layer on top of the sucrose cushion). At **step 7**, avoid any trace of the NP40-containing top layer to come in contact with the nuclear pellet (if required, nuclei can be rinsed with buffer III in an additional step, before proceeding with **step 8**).
4. When one assesses generalized DNase-I sensitivity in undifferentiated cells, we suggest addition of sodium butyrate (to a final concentration of 5 m*M*) to the buffers (I, II, and III) that are used for nuclei purification. This prevents artifacts due to the action of endogenous histone deacetylases (and therefore, possible changes in generalized nuclease sensitivity).

5. In the DNase-I digestion series on nuclei (Protocol A), DNase-I concentrations may need to be adjusted for chromosomal regions of interest.
6. Protocol B is designed to study nuclease sensitivity in small numbers of cells, but can be used for the analysis of early mammalian embryos as well *(24)*. However, post-blastocyst-stage embryos seem unsuitable because with excessive clumping of cells, the nucleases will not penetrate all nuclei uniformly.
7. DNase-I (and restriction endonuclease) hypersensitive sites are most readily detected when small (~4 kb) restriction fragments are analyzed in the Southern hybridization step (**Subheading 3.3.**).
8. Experiments involving PCR require extremely careful technique to prevent contamination *(30)*.
9. Instead of adding radioactive dNTPs α^{32}P-dCTP to the PCR reactions for SSCP analysis, one can radioactively end-label the PCR primers (forward and reverse) using T4 polynucleotide kinase and γ^{32}P-dATP.
10. In most cases, SSCP separates 150 to 300-bp single-stranded DNA molecules that have one or more nucleotide difference *(25–27)*. However, the migration of single-stranded fragments in the gel is strongly temperature-dependent. Ideally, therefore, samples to be compared should be run on the same gel. In addition, SSCP is more efficient for DNA with a relatively high G+C content. SSCP analysis of fragments with a lower G+C content can be enhanced by electrophoresis at 4°C.
11. The PCR-SSCP methodology was originally designed to analyze allele-specific DNase-I sensitivity in limited numbers of cells *(24)*. However, it is suitable also to analyze DNA samples obtained with Protocols A, C, and D (*see* **Fig. 2**).

Acknowledgments

The Human Frontier Science Program and the U.K. Ministry of Agriculture, Fisheries and Food supported this work.

References

1. Constância, M., Pickard, B., Kelsey, G., and Reik, W. (1998) Imprinting mechanisms. *Genet. Res.* **8,** 881–900.
2. Brannan, C. I. and Bartolomei, M. S. (1999) Mechanisms of genomic imprinting. *Curr. Opin. Genet. Dev.* **9,** 164–170.
3. Feil, R. and Khosla, S. (1999) Genomic imprinting in mammals: an interplay between chromatin and DNA methylation? *Trends Genet.* **15,** 431–435.
4. Schweizer, J., Zynger, D., and Francke, U. (1999) *In vivo* nuclease hypersensitivity studies reveal multiple sites of parental-origin-dependent differential chromatin conformation in the 150 kb *SNRPN* transcription unit. *Hum. Mol. Genet.* **8,** 555–566.
5. Hark, A. T. and Tilghman, S. M. (1998) Chromatin conformation of the *H19* epigenetic mark. *Hum. Mol. Genet.* **7,** 1979–1985.
6. Khosla, S., Aitchison, A., Gregory, R., Allen, N. D., and Feil, R. (1999) Parental allele-specific chromatin configuration in a boundary-imprinting-control element upstream of the mouse *H19* gene. *Mol. Cell. Biol.* **19,** 2556–2566.

7. Zaret, K. S. (1997) *In vivo* footprinting analysis of protein-nucleic acid interactions. *Methods* **11,** 149–263.
8. Szabó, P. E., Pfeifer, G. P., and Mann, J. R. (1998) Characterization of novel parent-specific epigenetic modifications upstream of the imprinted *H19* gene. *Mol. Cell. Biol.* **18,** 6767–6776.
9. Weintraub, H. and Groudine, M. (1976) Chromosomal subunits in active genes have an altered conformation. *Science* **193,** 848–856.
10. Gross, D. S. and Garrard, W. T. (1988) Nuclease hypersensitive sites in chromatin. *Annu. Rev. Biochem.* **57,** 159–197.
11. Boyes, J. and Felsenfeld, G. (1996) Tissue-specific factors additively increase the probability of the all-or-none formation of a hypersensitive site. *EMBO J.* **15,** 2496–2507.
12. Hebbes, T. R., Clayton, A. L., Thorne, A. W., and Crane-Robinson, C. (1994) Core histone hyperacetylation co-maps with generalized DNase-I sensitivity in the chicken α-globin chromosomal domain. *EMBO J.* **13,** 1823–1830.
13. Feil, R., Boyano, M. D., Allen, N. D.m and Kelsey, G. (1997) Parental chromosome-specific chromatin conformation in the imprinted *U2af1-rs1* gene in the mouse. *J. Biol. Chem.* **272,** 20,893–20,900.
14. Feil, R. and Kelsey, G. (1997) Insights from model systems. Genomic imprinting: a chromatin connection. *Am. J. Hum. Genet.* **61,** 1213–1219.
15. Wu, C. (1989) Analysis of hypersensitive sites in chromatin. *Methods Enzymol.* **170,** 269–289.
16. Bellard, M., Dretzen, G., Giangrande, A., and Ramain, P. (1989) Nuclease digestion of transcriptionally active chromatin. *Methods Enzymol.* **170,** 317–346.
17. Sasaki, H., Jones, P. A., Chaillet, J. R., Ferguson-Smith, A. C., Barton, S. C., Reik, W., and Surani, M. A. (1992) Parental imprinting: potentially active chromatin of the repressed maternal allele of the mouse insulin-like growth factor II (*Igf2*) gene. *Genes Dev.* **6,** 1843–1856.
18. Feil, R., Handel, M-A., Allen, N. D., and Reik, W. (1995) Chromatin structure and imprinting: developmental control of DNase-I sensitivity in the mouse insulin-like growth factor 2 gene. *Dev. Genet.* **17,** 240–252.
19. Stewart, A. F., Reik, A., and Schütz, G. (1991) A simpler and better method to cleave chromatin with Dnase-I for hypersensitive site analyses. *Nucleic Acids Res.* **19,** 3157.
20. Dean, W., Bowden, L., Aitchison, A., Klose, J., Moore, T., Meneses, J. J., Reik, W., and Feil, R. (1998) Altered imprinted gene methylation and expression in completely ES cell-derived mouse fetuses: association with aberrant phenotypes. *Development* **125,** 2273–2282.
21. Bartolomei, M. S., Webber, A. L., Brunkow, M. E., and Tilghman, S. M. (1993) Epigenetic mechanisms underlying the imprinting of the mouse *H19* gene. *Genes Dev.* **7,** 1663–1673.
22. Kramer, J. A., McCarrey, J. R., Djakiew, D., and Krawetz, S. A. (1998) Differentiation: the selective potentiation of chromatin domains. *Development* **125,** 4749–4755.

23. McCabe, V., Formstone, E. J., O'Neill, L. P., Turner, B. M., and Brockdorff, N. (1999) Chromatin structure analysis of the mouse *Xist* gene. *Proc. Natl. Acad. Sci. USA* **96,** 7155–7160.
24. Gregory, R. I. and Feil, R. (1999) Analysis of chromatin in limited numbers of cells: a PCR-SSCP based assay of allele-specific nuclease sensitivity. *Nucleic Acids Res.* **27,** e32i–iv.
25. Orita M., Iwahana, H., Kanazawa, K., and Sekiya, T. (1989) Detection of polymorphisms of human DNA by gel electrophoresis as single-strand conformation polymorphisms. *Proc. Natl. Acad. Sci. USA* **86,** 2766–2770.
26. Orita, M., Suzuki, Y, Sekiya, T., and Hayashi, K. (1989) A rapid and sensitive detection of point mutations and DNA polymorphisms using the polymerase chain reaction. *Genomics* **5,** 874–879.
27. Warren, W., Hovig, E., Smith-Sorensen, B., Borresen, A-L., Fujimura, F. K., Liu, Q., Feng, J., and Sommer, S. S. (1997) Detection of mutations by single-strand conformation polymorphism (SSCP) analysis and SSCP-hybrid methods, in *Current Protocols in Human Genetics* (Boyle, A. L., ed.), Wiley, New York, pp. 7.41–7.4.23.
28. Cartwright, I. L. and Elgin, S. C. R. (1989) Nonenzymatic cleavage of chromatin. *Methods Enzym.* **170,** 359–369.
29. Drew, H. R. (1984) Structural specificities of five commonly used DNA nucleases. *J. Mol. Biol.* **176,** 535–557.
30. Erlich, H. A. (1989) *PCR Technology*, Stockton, New York.

22

Examining Histone Acetlylation at Specific Genomic Regions

Ji-Fan Hu and Andrew R. Hoffman

1. Introduction

The acetylation of core histones can modulate the expression of numerous genes. In general, the deacetylation of histones results in transcriptional repression, whereas increases in histone acetylation lead to the enhancement of gene transcription. Since histone acetylation is maintained during mitosis, the acetylation pattern may contribute a heritable epigenetic imprint that can continue to influence gene transcription *(1)*. These findings suggest that the degree of histone acetylation may represent one of several nonexclusive mechanisms that can initiate or maintain the allele-specific silencing of genomic imprinting. We have recently shown that the inhibition of histone deacetylation by Trichostatin A (TSA) induces the expression of the normally imprinted maternal *IGF2* allele *(2)*, leading to biallelic expression in both human and murine cells. Partial loss of imprinting of both sense and antisense *Igf2r* is also observed after TSA treatment. It has also been shown that exposure of mouse conceptuses to TSA leads to loss of *H19* imprinting *(3)*. In X-chromosome inactivation, the inactive chromosome is associated with relatively underacetylated histones *(4)*. Changes in histone acetylation become apparent after the expression of *Xist*, the gene responsible for X-chromosome gene silencing, and after the downregulation of the inactivated X-chromosome genes *(5)*, suggesting that decreased histone acetylation may mark or otherwise stabilize transcriptional repression rather than actually initiate it.

From: *Methods in Molecular Biology, vol. 181: Genomic Imprinting: Methods and Protocols*
Edited by: A. Ward © Humana Press Inc., Totowa, NJ

2.1. Histone Acetylation and DNA Methylation

The interaction of histones with DNA methylation has been intensively studied. The inhibition of transcription after DNA methylation is seen only after chromatin formation *(6)*. DNA that is rich in methylated CpGs is associated with underacetylated histones, while DNA containing CpG islands that are unmethylated is associated with chromatin enriched in hyperacetylated histone cores *(7)*. DNA methylation may lead to repression of transcription by preventing access to and binding of transcription factors or by enhancing the binding of transcription-inhibitory proteins *(8)*. Methylation of DNA also generates an inactive DNase-resistant local chromatin structure with underacetylated core histones. It has been demonstrated that the inactive chromatin structure generated by the methylated CpGs can spread to adjacent nonmethylated DNA, thereby potentially inhibiting gene transcription over a larger segment of the chromosome *(9)*. The production of this inactive chromatin structure may lead to the creation of regions containing clusters of imprinted genes (such as human chromosome 11p15.5). The methyl-CpG-binding protein MeCP2 is found in a complex with histone deacetylase and other proteins that might regulate transcription. Moreover, it has recently been shown that DNA methyltransferase 1 (Dnmt1), the enzyme responsible for CpG methylation, is associated with histone deacetylase activity in vivo. In vitro, the histone deacetylating enzyme HDAC1 binds to Dnmt1 *(10)*.

Inhibitors of histone deacetylase can overcome DNA methylation-induced transcriptional repression, indicating a linkage between these two known mechanisms of transcriptional repression *(11,12)*. However, changes in histone acetylation appear before the increases in DNA methylation, and histone underacetylation is also correlated with X-inactivation in marsupials, animals in which CpG methylation does not occur *(13)*. In conjunction with DNA methylation, histone acetylation may represent a crucial molecular mechanism for initiating, maintaining and/or transmitting the genomic imprint.

2.2. Methods for Assessing Histone Acetylation in Vitro

The role of acetylated histones has traditionally been examined by using inhibitors of histone deacetylase. Application of these inhibitors leads to the development of hyperacetylated histones. Sodium butyrate is a reversible but nonspecific inhibitor of histone deacetylase; its effects are thought to be mediated through the activation of a protein phosphatase *(14)*. Trichostatin A is considered to be a specific inhibitor of histone deacetylase. In most experiments, TSA (0.3–3 μ*M*) or sodium butyrate (1–5 m*M*) is added to the culture medium for 48 h, and the cells are then harvested for allelic expression analysis.

Although several groups have shown that incubation of cells with TSA leads to changes in gene expression, it is necessary to show directly that histones that are associated with particular regions of DNA (e.g., those containing CpG islands) undergo changes in the amount of histone acetylation that correlate with the allele-specific gene expression. Since histone acetylation is associated with increased gene transcription, it would be predicted that the expressed allele would be associated with acetylated histones while the imprinted, nonexpressed allele would be associated with nonacetylated histones. It has recently been shown that the expressed maternal *H19* allele is in fact preferentially associated with acetylated histones ***(15)***.

The method described here will focus on the assessment of histone acetylation in livers collected from interspecific mice that are derived from *Mus spretus* crossed with *Mus musculus* (C57BL/6). The same method can be easily adopted for the quantitation of histone acetylation in samples collected from cultured cells and other tissues. In order to determine if there is allele-specific histone acetylation, it is necessary to use tissue that is informative for one or more polymorphisms in the gene of interest. In our model, polymorphisms have readily been found between these two strains of mice in most genes.

Allelic histone acetylation in tissues or cells is assessed by a four-step procedure: (1) oligonucleosomes are prepared from mild digestion of chromatin using micrococcal nuclease ***(16)***; (2) acetylated-H4 linked oligonucleosomes are separated from unacetylated-H4 linked oligonucleosomes by immunoprecipitation with antiserum specific for acetylated histone ***(16)***; (3) genomic DNA is extracted from the isolated acetylated-H4 oligonucleosomes and is then amplified with PCR primers that are specific for the genomic DNA region of interest ***(2)***; and (4) PCR products are digested with parent-specific polymorphic restriction enzymes and are separated on polyacrylamide-urea or agarose gel to distinguish and to quantitate the status of histone acetylation between two parental alleles ***(2)***.

2. Materials

2.1. Tissue Collection

1. Ketamine HCl (Ketaset, 100 mg/mL, Burns Veterinary Supply, Rockville Centre, NY).
2. Xylazine, (Gemini SA, 20 mg/mL, Burns Veterinary Supply, Rockville Centre, NY).
3. Phosphate-buffered saline (PBS): Dissolve 8.0 g NaCl, 0.2 g KCl, 1.44 g Na_2HPO_4, and 0.24 g KH_2PO_4 in 800 mL distilled H_2O. Adjust pH to 7.4, add H_2O to 1 L, and autoclave for use.
4. 35 × 10 mm Petri dish.

5. 1-mL syringe.
6. Scissors and other surgical materials.

2.2. Cell Preparation

1. Dounce tissue grinders with loose and tight pestles (Daigger, Lincolnshire, IL).
2. Phosphate-buffered saline (PBS).
3. Nylon gauze (200 μm).
4. Beckman centrifuge.

2.3. Nucleus Preparation

1. Tris-buffered saline (TBS): 0.01 *M* Tris-HCl, pH 7.5, 3 m*M* $CaCl_2$, 2 m*M* $MgCl_2$, 5 m*M* sodium butyrate.
2. Buffer A (suspension buffer): 0.01 *M* Tris-HCl, pH 7.5, 3 m*M* $CaCl_2$, 2 m*M* $MgCl_2$, 5 m*M* sodium butyrate, 0.5% Tween-40, 0.5 m*M* phenylmethylsulfonyl fluoride (PMSF).
3. 25% sucrose.
4. 50% sucrose.
5. Buffer B (digestion buffer): 0.32 *M* sucrose, 50 m*M* Tris-HCl, pH 7.5, 4 m*M* $MgCl_2$, 1 m*M* $CaCl_2$, 0.1 m*M* PMSF, 5 m*M* sodium butyrate.
6. Microscope.

2.4. Oligonucleosome Preparation

1. Micrococcal nuclease (Amersham Pharmacia, Piscataway, NJ).
2. 250 m*M* ethylenediaminetetraacetic acid (EDTA).
3. Buffer C (lysis buffer): 1 m*M* Tris-HCl, pH 7.5, 0.2 m*M* Na_2-EDTA, 0.2 m*M* PMSF, 5 m*M* sodium butyrate.
4. Dialysis bag.
5. Phenol/chloroform.

2.5. Immunoprecipitation

1. Buffer D (incubation buffer): 50 m*M* NaCl, 20 m*M* Tris-HCl, pH 7.5, 20 m*M* sodium butyrate, 5 m*M* Na_2-EDTA, 0.1 m*M* PMSF.
2. Acetylated histone antiserum (Serological, Inc., Raleigh, NC).
3. Protein-Sepharose CL-4B (Amersham Pharmacia, Piscataway, NJ): Weigh 0.2 g of protein-Sepharose and wash with 20 mL distilled H2O. Centrifuge at 2000 rpm for 10 min and keep the gel at 4°C for immunoprecipitation.
4. Buffer E (washing buffer): 50 m*M* Tris-HCl, pH 7.5, 10 m*M* EDTA, 5 m*M* sodium butyrate.
5. 1 *M* NaCl.
6. 10% sodium dodecyl sulfate (SDS).

2.6. Genomic DNA Extraction

1. Phenol/chloroform.
2. 100% ethanol.

3. 75% ethanol.
4. 7.5 *M* ammonium acetate.

2.7. PCR Amplification

1. dNTP mix (2.5 m*M*): Dilute 10 m*M* dNTP (dATP, dTTP, dCTP, dGTP) mix (Gibco Life Technologies, Rockville, MD) to 2.5 m*M* dNTP with distilled water.
2. Tfl DNA polymerase (Epicentre Technologies, Madison, WI) (*see* **Note 1**).
3. $\gamma[^{32}P]$-ATP (Amersham Life Science, Arlington Heights, IL).
4. T4 polynucleotide kinase (10U/µL, BioLabs, Beverly, MA).
5. Liquid wax (Chill-out 14), (J Research, Boston, MA).
6. 0.2-mL 8-strip PCR tube.
7. PCR primer (20 µ*M*).
8. 3× PCR master mixture (for a final 100-µL reaction):
   Mix 10 µL 10× Tfl DNA polymerase buffer, 2 µL 2.5 mM 4-dNTP, 2.5 U Tfl Taq DNA polymerase, and add distilled H2O to make a final volume of 33 µL. The prepared PCR master mixture can be stored at 4°C for as long as 1 mo.
9. 96-well Programmable Thermal Controller (MJ Research, Watertown, MA).

2.8. Quantitation of Allelic Histone Acetylation

1. 2× restriction enzyme digestion mixture: For a single 6-µL reaction, add 0.6 µL of 10× restriction enzyme digestion buffer, 1 U of polymorphic restriction enzyme, and add H_2O to make a total of 3 µL. Diluted PCR product (3 µL) will be added later (**Subheading 3.8., step 3**) to make 6 µL. The digestion mixture can be made for multiple samples by using the same ratio.
2. 5% Polyacrylamide-urea gel or 2% agarose gel.
3. Electrophoresis apparatus.
4. PhosphoImager scanner or agarose gel scanner.
5. X-ray film.

3. Methods

3.1. Liver Tissue Collection

1. Place PBS on ice for 30 min.
2. Add 2 mL PBS to 35 × 10 mm Petri dishes, and put the dishes on ice.
3. Prepare ketamine/xylazine anaesthetizing solution by mixing 0.4 mL ketamine, 0.3 mL xylazine, and 0.7 mL distilled water in a 1.5-mL tube.
4. Anaesthetize animals by injecting 30 µL of the ketamine/xylazine mixed solution intraperitoneally.
5. After animals are completely unconscious, make an abdominal incision with scissors.
6. Remove the liver and place it in the 35 × 10 mm Petri dish containing ice-cold PBS.

3.2. Liver Cell Preparation

1. Mince the liver into small pieces with the scissors.
2. Transfer the minced liver pieces into a glass homogenizer.
3. Wash the Petri dish with 9 mL ice-cold PBS and combine this wash with the samples in the glass homogenizer.
4. Gently homogenize the liver pieces with the loose pestle, up and down 20 times, to release single cells.
5. Pass the cells through the nylon layer to remove fibrous connective tissues.
6. Centrifuge the samples at 600*g* at 4°C, for 15 min, to collect the cells.
7. Wash the cells two times with 10 mL ice-cold PBS.

If cultured cells are being used instead of whole tissues for the measurement of histone acetylation, detach the cells using an EDTA-trypsin solution, and then proceed from **step 5**.

3.3. Nucleus Preparation

1. Suspend the cells in 5 mL 1× TBS and transfer them to another glass homogenizing tube. Save 30 µL of the cell suspension in a 0.6-mL tube on ice at this step for comparison with the nuclei that will be collected later (**Subheading 3.3., step 8**).
2. Add 5 mL suspension buffer (buffer A). Homogenize the cells with a tight pestle up and down 20 times to release the nuclei from the cells.
3. Centrifuge the homogenates at 600*g* at 4°C, for 20 min.
4. Suspend the pellet containing the nuclei in 25% sucrose solution and gently transfer the suspension to a polynylon tube that has been underlayered with 50% sucrose for the gradient separation of the nuclei.
5. Centrifuge the suspension at 1500*g* at 4°C for 20 min to collect the nuclei in the pellet.
6. Wash the pellet once with 10 mL 25% sucrose solution and centrifuge again at 1500*g* at 4°C for 20 min.
7. Suspend the pellet in 0.5–1.0 mL digestion buffer (buffer B).
8. Take a small drop of this suspension of nuclei and compare it with the cells collected at **Subheading 3.3., step 1**, under a microscope. Nuclei will look like small shiny particles under the microscope, and they should differ from the shape of cells collected at **step 1** (*see* **Notes 2** and **3**).
9. Measure the DNA content of the nuclear extracts at 260 nm. The extracts collected at this step can be aliquoted and saved at –80°C for future use.

3.4. Oligonucleosome Preparation

1. Take a portion of the suspension of nuclei (20 µg of equivalent DNA) and dilute it to a concentration of 0.5 µg/µL with digestion buffer (buffer B). The total reaction volume will be 400 µL.

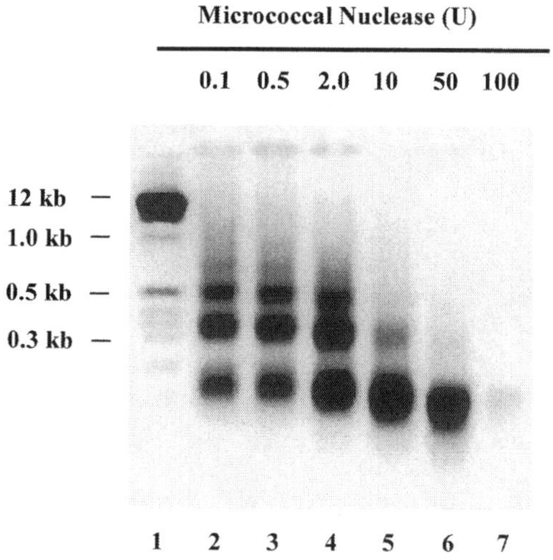

Fig. 1. Nucleosome DNA following digestion with micrococcal nuclease. Nuclei (equivalent to 20 µg DNA) were digested with various amounts of micrococcal nuclease, as indicated (0.1–100 U). After phenol/chloroform extraction, DNA was precipitated with ethanol and analyzed on a 2% agarose gel. Nucleosomes that were treated with the optimal nuclease concentration (0.5–2.0 U, lanes 3 and 4) were used for immunoprecipitation.

2. Add 0.5–2.0 U of micrococcal nuclease, and then incubate at 37°C for 5 min (*see* **Note 4**).
3. Terminate the nuclease reaction by adding 8 µL of 250 m*M* EDTA.
4. Centrifuge at 12,000*g* at 4°C, for 10 min.
5. Suspend the pellet in 100 µL lysis buffer (buffer C).
6. Transfer the suspension into a small dialysis bag and dialyze it against 2 L of lysis buffer (buffer C) at 4°C overnight.
7. Centrifuge at 12,000*g* and 4°C for 10 min. Aliquot and save the oligonucleosome supernatants at –80°C.
8. Take one aliquot of the supernatant to check the distribution of oligonucleosomes following **steps 9–11**.
9. Extract the oligonucleosome supernatant with an equal volume of phenol/chloroform.
10. Precipitate DNA by adding one-half volume of 7.5 *M* ammonium acetate and three volumes of absolute ethanol, then place at –20°C for 1 h. Centrifuge at 12,000*g* in a microcentrifuge and wash the pellet once with 75% ethanol.
11. Check DNA size on a 2% agarose gel. A proper nuclease digestion should produce an oligonucleosome DNA ladder, varying from 123 bp (mononucleosome) to 700 bp (pentonucleosomes) (**Fig. 1**).

3.5. Immunoprecipitation

1. Place aliquots of nuclease-digested oligonucleosomes (5 μg) into 0.6-mL tubes and dilute to 100 μL with incubation buffer (buffer D).
2. Add 0.0–4.0 μL acetylated histone antiserum (1 μg/μL). Mix and incubate on a rotation plate at 4°C overnight. (The sample without H4 antiserum will serve as the experimental control, which should give equal amounts of the two parental alleles in the PCR amplification).
3. Add 10 μL protein A-Sepharose gel solution and incubate at room temperature for 3 h.
4. Centrifuge at 12,000*g* at 4°C for 10 min to separate bound (acetylated) and unbound (unacetylated) nucleosomes.
5. Suspend the bound nucleosome pellet (acetylated form) in 100 μL washing buffer (buffer E) containing 50 m*M* NaCl.
6. Layer the suspension onto 900 μL buffer E and centrifuge at 600*g* and 4°C.
7. Wash the pellet with 500 μL buffer E containing 100 m*M* NaCl and then with 500 μL buffer E containing 150 m*M* NaCl.
8. Elute DNA twice with 60 μL incubation buffer (buffer D) containing 1% SDS. Combine eluates for PCR analysis. The eluates contain oligonucleosomes in which histone H4 is acetylated.

3.6. Genomic DNA Extraction

1. Extract eluates with an equal volume of phenol/chloroform.
2. Add one-half volume of 7.5 *M* ammonium acetate and precipitate DNA with three volumes of absolute ethanol using 5 μg glycogen as carrier.
3. Suspend the DNA pellet into 100 μL water in preparation for PCR amplification.

3.7. PCR Amplification of Parental Alleles

3.7.1. Primer End-Labeling

1. Into a 0.6-mL microcentrifuge tube, add 12 μL chill-out 14 liquid wax, 3 μL 20 m*M* 5′ or 3′ primer, 5 μL γ[^{32}P]-ATP, 1 μL 10× T4 polynucleotide kinase buffer, 1 μL 0.1 *M* DTT, and 1.0 μL T4 polynucleotide kinase (*see* **Notes 5** and **6**).
2. Mix, then spin for 5 s in microcentrifuge. Incubate the tube at 37°C for 15–30 min.
3. Inactivate T4 polynucleotide kinase by heating the tube at 99°C for 3 min.
4. Add 20 μL H_2O to dilute the end-labeled PCR primer. Mix the end-labeled PCR primer with an equal amount of the other cold (unlabeled) primer as the primer set for the following PCR amplification.

3.7.2. PCR Amplification

1. Into 0.2-mL 8-strip tubes, add 12 μL Chill-out 14 liquid wax, 1 μL PCR primer set, and 1 μL diluted DNA template.

2. Set the PCR program for an initial denature at 96°C for 1.5 min, followed by 30–35 cycles of amplification: denaturing at 95°C × 12 s, annealing at 65°C × 40 s, and extension at 72°C × 30 s.
3. Start the PCR program and pause PCR amplification at the first annealing stage (65°C) by using the PCR PAUSE function key.
4. Add 1 μL 3× PCR master mixture.
5. Release the PAUSE function key to continue PCR amplification.

3.8. Quantitation of Allelic Histone Acetylation

1. After PCR amplification, remove as much of the wax that has been added in PCR amplification as possible.
2. Add 8 μL distilled water to dilute the PCR products for restriction enzyme digestion.
3. Remove 3 μL of the diluted PCR-derived DNA and add 3 μL of restriction enzyme digestion mixture.
4. Incubate the reaction tubes at 37°C for 2–3 h in order to obtain complete digestion of the PCR products.
5. Add 12 μL of sequencing loading buffer.
6. Heat the samples at 96°C for 1.5 min and immediately put the samples on ice.
7. Load 3–5 μL of the sample solution onto a polyacrylamide-urea gel.
8. Run the electrophoresis at 1000 V until the first dye reaches the bottom of the gel.
9. Unload the gel, attach it to 3MM filter paper, and cover it with Saran wrap.
10. Expose the gel to X-ray film or to a PhosphoImager screen.
11. Scan the PCR bands and quantitate the relative densities of the two parental alleles that have been separated by the digestion of the polymorphic restriction enzyme (*see* **Note 6**). An example of differential histone acetylation in the insulin-like growth factor II receptor (*Igf2r*) promoter region is shown in **Fig. 2**.

4. Notes

1. Other PCR enzymes and PCR programs can also be used to amplify genomic DNA, depending on the preference of the PCR method and availability of DNA polymerase in each laboratory.
2. Under the microscope, nuclei should look like small uniform particles. Nuclei should be round and shiny when the focus is slightly changed. Tissue and cell debris will appear as irregular shapes and should not be present if pure suspensions of nucleus extracts have been achieved.
3. Other methods that stain specifically for nuclei, such as Hoechst Dye (Molecular Probes, Eugene, OR), can also be used to examine the nuclei collected.
4. Complete digestion of chromatin into mononucleosomes by micrococcal nuclease may not suitable for PCR amplification, especially when PCR primers cover a relatively long fragment. Thus, it will be necessary to achieve the optimal micrococcal nuclease digestion to produce oligonucleosomes varying from 1 to 5 nucleosomes, such that the interested region can be amplified easily by PCR.

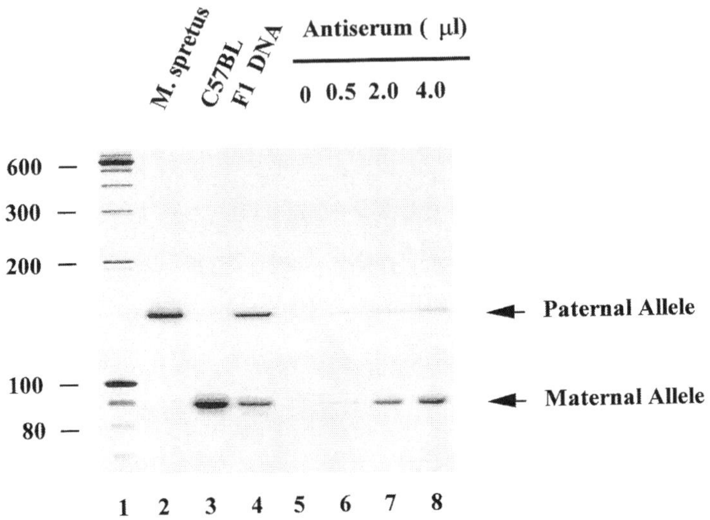

Fig. 2. Differential allelic histone acetylation of *Igf2r* promoter. Oligonucleosomes, prepared from the liver of backcross mice (F1 female × C57BL male), were immunoprecipitated with acetylated histone antiserum, as indiated (0.0–4.0 μL). DNA was extracted from immunoprecipitated (acetylated) nucleosomes and *Igf2r* sense RNA promoter regions were amplified with PCR primers. Two parental alleles were separated following digestion at the polymorphic Dde1 restriction enzyme site that is present in *M. spretus* but not in C57BL. Samples were: 100-bp DNA ladder (lane 1); PCR products amplified from genomic DNA isolated from *M. spretus* (lane 2); C57BL (lane 3); and F1 mice (*M. spretus* male × C57BL female) (lane 4); PCR products derived from F1 DNA incubated with increasing amounts of acetylated histone antiserum (lanes 5–8). Most, but not all, of the acetylated histone DNAs are associated with the expressed maternal *Igf2r* allele.

5. For primer end-labeling, the total reaction volume can be scaled up or down using appropriate ratios, depending on the volume of PCR reactions.
6. PCR amplification can also be performed using random incorporation of α[^{32}P]dCTP (which can be included in the PCR master mixture) in place of the labeled primer. Furthermore, PCR amplification can be performed without radioisotopes when PCR products are separable on an agarose gel. In this case, an agarose-gel scanner can be used for the quantitation of the two parental alleles. However, the relative amounts of the two parental alleles cannot be quantitated as accurately as when end-labeled primer PCR is used. The signal density from the unlabeled PCR on agarose gel or from random incorporation of α[^{32}P]dCTP is closely related to the length of the PCR products. In the end-labeled primer reaction, the PCR products are quantitated strictly on the basis of molecule number.

References

1. Jeppesen, P. (1997) Histone acetylation: a possible mechanism for the inheritance of cell memory at mitosis. *Bioessays* **19,** 67–74.
2. Hu, J.-F., Oruganti, H., Vu, T. H., and Hoffman, A. R. (1998) The role of histone acetylation in the allelic expression of the imprinted human insulin-like growth factor II gene. *Biochem. Biophys. Res. Commun.* **251,** 403–408.
3. Svensson, K., Mattsson, R., James, T. C., Wentzel, P., Pilartz, M., MacLaughlin, J., Miller, S. J., Olsson, T., Eriksson, U. J., and Ohlsson, R. (1998) The paternal allele of the H19 gene is progressively silenced during early mouse development: the acetylation status of histones may be involved in the generation of variegated expression patterns. *Development* **125,** 61–69.
4. Jeppesen, P. and Turner, B. M. (1993) The inactive X chromosome in female mammals is distinguished by a lack of histone H4 acetylation, a cytogenetic marker for gene expression. *Cell* **74,** 281–289.
5. Keohane, A. M., O'Neill, L. P., Belyaev, N. D., Lavender, J. S., and Turner, B. M. (1996) X-inactivation and histone H4 acetylation in embryonic stem cells. *Dev. Biol.* **180,** 618–630.
6. Buschhausen, G., Wittig, B., Graessmann, M., and Graessman, A. (1987) Chromatin structure is required to block transcription of the methylated herpes simplex virus thymidine kinase gene. *Proc. Natl. Acad. Sci. USA* **84,** 1177–1181.
7. Tazi, J. and Bird, A. (1990) Alternative chromatin structure at CpG islands. *Cell* **60,** 902–920.
8. Boyes, J. and Bird, A. (1991) DNA methylation inhibits transcription indirectly via a methyl-CpG binding protein. *Cell* **64,** 1123–1134.
9. Kass, S. U., Goddard, J. P., and Adams, R. L. P. (1993) Inactive chromatin spreads from a focus of methylation. *Mol. Cell Biol.* **13,** 7372–7379.
10. Fuks, F., Burgers, W. A., Brehm, A., Hughes-Davies, L., and Kouzarides, T. (2000) DNA methyltransferase Dnmt1 associates with histone deacetylase activity. *Nature Genet.* **24,** 88–91.
11. Nan, X., Ng, H. H., Johnson, C. A., Laherty, C. D., Turner, B. M., Eisenman, R. N., and Bird, A. (1998) Transcriptional repression by the methyl-CpG-binding protein MeCP2 involves a histone deacetylase complex. *Nature* **393,** 386–389.
12. Jones, P. L., Veenstra, G. J. C., Wade, P. A., Vermaak D., Kass, S. U., Landsberger, N., Strouboulis, J., and Wolffe, A. P. (1998) Methylated DNA and MeCP2 recruit histone deacetylase to repress transcription. *Nature Genet.* **19,** 187–191.
13. Wakefield, M. J., Keohane, A. M., Turner, B. M., and Graves, J. A. M. (1997) Histone underacetylation is an ancient component of mammalian X chromosome inactivation. *Proc. Natl. Acad. Sci. USA* **94,** 9665–9668.
14. Cuissett, L., Tichonicky, L., Jaffray, P., and Delpech, M. (1997) The effects of sodium butyrate on transcription are mediated through activation of a protein phosphatase. *J. Biol. Chem.* **272,** 24,148–24,153.
15. Pedone, P. V., Pikaart, M. J., Cerrato, F., Vernucci, M., Ungaro, P., Bruni, C. B., and Riccio, A. (1999). Role of histone acetylation and DNA methylation in

the maintenance of the imprinted expression of the *H19* and *Igf2* genes. *FEBS Lett.* **458,** 45–50.
16. O'Neill, L. P. and Turner, B. M. (1995) Histone H4 acetylation distinguishes coding regions of the human genome from heterochromatin in a differentiation-dependent but transcription-independent manner. *EMBO J.* **14,** 3946–3957.

23

Purification of the MeCP2/Histone Deacetylase Complex from *Xenopus laevis*

Peter L. Jones, Paul A. Wade, and Alan P. Wolffe

1. Introduction

DNA methylation has long been associated with stable transcriptional silencing and a repressive chromatin structure (reviewed in **refs.** *1,2*). Differential methylation is associated with imprinting, carcinogenesis, silencing of repetitive DNA, and allows for differentiating cells to efficiently shut off unnecessary genes. In vertebrates, where 60–90% of genomic CpG dinucleotides are methylated, methylation-dependent repression is vital for proper embryonic development *(3)*. Microinjection experiments using methylated DNA templates implicate chromatin structure as an underlying mechanism of methylation-dependent silencing *(4,5)*. Methyl-specific transcriptional repression requires chromatin assembly, and can be partially relieved by the histone deacetylase inhibitor Trichostatin A. In addition, several proteins have been identified that specifically bind to methylated DNA *(6–8)*. Two of these methyl-DNA binding proteins, MeCP1 and MeCP2, have been shown to mediate transcriptional repression *(6,7)*. MeCP1 is a relatively uncharacterized complex that requires at least 12 symmetrical methyl-CpGs for DNA binding *(6)*. MeCP2 is a single polypeptide containing a methyl-binding domain capable of binding a single methyl-CpG, and a transcriptional repression domain *(9)*. Recently MeCP2 was shown to interact with the Sin3 corepressor and histone deacetylase *(10,11)*. Changes in the acetylation state of the core histone tails correlates with changes in transcription (reviewed in **refs.** ***12,13)***, and several transcriptional repression complexes containing histone deacetylases have recently been described ***(10,14,15)***. These data provide a direct link between methyl-dependent transcriptional repression and the modification of chromatin

From: *Methods in Molecular Biology, vol. 181: Genomic Imprinting: Methods and Protocols*
Edited by: A. Ward © Humana Press Inc., Totowa, NJ

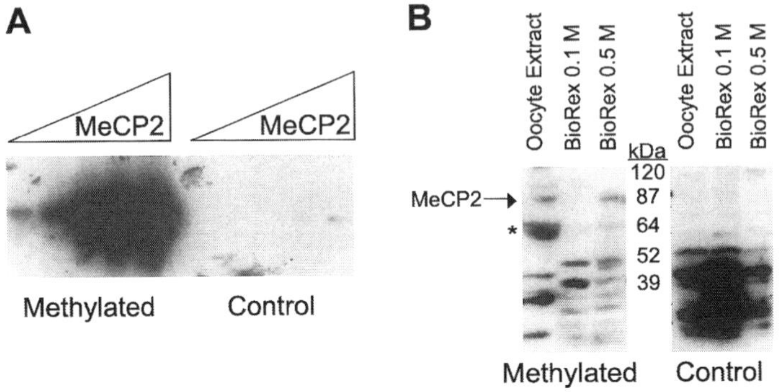

Fig. 1. Southwestern blot analysis of (**A**) recombinant *X. laevis* MeCP2 and (**B**) endogenous *X. laevis* MeCP2 shows the preference of MeCP2 binding for methylated DNA. (**A**) Increasing amounts of recombinant xMeCP2 (75 to 900 ng) hybridized with either a methylated (left) or control unmethylated (right) probe. (**B**) Fractionation of oocyte extract over BioRex70 resin hybridized with either a methylated (left) or control unmethylated (right) probe. *Xenopus* MeCP2 migrates at 87 kDa (* indicates a degradation product).

structure. Here, we describe techniques for purifying the MeCP2-contining histone deacetylase complex from *Xenopus laevis* oocytes.

Purification schemes for DNA-binding proteins often utilize the specific DNA-binding site sequences for the protein of interest in a DNA-binding assay. Southwestern analysis allows for the separation of multiple peptides in a sample that may bind to the same probe. Southwestern analysis is based on the ability of many proteins to be denatured with guanidine hydrochloride (G-HCl) and renatured such that the protein, or a portion thereof, refolds such that the DNA-binding activity is retained ***(16,17)*** (**Note 1**). Protein samples are separated by size on sodium dodecyl sulfate polyacrylamide gel electrophoresis (SDS-PAGE), immobilized to a membrane, denatured with G-HCl followed by renaturation, and then hybridized with a radiolabeled nucleic acid probe. MeCP2 was originally characterized from rat by the Southwestern technique *(7)* and was cloned from *X. laevis* (migrating at a molecular weight of 87 kDa on SDS-PAGE) *(10)*. Thus, MeCP2 can be accurately monitored by Southwestern assay by its very high binding preference for methylated DNA (**Fig. 1A,B**).

Due to the association of MeCP2 with a histone deacetylase complex, it is useful to follow histone deacetylase enzymatic activity. The purification of histone deacetylase complexes depends in large part on having a sensitive and reliable assay. We outline an assay that utilizes purified recombinant histone acetyltransferase and purified chicken erythrocyte histones to specifically label

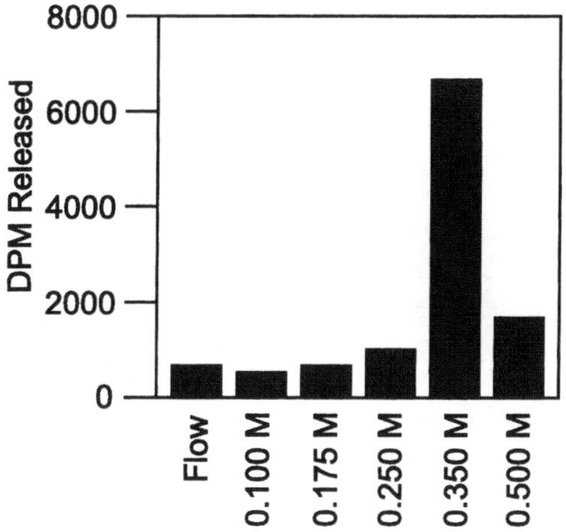

Fig. 2. Equal volumes of fractions from BioRex70 step-elutions of oocyte extract were assayed for deacetylase activity as described.

a desired histone to a high specific activity. This assay allows for monitoring the histone deacetylase activity of the MeCP2 complex during purification from oocyte extract.

The major advantage in using *X. laevis* oocyte extracts is that many chromatin components including MeCP2 are present in large quantities in storage forms, and these chromatin components can be extracted in low salt, preserving the integrity of the complexes *(18)*. Oocyte extracts prepared by this method contain robust histone deacetylase activity (**Fig. 2**; **refs.** ***10,14***). Using the assays described to monitor MeCP2 and histone deacetylase activities through the following chromatography protocols, the MeCP2-histone deacetylase complex can be purified.

2. Materials

2.1. Southwestern Oligo Preparation

1. Complementary oligonucleotides were synthesized (Operon Technologies) either with (oligos 3 and 4) or without (oligos 1 and 2) 5-methylcytosine (M) at each CpG residue for the following sequences *(7)*:
   Oligo 1: GATC$(CGA)_{12}$TC
   Oligo 2: GA$(TCG)_{12}$GATC
   Oligo 3: GATC$(MGA)_{12}$TC
   Oligo 4: GA$(TMG)_{12}$GATC
2. Elution buffer: 0.5 *M* ammonium acetate and 1 m*M* ethylenediaminetetraacetic acid (EDTA).
3. Kinase reagents: $[^{32}P]\gamma$-ATP (3000 Ci/mmol), and T4 polynucleotide kinase.

2.2. Southwestern Assay

1. 8% SDS gel with 4% stacking gel.
2. 2× SDS-PAGE loading buffer: 30 m*M* Tris-HCl, pH 6.8, 1% SDS, 5% glycerol, 2.5% β-mercaptoethanol, 0.1% bromophenol blue.
3. Nitrocellulose membrane (*see* **Note 2**).
4. SW transfer buffer: 25 m*M* Tris base and 190 m*M* glycine.
5. SW buffer: 20 m*M* HEPES (pH 7.9), 3 m*M* $MgCl_2$, 40 m*M* KCl, and 10 m*M* 2-mercaptoethanol.
6. SW buffer + 6 *M* guanidine hydrochloride (G-HCl) (avoid contact with skin).
7. Blocking buffer: SW buffer + 2% nonfat dried milk.
8. Binding buffer: SW buffer, 25-µg/mL sonicated native *Escherichia coli* DNA, 2-µg/mL denatured *E. coli* DNA, and 0.1% Triton X-100.
9. SW washing buffer: SW buffer + 0.01% Triton X-100.

2.3. In-Vitro Histone Acetylation

1. Purified chicken erythrocyte histones ***(19)***.
2. Acetylation buffer (1×): 25 m*M* Tris-HCl pH 8.0, 100 m*M* NaCl, 0.1 m*M* EDTA, 0.2% pheylmethylsulfonyl fluoride (PMSF), and 10% glycerol.
3. Recombinant Hat1p (*see* **Note 3**).
4. [^{3}H]Acetyl-coenzyme A (4.90 Ci/mmol) (Amersham Life Science).
5. Buffer A(200): 10 m*M* Tris-HCl (pH 8.0), 0.1 m*M* EDTA, and 200 m*M* NaCl.
6. Buffer A(2000): 10 m*M* Tris-HCl (pH 8.0), 0.1 m*M* EDTA, and 2 *M* NaCl.
7. BioRex70 resin (Bio-Rad, Inc).

2.4. Histone Deacetylase Assay

1. Deacetylase buffer (1×): 25 m*M* Tris-HCl pH 8.0, 10 % glycerol, 50 m*M* NaCl, and 1 m*M* EDTA.
2. [^{3}H] histone octamers (see above).
3. Deacetylase stop solution: 0.1 *M* HCl and 0.16 *M* acetic acid.
4. Ethyl acetate.

2.5. Oocyte Extract Preparation

2.5.1. Equipment

1. Female *Xenopus laevis*.
2. Dissection scissors and forceps.
3. SW-41Ti ultracentrifuge rotor and 12-mL tubes.

2.5.2. Reagents

1. OR-2 buffer: 5 m*M* HEPES (pH 7.9), 1 m*M* Na_2HPO_4, 82.5 m*M* NaCl, 2.5 m*M* KCl, and 1 m*M* $MgCl_2$.
2. Extraction buffer: 20 m*M* HEPES (pH 7.5), 5 m*M* KCl, 1.5 m*M* $MgCl_2$, 1 m*M* EGTA, 10% glycerol, 10 m*M* β-glycerophosphate, 0.5 m*M* dithiothreitol (DTT), 1 m*M* PMSF, 2-µg/mL pepstatin A, and 1-µg/mL leupeptin.

2.6. Chromatography

2.6.1. Equipment

1. BioRex 70 resin 100–200 mesh (Bio-Rad) equilibrated to Na^+ form.
2. Superose 6 HR 10/30 FPLC column (Pharmacia Biotech).
3. MonoQ Sepharose HR 5/5 or HR 10/10 FPLC column (Pharmacia Biotech).
4. HiTrap Heparin 1-mL column (Pharmacia Biotech).

2.6.2. Reagents

All buffers are at 4°C. DTT and protease inhibitors are added just prior to use. All buffers must be filtered through a 0.45-μm filter before use with the FPLC.

1. Buffer B(0): 20 m*M* HEPES, pH 7.5, 1.5 m*M* $MgCl_2$, 1 m*M* EGTA, 10 m*M* β-glycerophosphate, 10% glycerol, 0.5 m*M* DTT, 1 m*M* PMSF, 2-μg/mL pepstatin A, and 1-μg/mL leupeptin.
2. Buffer B(100): 100 m*M* NaCl, 20 m*M* HEPES, pH 7.5, 1.5 m*M* $MgCl_2$, 1 m*M* EGTA, 10 m*M* β-glycerophosphate, 10% glycerol, 0.5 m*M* DTT, 1 m*M* PMSF, 2-μg/mL pepstatin A, 1-μg/mL leupeptin.
3. Buffer B(500): 500 m*M* NaCl, 20 m*M* HEPES, pH 7.5, 1.5 m*M* $MgCl_2$, 1 m*M* EGTA, 10 m*M* β-glycerophosphate, 10% glycerol, 0.5 m*M* DTT, 1 m*M* PMSF, 2-μg/mL pepstatin A, 1-μg/mL leupeptin.
4. Buffer B(1000): 1 *M* NaCl, 20 m*M* HEPES, pH 7.5, 1.5 m*M* $MgCl_2$, 1 m*M* EGTA, 10 m*M* β-glycerophosphate, 10% glycerol, 0.5 m*M* DTT, 1 m*M* PMSF, 2-μg/mL pepstatin A, 1-μg/mL leupeptin.

3. Methods

3.1. Probe Preparation

1. Mix equimolar quantities of oligos 1 and 2 (20 μ*M* final) and make to 1× TE + 0.1 *M* KCl in a 500-μL Eppendorf tube.
2. Heat the oligos to 100°C for 2 min in a thermocyler and bring back to 30°C over 60 min, then immediately cool to 4°C.
3. Radiolabel using 4 pmol of duplex oligo (8-pmol ends) with 16 pmol ^{32}P-γ-ATP and T4 polynucleotide kinase for 60 min at 37°C.
4. Gel-purify the kinase reactions on a 5% native polyacrylamide gel run at 200 V for 2 h (*see* **Note 4**).
5. Excised the labeled duples from the gel with a razor blade and elute overnight in 0.5 mL elution buffer at 37°C with shaking.
6. Ethanol-precipitate the probe and resuspend in 1× TE (pH 8.0). This purification step removes all unincorporated label as well as any single-strand oligonucleotides.

3.2. Southwestern Blotting

Depending on the concentration of MeCP2 in the sample, a TCA precipitiation may have to be performed before running the SDS-PAGE (*see* **Note 5**).

1. The protein sample is made 1× in SDS-PAGE loading buffer and loaded directly onto an 8% SDS-PAGE gel with a 4% stacking gel and electrophoresced at 100 V in 4°C until the bromphenol blue reaches the botttom of the gel.
2. Transfer the proteins to nitrocellulose in 1 L SW transfer buffer at 4°C in a Mini Trans-blot transfer cell (Bio-Rad) for 5 h at 350 mA.
3. Remove the membranes from the cell and soak in SW buffer + 6 *M* G-HCl for 5 min with shaking at 4°C.
4. Renature the filters by four twofold dilutions with SW buffer for 5 min each at 4°C with shaking.
5. Wash once with straight SW buffer.
6. Block the filters for 10 min at room temperature with SW blocking buffer.
7. Rinse the filters once in SW buffer.
8. Hybridize in binding buffer with 2×10^5 CPM/mL labeled probe for 1 h at room temperature (*see* **Note 6**).
9. Wash the filters twice with SW buffer for 5 min at room temperature, air-dry on 3MM paper, and expose to film overnight with an intensifying screen.

3.3. In-Vitro Histone Acetylation

1. Incubate 1 mg core histones with 100 µg rHat1p and 100 µL [^{3}H] acetyl coenzyme A (Amersham) in 1× acetylation buffer in a final volume of 8.8 mL for 30 min at 37°C.
2. Chase by adding 100 nmol unlabeled acetyl coenzyme A and incubating for 30 min at 37°C.
3. Equilibrate a 1-mL BioRex70 column with Buffer A(200).
4. Load the acetylation reaction and allow it to enter the column by gravity.
5. Wash the column with 5 mL buffer A(200).
6. Elute the histones with careful addition of 5 mL buffer A(2000).
7. Collect the eluate in 0.5-mL fractions and assay by liquid scintillation (*see* **Note 7**).

3.4. Deacetylase Assay

1. Incubate a small volume of sample (1–10 µL) in a 200-µL reaction (made to 1× deacetylase buffer) with 1 µg acetylated histones, for 30 min at 30°C.
2. Add 50 µL stop solution.
3. Extracted the acetate from the reaction with 600 µL ethyl acetate.
4. Remove 450 µL of the organic phase to a liquid-scintillation vial.
5. Add liquid-scintillation fluid and count the samples in a liquid-scintillation counter.

3.5. Preparation of Oocyte Extracts (*see* Note 8)

1. Dissect ovaries from mature female *X. laevis* and rinse in OR-2 buffer.
2. Cut ovaries into small pieces and put into 50-mL conical tubes (15–20 mL ovary tissue per tube).

3. Rinse several times with OR-2.
4. Add fresh OR-2 is to a final volume of about 35 mL.
5. Add collagenase type II (Sigma Chemical) to 0.75 mg/mL.
6. Place the ovaries on a platform shaker and agitate for 60–90 min, until the oocytes are dispersed.
7. Wash the oocytes at least 10 times in OR-2 with rapid decanting to remove the immature oocytes and follicle cells.
8. Transfer the oocytes to SW-41 tubes (6 mL per tube).
9. Wash twice with extraction buffer, and fill to 12 mL with extraction buffer.
10. Centrifuge in a SW-41Ti rotor at 38,000 rpm (250,000*g*) for 1 h at 4°C.
11. Carefully remove the clear supernatant using a 21-gage needle.

3.6. Extract Fractionation

All procedures are performed at 4°C. Fractions are stored at –70°C.

3.6.1. BioRex 70 Fractionation (*see* **Note 9**)

1. Equilibrate BioRex 70 column (1 mL packed bed volume per 10 mg extract to be applied) in buffer B(100).
2. Apply extract and wash with three column volumes (cv) buffer B(100)
3. Elute bound protein with buffer B(500).

3.6.2. Superose 6 Gel Filtration

1. Equilibrate the Superose 6 HR 10/30 (Pharmacia) FPLC column in buffer B(500).
2. Filter BioRex 70 B(500) fraction through 0.45-μm syringe filter.
3. Load onto column at 2 mg protein in 500 μL volume.
4. Run FPLC at 0.1 mL/min and collect 250-μL fractions.
5. Assay fractions by Southwestern analysis for MeCP2 and by Western analysis for Sin3 (**Fig. 3**).

3.6.3. MonoQ Sepharose Fractionation

1. Dialyze the BioRex 500 m*M* elution against 200 vol buffer B(0) until the conductivity of the sample is <100 m*M* NaCl.
2. Centrifuge at 12,000*g* and 4°C for 20 min to remove insoluble material.
3. Equilibrate a MonoQ Sepharose (Pharmacia) HR10/10 (>20 mg protein to be applied) or HR5/5 (<20 mg protein to be applied) FPLC column in buffer B(100).
4. Apply the sample to the column, wash with 5 cv buffer B(100), and elute in a 20-cv linear gradient from buffer B(100) to buffer B(500).
5. Collect fractions of 0.5 cv (*see* **Note 10**).
6. Analyze fractions for histone deacetylase activity and by Western and/or Southwestern analysis for MeCP2. The MeCP2/histone deactylase complex elutes between 270 m*M* and 310 m*M* NaCl.

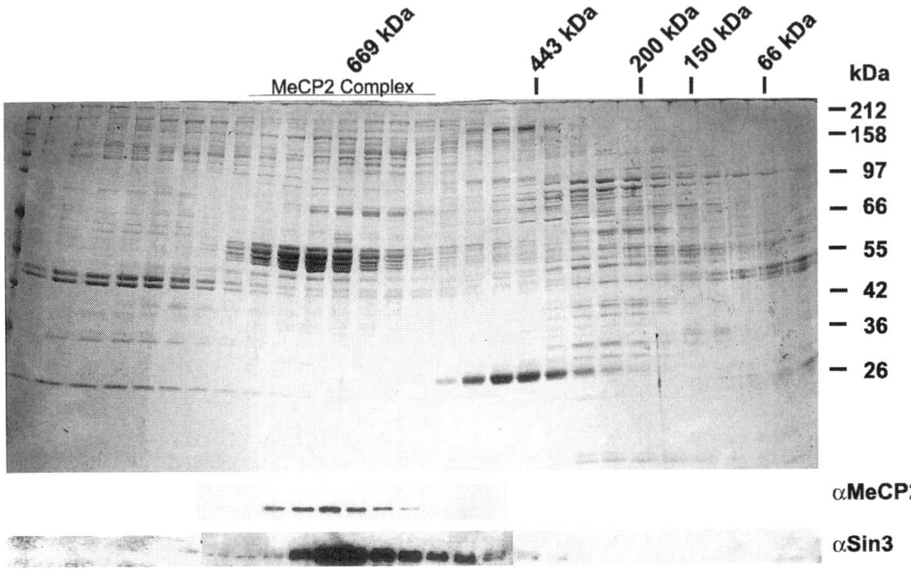

Fig. 3. Cofractionation of MeCP2 and Sin3 by gel filtration. Equal volumes of fractions from the BioRex70 high-salt (0.5 *M*) step elution separated over a Superose 6 HR 6/30 gel filtration column were assayed by Western blot for MeCP2 and Sin3 (lower), or stained with Coomassie blue (upper).

3.6.4. Heparin Fractionation

The fraction(s) from the MonoQ fractionation containing MeCP2 are used for further purification as follows.

1. Dialyze the sample against 200 vol buffer B(0) until the conductivity of the sample is <100 m*M* NaCl.
2. Filter the sample through a 0.45-µm syringe filter.
3. Equilibrate a 1-mL HiTrap heparin FPLC column (Pharmacia) in buffer B(100).
4. Load the sample onto the column, wash with 5 cv buffer B(100), and elute in a 20-cv linear gradient from buffer B(100) to buffer B(1000).
5. Fractions are collected in 0.4-mL volumes and assayed for deacetylase activity and by Western/Southwestern blotting for MeCP2 (**Fig. 4A**).

Using this purification protocol, a complex containing MeCP2, Sin3, and histone deacetylase activity, as well as several currently unidentified proteins (**Fig. 4B**), is routinely isolated from *Xenopus* oocytes.

4. Notes

1. It is important to remember that for MeCP2, the binding preference is for the symmetrically modified 5-methylcytosine CpG dinucleotide with no other

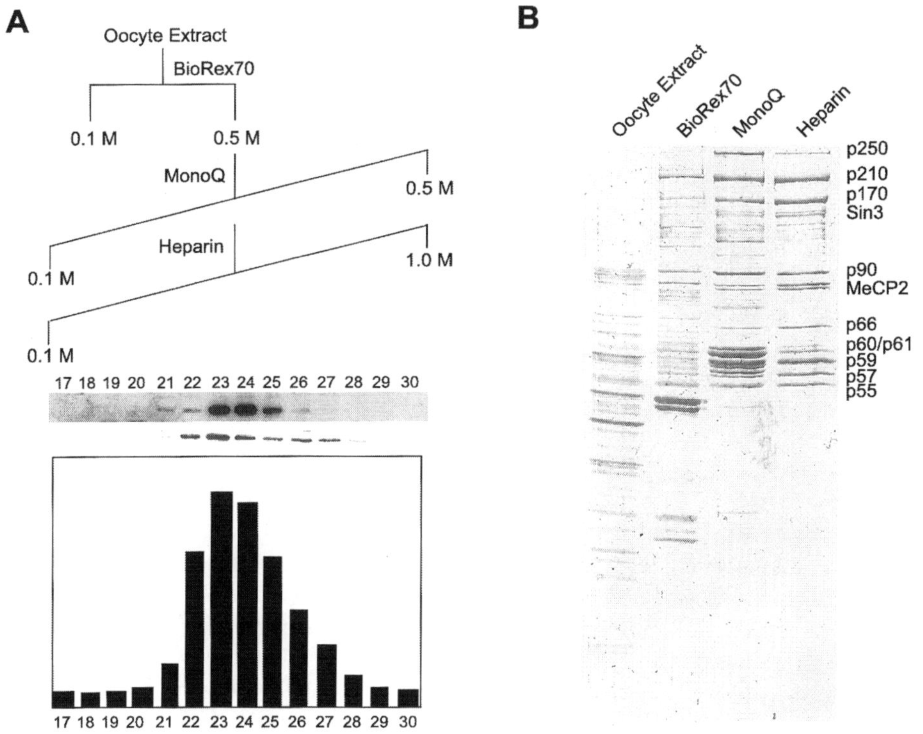

Fig. 4. MeCP2, Sin3, and histone deacetylase copurify. **(A)** Flow chart showing the fractionation of MeCP2 and Sin3 (upper). Fractions were assayed by Western blot for MeCP2 and Sin3 and for histone deacteylase activity. Assays for the final heparin fractionation are shown. **(B)** Coomassie blue stain of protein (7 µg) from the peak fractions at each step of purification shows that MeCP2 and Sin3 cofractionate with nine additional proteins.

sequence specificity. Therefore, assays must be done in duplicate with probes identical in sequences and differing only in their methylation status.

2. Alternatively, PVDF membrane can be used with an SDS-containing transfer buffer.
3. For the purification of the MeCP2-containing deacetylase complex, recombinant yeast Hat1p (generous gift of Dr. Mark. Parthun, Ohio State University) *(20)* was used to specifically acetylate histone H4. Other specificities can be obtained by using other histone acetyltransferases.
4. We have also successfully used G-25 spin columns (5Prime to 3Prime, Inc.) to purify the labeled oligonucleotides as per the manufacturer's instructions.
5. The sensitivity of the Southwestern assay depends on how efficiently the protein of interest is able to regain its DNA-binding ability after denaturation, immobilization, and renaturation. For MeCP2 from *Xenopus* oocyte extract, it is

necessary to precipitate a large protein sample (50 µg) in cold 20% trichloroacetic acid followed by a cold acidic acetone wash. For fractions, a smaller sample is needed because the MeCP2 becomes much more concentrated. The sensitivity of the assay for recombinant MeCP2 is about 25 ng.

6. The hybridization can easily be done in a plastic bag on a rotating platform to allow for smaller volumes of hybridization fluid and thus less probe is needed.
7. The extent of acetylation can be assayed by Triton acid urea gel ***(18)***. This method achieves a specific activity of 2000–5000 dpm per pmol histone H4.
8. We generally use 6 female *X. laevis* to do one oocyte extract preparation, yielding around 400 mg soluble protein. It is very important to wash the collagenase treated oocytes thoroughly. In addition to the collagenase itself being a protease, it is by no means a pure preparation and may contain additional protease activity. If proteolysis persists despite all efforts at protease inhibition, eggs may be used instead of oocytes and following the same extraction and purification scheme presented, starting with loading the eggs into the centrifuge tubes (**Subheading 3.5., step 8**). The reason for using oocytes is that they are easier than eggs to obtain in large numbers; however, eggs do not need the collagenase treatment.
9. All of the detectable MeCP2 elutes from the BioRex70 column between 250 m*M* NaCl and 375 m*M* NaCl. Performing this step elution increases the purification at this stage five- to eightfold and can be used for accommodating the smaller MonoQ HR5/5 at the next step.
10. The cv for MonoQ HR5/5 is 1 mL and for HR10/10 is 8 mL.

References

1. Kass, S. U., Pruss, D., and Wolffe, A. P. (1997) How does DNA methylation repress transcription? *Trends Genet.* **13,** 444–449.
2. Razin, A. (1998) CpG methylation, chromatin structure and gene silencing—a three-way connection. *EMBO J.* **17,** 4905–4908.
3. Tate, P., Skarnes, W., and Bird, A. P. (1996) The methyl-CpG binding protein MeCP2 is essential for embryonic development in the mouse. *Nature Genet.* **12,** 205–208.
4. Buschhausen, G., Wittig, B., Graessmann, M., and Graessmann, A. (1987) Chromatin structure is required to block transcription of the methylated herpes simplex virus thymidine kinase gene. *Proc. Natl. Acad. Sci. USA* **84,** 1177–1181.
5. Kass, S. U., Landsberger, N., and Wollfe, A. P. (1997) DNA methylation directs a time-dependent repression of transcription initiation. *Curr. Biol.* **7,** 157–165.
6. Meehan, R. R., Lewis, J. D., McKay, S., Kleiner, E. L., and Bird, A. P. (1989) Identification of a mammalian protein that binds specifically to DNA containing methylated CpGs. *Cell* **58,** 499–507.
7. Lewis, J. D., Meehan, R. R., Henzel, W. J., Maurer-Fogey, I., Jeppesen, P., Klein, F., and Bird, A. P. (1989) Purification, sequence, and cellular localization of a novel chromosomal protein that binds to methylated DNA. *Cell* **69,** 905–914.
8. Hendrich, B. and Bird, A. (1998) Identification and characterization of a family of mammalian methyl-CpG binding proteins. *Mol. Cell Biol.* **18,** 6538–6547.

9. Nan, X., Campoy, F. J., and Bird, A. P. (1997) MeCP2 is a transcriptional repressor with abundant binding sites in genomic chromatin. *Cell* **88,** 471–481.
10. Jones, P. L., Veenstra, G. J. C., Wade, P. A., Vermaak, D., Kass, S. U., Landsberger, N., Strouboulis, J., and Wolffe, A. P. (1998) Methylated DNA and MeCP2 recruit histone deacetylase to repress transcription. *Nature Genet.* **19,** 187–190.
11. Nan, X., Ng, H., Johnson, C. A., Laherty, C. D., Turner, B. M., Eisenman, R. N., and Bird, A. P. (1998) Transcriptional repression by the methyl-CpG-binding protein MeCP2 involves a histone deacetylase complex. *Nature* **393,** 386–389.
12. Struhl, K. (1998) Histone acetylation and transcriptional regulatory mechanisms. *Genes Dev.* **12,** 599–606.
13. Grunstein, M. (1997) Histone acetylation in chromatin structure and transcription. *Nature* **389,** 349–352.
14. Wade, P. A., Jones, P. L., Vermaak, D., and Wolffe, A. P. (1998) A multiple subunit Mi-2 histone deacetylase from *Xenopus laevis* cofractionates with an associated Snf2 superfamily ATPase. *Curr. Biol.* **8,** 843–846.
15. Lavinsky, R. M., Mullen, T. M., Soderstrom, M., Laherty, C. D., Torchia, J., et al., (1997) A complex containing N-CoR, mSin3 and histone deacetylase mediates transcriptional repression. *Nature* **387,** 43–48.
16. Miskimins, W. K., Roberts, M. P., McClelland, A., and Ruddle, F. H. (1985) Use of a protein-blotting procedure and a specific DNA probe to identify nuclear proteins that recognize the promoter region of the transferrin receptor gene. *Proc. Natl. Acad. Sci. USA* **32,** 6741–6744.
17. Vinson, C., LaMarco, K., Johnson, P., Landschulz, W., and McKnight, S. (1988) *In situ* detection of sequence-specific DNA binding activity specified by a recombinant bacteriophage. *Genes Dev.* **2,** 801–806.
18. Shimamura, A. and Worcel, A. (1989) The assembly of regularly spaced nucleosomes in the *Xenopus* oocyte S-150 extract is accompanied by deacetylation of histone H4. *J. Biol. Chem.* **264,** 14,524–14,530.
19. Chandler, S. P. and Wolffe, A. P. (1998) Analysis of linker histone binding to mono- and dinucleosomes, in *Methods in Molecular Biology: Chromatin Protocols* (Becker, P. B., ed.), Humana, London.
20. Parthun, M. R., Widom, J., and Gottschling, D. E. (1996) The major cytoplasmic histone acetyltransferase in yeast: links to chromatin replication and histone metabolism. *Cell* **87,** 85–94.

24

Reconstitution of Chromatin In Vitro

Kiyoe Ura and Yasufumi Kaneda

1. Introduction

1.1. Chromatin and DNA Methylation

It is now generally believed that DNA methylation is responsible for genomic imprinting in mammals *(1)*. Recent experimental evidence has provided an elegant mechanism for repression of gene expression by DNA methylation. This evidence suggests that proteins that recognize specifically methylated CpGs may contribute to the formation of inactive chromatin *(2–5)*. Nucleosomes are the basic unit of chromatin, consisting of a core of 146 bp of DNA wrapped around a histone octamer (two molecules of each of H2A, H2B, H3, and H4) and a stretch of linker DNA between adjacent nucleosomes. The binding of a fifth histone, known as a linker histone or H1, promotes the packaging of strings of nucleosomes into a 30 nm chromatin fiber *(6)*. Packaging of DNA into nucleosomes and the chromatin fiber greatly restricts the availability of the DNA for nuclear processes such as transcription.

It has been observed that a tight correlation exists between high levels of DNA methylation and specialized nuclease-resistant chromatin structure in cultured cells *(2,3)*. Recent studies demonstrating that the methylated DNA-binding protein MeCP2 is associated with histone deacetylases strongly suggests a functional link between DNA methylation and chromatin structure *(7,8)*. Control of DNA methylation may also be affected by higher-order chromatin structures *(9)*. Therefore, elucidation of structure–function relationships of DNA/protein interaction at the chromatin level is critical for the understanding of the regulatory events of gene expression.

A powerful approach for the analysis of structure–function relationships within chromatin has been the use of reconstituted model chromatin complexes.

From: *Methods in Molecular Biology, vol. 181: Genomic Imprinting: Methods and Protocols*
Edited by: A. Ward © Humana Press Inc., Totowa, NJ

For example, it has been reported that DNA methylation can alter the stability and the positioning of reconstituted mononucleosomes on the promoter of the fragile-X mental retardation gene ***(10)***, but not on the *Xenopus* 5S rRNA gene ***(11)***. Histone H1 binding to reconstituted chromatin is not affected by methylation of CpG using two different nucleosome assembly systems ***(11,12)***. However, under certain condition, a slight preference of this protein for naked methylated DNA can be detected ***(13)***.

1.2. Assembly of Chromatin in Vitro

There are several methods for assembly of chromatin on DNA fragments or plasmids. The simplest methods include: (1) the salt dialysis method, in which DNA and purified core histones are mixed and then dialyzed against a series of buffers starting from 2 *M* NaCl to low ionic strength; and (2) the histone octamer transfer method, in which histone octamers are redistributed from a purified donor chromatin to labeled specific DNA fragments during incubation in buffer with 1 *M* NaCl, followed by reduction of the salt concentration either by dialysis or stepwise dilution ***(14)***. In both methods the composition of the reconstituted chromatin is well defined and such chromatin is useful for analysis of nucleosome positioning due to intrinsic sequence-dependent DNA structure and assays of the binding of sequence-specific transcription factors or structural proteins to nucleosomal DNA. In vertebrates, core histones are highly conserved and, so far, any limited interspecies variation has had no functional significance at this level of analysis. Therefore, cultured cells or chicken erythrocytes are a convenient source of core histones or chromatin because of the low contamination of proteases and nucleases in these preparations.

Importantly, it should be noted that salt-mediated nucleosome assembly methods fail to space nucleosomes at the physiological nucleosomal repeat length of about 200 bp on general DNA fragments ***(14)***. Rather, multiple nucleosomes reconstituted by these methods are generally "closely packed," with one nucleosome every ~150 bp and little or no resulting linker DNA. However, salt dialysis methods can be used to produce physiologically spaced oligonucleosomes by employing DNA templates containing tandem repeats of strong nucleosome positioning sequences. For example, a template containing about 200-bp tandem repeats of a DNA fragment containing a *Lytechinus* 5S rRNA gene has been used to assemble properly spaced oligonucleosomes ***(15)***. Such oligonucleosomes have proven to be very useful systems for studying higher-order chromatin structures and transcription ***(16,17)***.

Alternative methods for assembly of chromatin employ crude cell extracts, or histone chaperones ***(14,18)*** at physiological ionic strength. We will not describe the detail of these methods here. It is notable that assembly of chromatin using cell extracts derived from *Xenopus* eggs ***(19)*** or *Drosophila* embryos ***(20)***, can produce regularly spaced chromatin in vitro, although the composition of the reconstitutes is complicated.

1.3. 5S Dinucleosome

A useful approach to investigate the relationship between chromatin structure and transcription has been to use short DNA fragments that are long enough to be competent for transcription but short enough for all aspects of their nucleoprotein organization to be determined. We established a physiologically spaced dinucleosome system in which both chromatin structural and transcriptional analyses were possible ***(21)***. Our dinucleosome DNA template is 424 bp long and is constructed from two tandem repeats of a DNA fragment containing a *Xenopus borealis* somatic 5S rRNA gene. This DNA contains intrinsic DNA structure sufficient to direct the translational and rotational positioning of a histone octamer with respect to the DNA sequence. The 5S gene encodes a short transcript (120 bp) and has an internal promoter located near one end of the nucleosome position. Linker histones can be incorporated onto the 5S dinucleosome without aggregation. The reconstituted dinucleosome system was used to demonstrate that spaced histone octamers could repress transcription partially and the addition of linker histone H1 or H5 established a completely repressive chromatin state by fixing histone-DNA contacts over essential promoter elements ***(21,22)***. Model dinucleosome templates showed no significant difference of nucleosome positioning or stability as a result of acetylation of the N-terminal core histone tails. However, transcriptional activity was significantly induced in response to acetylation ***(23)***.

This chapter describes the methods used for reconstitution of chromatin and characterization of reconstitutes in vitro, using the 5S dinucleosome as a model chromatin system.

2. Materials

2.1. Isolation and Labeling of DNA Fragments

1. Plasmid DNA (pX5S197-2, which contains the 424-bp *Xba* I-*Xho* I dinucleosome template).
2. Restriction enzymes (*Xba* I, *Xho* I).
3. Calf intestinal alkaline phosphatase.
4. T4 polynucleotide kinase.
5. [γ-^{32}P]ATP.
6. Sephadex G-50 spin column (e.g., Amersham Pharmacia).

7. 19/1 acrylamide/bisacrylamide, 6% (w/v) polyacrylamide gel containing 1× TBE (20 × 20 × 0.1 cm gel size).
8. 1× TBE: 90 m*M* Tris-HCl, 90 m*M* boric acid, 2.5 m*M* ethylenediaminetetraacetic acid (EDTA), pH 8.3.
9. Vertical DNA gel electrophoresis apparatus and power supply.
10. Equipment for autoradiography.
11. Extraction buffer: 10 m*M* Tris-HCl pH 8.0, 1 m*M* EDTA, 0.1% sodium dodecyl sulfate (SDS).
12. Micropure separator (Amicon, 0.22 µm pore size).
13. TE buffer: 10 m*M* Tris-HCl pH 8.0, 1 m*M* EDTA.

2.2. Preparation of Nuclei from HeLa Cells

1. Phosphate-buffered saline (PBS), pH 7.4.
2. Nuclei isolation buffer: 10 m*M* Tris-HCl pH 7.5, 1.5 m*M* $MgCl_2$, 1.0 m*M* $CaCl_2$, 0.25 *M* sucrose, 0.1 m*M* phenylmethylsulfonyl fluoride (PMSF).
3. 20% (v/v) Triton X-100.
4. Saturated NaCl-Urea: 5 *M* NaCl, 8 *M* urea, stir overnight and store at room temperature.
5. Dounce homogenizer (B pestle).

2.3. Preparation of Histone H1-Depleted Chromatin

1. Nuclei isolation buffer: 10 m*M* Tris-HCl pH 7.5, 1.5 m*M* $MgCl_2$, 1.0 m*M* $CaCl_2$, 0.25 *M* sucrose, 0.1 m*M* PMSF.
2. Micrococcal nuclease (Worthington). Dissolve at 15,000 U/mL in water and store frozen at –20°C.
3. 100 m*M* ethyleneglycol-bis-(2-amino ethylether-tetraacetic acid (EGTA).
4. Lysis buffer: 10 m*M* Tris-HCl, pH 6.85, 5 m*M* EDTA, 0.1 m*M* PMSF.
5. Protease inhibitor cocktail (Boehringer).
6. Dialysis bag (Spectrum, molecular weight cutoff (MWCO) 12,000–14,000).

Protocol A

7. Buffer A: 0.1 m*M* PMSF.
8. Buffer B: 1 *M* K-phosphate pH 6.8, 0,1 m*M* PMSF.
9. 20 mL hydroxyapatite column (Bio-Rad, Bio-Gel HTP gel).
10. FPLC system (Amersham Pharmacia).
11. TEP buffer: 10 m*M* Tris-HCl, pH 7.5, 1 m*M* EDTA, 0.1 m*M* PMSF.
12. Concentrator (e.g., Filtron or Amicon YM membrane, MWCO 30 K).

Protocol B

7. Phosphate buffer: 50 m*M* Na-phosphate pH 7.0, 0.15 *M* NaCl, 0.1 m*M* PMSF.
8. Cation-exchange resin AG 50W-X2 (Bio-Rad).
9. TEP buffer: 10 m*M* Tris-HCl, pH 7.5, 1 m*M* EDTA, 0.1 m*M* PMSF.

2.4. Preparation of Core Histones

2.4.1. Preparation of Core Histones Using Fast Performance Liquid Chromatography (FPLC) System

1. Buffer A: 50 m*M* Na-phosphate, pH 6.8, 0.1 m*M* PMSF.
2. Buffer B: 50 m*M* Na-phosphate, pH 6.8, 2 *M* NaCl, 0.1 m*M* PMSF.
3. Dialysis bag (Spectrum, MWCO 12,000–14,000).
4. Concentrator (e.g., Filtron or Amicon YM membrane, MWCO 30K and MWCO 10K).
5. 20 mL Hydroxyapatite column (Bio-Rad, Bio-Gel HTP Gel).
6. FPLC system (Amersham Pharmacia).

2.4.2. Further Purification and Concentration of Core Histones

1. Dialysis buffer: 0.2 *M* NaCl, 10 m*M* Tris-HCl, pH7.5, 1 m*M* EDTA, 0.1 m*M* PMSF.
2. Elution buffer: 2 *M* NaCl, 10 m*M* Tris-HCl, pH 7.5, 1 m*M* EDTA, 0.1 m*M* PMSF.
3. 2-mL CM52 column (Whatman) (5 mg histone/mL bed volume).
4. Dialysis bag (Spectrum, MWCO 6000–8000).

2.5. Reconstitution of Chromatin

2.5.1. Reconstitution of Chromatin with Purified Core Histones

1. Radiolabeled DNA fragments.
2. Carrier DNA (cold DNA fragment or plasmid).
3. Core histones.
4. TEMP buffer: 10 m*M* Tris-HCl, pH 7.5, 1 m*M* EDTA, 1 m*M* 2-mercaptoethanol, 0.1 m*M* PMSF.
5. 2 *M*, 1.5 *M*, 1 *M*, and 0.75 *M* NaCl in TEMP buffer.
6. Dialysis bag (Spectrum, MWCO 6000–8000, flat width 1 cm).

2.5.2. Reconstitution of Chromatin by the Histone Octamer Exchange Method

1. Radiolabeled DNA fragments.
2. Histone H1-depleted chromatin.
3. 5 *M* NaCl.
4. TEMP buffer: 10 m*M* Tris-HCl, pH 7.5, 1 m*M* EDTA, 1 m*M* 2-mercaptoethanol, 0.1 m*M* PMSF.
5. Dialysis bag (Spectrum, MWCO 6000–8000, flat width 1 cm).

2.6. Agarose Nucleoprotein Gel Analysis

1. 10% (v/v) glycerol.
2. 0.7% agarose gel containing 0.5× TBE (20 × 25 × 0.5 cm gel size).
3. TBE 1×: 90 m*M* Tris-borate, 2.5 m*M* EDTA, pH 8.3.

4. Horizontal DNA gel electrophoresis apparatus and power supply.
5. 3MM paper (Whatman).
6. Gel dryer.
7. Equipment for autoradiography.

2.7. Purification of Reconstituted Oligonucleosomes Using Sucrose Gradient

1. Sucrose solutions: 5% and 20% (w/v) in 10 m*M* Tris-HCl, pH 7.5, 1 m*M* EDTA, 0.1 m*M* PMSF.
2. Gradient maker: pump, gradient former, magnetic stirrer, and capillary or gradient master (BioComp).
3. Ultracentrifuge, swing rotor (e.g., Beckman SW41Ti) and tubes.
4. Equipment for fractionation: fraction collector, pump.
5. Nucleoprotein agarose gel (as in under **Subheading 2.6.**).
6. Microcon-30 (Amicon).

2.8. Binding of Linker Histones to Reconstituted Chromatin

1. Binding buffer: 10 m*M* Tris-Cl, pH 7.5, 50 m*M* NaCl, 0.1 m*M* EDTA, 3% (v/v) glycerol.
2. Nucleoprotein agarose gel (as under **Subheading 2.6.**).

2.9. DNaseI Footprinting Analysis

1. DNase I solution: Dilute DNase I stock solution (Boehringer, RNase free, 10 U/μL) to 10 U/mL in water, just before use.
2. 40 m*M* MgCl.
3. Stop solution: 50 m*M* EDTA, 30% (v/v) glycerol.
4. Nucleoprotein agarose gel (as under **Subheading 2.6.**).
5. Formamide loading buffer: 90% (v/v) formamide, 1× TBE, 10 m*M* EDTA, 0.005% bromophenol blue, 0.05% xylene cyanol.
6. 19/1 acrylamide/bisacrylamide, 6% (w/v) denaturing polyacrylamide gel containing 7 *M* urea, 1× TBE (31 × 38.5 × 0.04 cm gel size).
7. 1× TBE: 90 m*M* Tris-borate, 2.5 m*M* EDTA, pH 8.3.
8. Sequencing gel electrophoresis apparatus and power supply.
9. Gel dryer.
10. Equipment for autoradiography.

3. Methods

3.1. Isolation of 5′-End Radiolabeled DNA Fragments (see Note 1)

The 424-bp fragment from pS5S197-2 is long enough to accommodate two nucleosomes and contains two tandem repeats of the *Xenopus* 5S nucleosome positioning element. Templates to be used for nucleosome reconstitution are typically radiolabeled to allow easy monitoring of the extent of reconstitution on nucleoprotein gels, isolation of particles from sucrose gradients, and for subsequent footprinting analysis.

1. Digest ~20 µg pX5S197-2 plasmid DNA with *Xba* I and subsequently treat with calf intestinal alkaline phosphatase at 37°C for 60 min. Purify the DNA by ethanol precipitation and remove excess salt by gently rinsing the pellet with ice-cold 70% ethanol.
2. Radiolabel the 5′-end of restriction endonuclease cleavage site using T4 polynucleotide kinase and [γ- ^{32}P]ATP by standard methods. Remove unincorporated labeled nucleotide using a Sephadex G-50 microcolumn, followed by ethanol precipitation as above.
3. Digest radiolabeled DNA with *Xho* I, liberating the 424-bp end-labeled DNA fragment.
4. Isolate the end-labeled DNA fragment on a preparative 6% nondenaturing polyacrylamide gel in 1× TBE.
5. After autoradiography of the wet gel, excise the radiolabeled band and crush the gel slice in an Eppendorf tube with a siliconized glass rod or a pestle.
6. Add ~300 µL elution buffer to the crushed gel, resuspend, and incubate at 37°C to elute the end-labeled DNA fragment.
7. Filter elution buffer using a micropure separator to remove gel pieces, followed by extraction with phenol/chloroform/isoamyl alcohol and ethanol precipitation of the labeled DNA fragment. Resuspend in TE buffer.

3.2. Preparation of Nuclei from HeLa Cells

The following method is used to isolate nuclei not only from HeLa but also from other mammalian cell lines.

1. Harvest HeLa S3 cells from a 2-L spinner flask culture at about 5 × 10^5 cells/mL by centrifugation at 500*g* for 10 min at 4°C.
2. Wash the cell pellet by gentle resuspension in ice-cold PBS and repeat the centrifugation as above.
3. Wash the cell pellet again with ice-cold nuclei isolation buffer and resuspend the cell pellet in 50 mL ice-cold nuclei isolation buffer plus 0.5% (v/v) Triton X-100 with gentle agitation. Allow the cells to swell for 10 min on ice.
4. Homogenize the swollen cells on ice with 15 strokes of a Dounce homogenizer on ice. The homogenized cells (about 60 mL) are poured into two 50-mL conical tubes and centrifuged at 500*g* for 5 min at 4°C.
5. Discard the cloudy supernatants slowly and gently resuspend each nuclear pellet in about 5 mL of nuclei isolation buffer without Triton X-100. Add nuclei isolation buffer to the bring to the original volume (about 30 mL each) and spin down the nuclei again at 500*g* for 5 min at 4°C.
6. Repeat **step 5** at least twice, until the nuclei are pure white and supernatant becomes clear.

7. Resuspend each nuclear pellet gently in about 5 mL of nuclei isolation buffer and combine in one conical tube. Adjust the total volume to approximately 20 mL with nuclei isolation buffer. Remove a 5-µL aliquot and measure the absorbance at 260 nm in 1 mL saturated NaCl-urea using saturated NaCl-urea as a blank to quantify the amount of chromatin DNA (20 OD_{260nm} = 1 mg/mL DNA). The total yield should be about 30 mg of chromatin as DNA. Spin down the nuclei and carefully remove the supernatant. Freeze the nuclear pellet in liquid nitrogen and store at –80°C.

3.3. Preparation of Histone H1-Depleted Chromatin

In order to solubilize chromatin of isolated nuclei, the most efficient method is micrococcal nuclease digestion (*see* **Note 2**). Linker histones are removed from solubilized chromatin by chromatography on a hydroxyapatite FPLC column (protocol A) or by a batch method with a cation-exchange resin (protocol B) *(24)*.

1. Thaw the nucler pellet quickly at 37°C with gentle agitation.
2. Dilute the nuclear pellet to 2 mg/mL DNA with nuclei isolation buffer and preincubate at 35°C for 5 min with constant gentle agitation.
3. Add micrococcal nuclease (25 U/mg DNA) and digest for 10 min at 35°C with constant gentle agitation to break chromatin in the nuclei.
4. Stop the digestion by adding 100 m*M* EGTA to 2 m*M* final concentration. Spin down the nuclei at 500*g* for 5 min at 4°C and remove the supernatant.
5. To lyse the nuclei, resuspend the nuclear pellet in 10 mL of lysis buffer and vortex vigorously.
6. Transfer the lysed nuclei to a dialysis bag and dialyze against lysis buffer overnight at 4°C.
7. Spin out the nuclear debris at 10,000*g* for 10 min at 4°C and save the supernatant, which contains the solubilized chromatin (the chromatin ranges in size from 1 to 10 nucleosomes). Measure the absorbance of a small aliquot in water at 260 nm to determine the concentration of chromatin DNA. The yield should be about 50% of the original total DNA (*see* **Note 3**). Add the protease inhibitor cocktail to the solubilized chromatin.

Protocol A

8. Equilibrate hydroxyapatite in buffer A, after removing fine particles. Prepare a 20-mL hydroxyapatite FPLC column. Equilibrate at 4°C and wash the column with 200 mL of 1% buffer B (10 m*M* K-phosphate) at a flow rate of 8 mL/min using the FPLC system.
9. Apply no more than 15 mg (DNA content) of solubilized chromatin to the hydroxyapatite column at a flow rate of 4 mL/min in 1% buffer B. Wash the column with the same buffer over 20 min.
10. Use the following elution conditions: linear gradient 1–9% buffer B (10–90 m*M* K-phosphate) over 5 min, linear gradient 9–35% (90–350 m*M* K-phosphate) over

70 min (H1-depleted chromatin is eluted), linear gradient 35–100% (0.35–1 *M* K-phosphate) over 30 min (H1 is eluted), at a flow rate of 4 mL/min. The fraction collector is programmed to collect 8-mL fractions.

11. Analyze proteins in each fraction by an 18% (w/v) SDS/polyacrylamide gel electrophoresis (*see* **Note 4**). Collect H1-depleted chromatin.
12. Dialyze against TEP buffer overnight. Determine the concentration of H1-depleted chromatin DNA from the OD_{260} (20 OD_{260} = 1 mg/mL DNA). Concentrate chromatin to 0.3 mg/mL (DNA content) using a concentrator (MWCO 30K) for chromatin reconstitution, if necessary. Freeze the chromatin sample rapidly and store at –80°C (*see* **Note 5**).

Protocol B

13. Dialyze the solubilized chromatin against phosphate buffer at 4°C.
14. Add 4 mL cation-exchange resin AG 50W-X2 equilibrated with phosphate buffer to 20 mL of solubilized chromatin. Stir or shake gently at 4°C for 90 min.
15. Filter or decant the H1-depleted chromatin from the resin. Check proteins in an 18% (w/v) SDS/polyacrylamide gel (*see* **Note 4**).
16. Dialyze against TEP buffer overnight. Determine the concentration of H1-depleted chromatin DNA from the OD_{260} (20 OD_{260} = 1 mg/mL DNA). Freeze the chromatin sample rapidly and store at –80°C (*see* **Note 5**).

3.4. Preparation of Core Histones

3.4.1. Preparation of Core Histones Using FPLC System (see **Note 6**)

The most defined and useful source of material for reconstitution of chromatin is purified core histones. These can be prepared from either solubilized chromatin or H1-depleted chromatin using hydroxyapatite column chromatography. To keep the correct stoichiometry of all histones, it is recommended to prepare from H1-depleted chromatin by a single step elution with 2 *M* NaCl (*see* **Note 7**).

1. Dialyze the pooled H1-depleted chromatin against buffer A overnight at 4°C.
2. Equilibrate a 20-mL hydroxyapatite column with buffer A and apply H1-depleted chromatin at a flow rate of 2 mL/min, 0% buffer B (*see* **Note 8**). Wash the column with the same buffer conditions for 30 min.
3. Elute nonhistone chromatin proteins with 25% buffer B (0.5 *M* NaCl), at a flow rate of 2 mL/min over 30 min. Elute the core histones as octamers with 100% buffer B (2 *M* NaCl) over 15 min (collect 2-mL fractions). Determine the concentration of core histones in each fraction from the OD_{230} (OD_{230} = 4.2 for 1 mg/mL). Analyze the proteins of each fraction in an 18%(w/v) SDS/polyacrylamide gel (*see* **Note 4**).
4. Pool fractions containing core histones. Concentrate to more than 0.1 mg/mL for reconstitution of chromatin, if necessary, using a concentrator (MWCO 10K) or a CM52 column (**Subheading 3.4.2.**). Store at –20°C in siliconized tubes.

3.4.2. Further Purification and Concentration of Core Histones

1. Dialyze core histones against dialysis buffer at 4°C.
2. Prepare a 2-mL CM52 column (5 mg histone/mL bed volume) and equilibrate with dialysis buffer.
3. Apply core histones to the column, wash thoroughly with 15 mL dialysis buffer, and elute with 5 mL elution buffer. Measure the absorbance of a small aliquot in water at 230 nm to quantify the amount of histones. Store at –20°C in siliconized tubes.

3.5. Reconstitution of Chromatin

Dinucleosomes can be reconstituted onto 424-bp radiolabeled DNA fragments (*see* **Notes 9** and **10**) either by salt dialysis using purified core histones or by histone octamer transfer from donor H1-depleted chromatin. Usually, reconstitutes contain naked DNA, mono-, di-, and trinucleosome cores.

3.5.1. Salt Dialysis Using Purified Core Histones

1. Mix 5 μg of DNA (50–500 ng radiolabeled DNA and unlabeled carrier DNA) and 5 μg core histones in 75–100 μL of 2 *M* NaCl, TEMP buffer on ice, to give a histone-to-DNA ratio (w/w) of 1/1 (*see* **Notes 11** and **12**).
2. Transfer the mixture into a dialysis bag and dialyze at 4°C against 1 L TEMP buffer containing NaCl as follows: 2 *M* NaCl, overnight; 1.5 *M* NaCl, 1 h; 1 *M* NaCl, 4 h; 0.75 *M* NaCl, 4 h; and without NaCl, overnight.
3. Change to another 1 L TEMP buffer and dialyze for 4 h. Keep reconstituted chromatin samples in siliconized tubes on ice until use (*see* **Note 13**).

3.5.2. Octamer Transfer from Donor Chromatin (*see* **Note 14**)

1. Mix 50–500 ng radiolabeled DNA fragments with H1-depleted donor chromatin to give a chromatin DNA-to-fragment DNA ratio (w/w) of 50/1–100/1) in 80 μL TEME buffer followed by slow adjustment of NaCl concentration to 1 *M* with 20 μL 5 *M* NaCl.
2. Incubate at 37°C for 20 min and transfer the mixture into a dialysis bag. Dialyze at 4°C against 1 L TEMP buffer containing 1 *M* NaCl for 4 h and then dialyze against TEMP buffer containing 0.75 *M* NaCl for 4 h.
3. Finally, the samples are dialyzed against TEMP buffer overnight.
4. Change to another 1 L TEMP buffer and dialyze for 4 h. Keep reconstituted chromatin samples in siliconized tubes on ice until use (*see* **Note 13**).

3.6. Nucleoprotein Agarose Gel Analysis

The reconstitution efficiency and the number of nucleosome cores on a labeled DNA fragment can be analyzed by agarose gel electrophoresis.

1. Load samples in 3% (v/v) glycerol without dye onto a 0.7% agarose gel in 2 L 0.5X TBE.

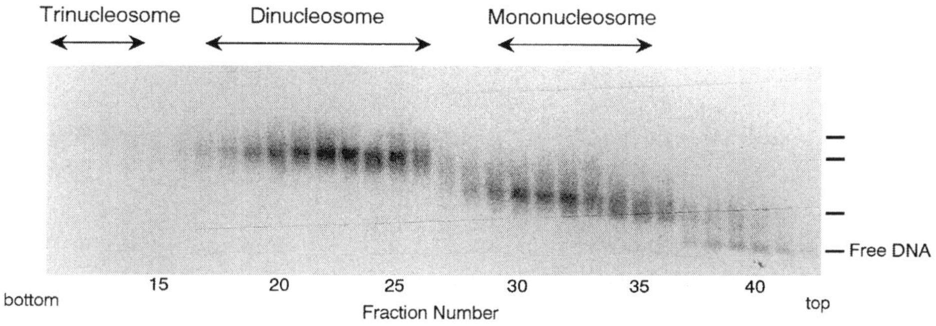

Fig. 1. Purification of reconstituted dinucleosomes using sucrose gradient. Reconstituted oligonucleosomes on dinucleosome DNA were fractionated on 5–20% sucrose gradients, and each fraction was analyzed by 0.7% nucleoprotein agarose gel electrophoresis (left is bottom fraction). The reconstitutes were separated into four bands as indicated by the bars at the right-hand side. The bottom band represents free DNA. Fractions containing mono-, di-, or trinucleosomes are indicated above.

2. Run the gel at room temperature at 130 V, for 3–4 h (less than 60 mA [*see* **Note 15**]). After electrophoresis, put the gel on two sheets of 3MM paper and start drying the gel without heating. Turn on the heater of the gel dryer after flattening the gel.
3. Subject to autoradiography.

3.7. Purification of Reconstituted Chromatin Using Sucrose Gradient (25)

Reconstituted oligonucleosomes are separated by sucrose gradient centrifugation depending on the number of histone octamers bound to the DNA fragment. We use 5–20% (w/v) sucrose linear gradients to isolate dinucleosomes from unreconstituted DNA and histones, mono-, and trinucleosomes.

1. Prepare 12 mL 5–20% (w/v) sucrose linear gradients using a gradient former or gradient master (*see* **Note 16**) at room temperature. Cool down gradients at 4°C and keep up to a few hours until use.
2. Load samples (less than 300 µL) on gradients and centrifuge at 36,000 rpm for 18 h at 4°C in a Beckman SW41Ti rotor.
3. Fractionate the gradients by pumping out from the bottom. The fraction collector is programmed to collect about 300 µL fractions. Analyze each fraction (10 µL each) by 0.7% nucleoprotein agarose gel electrophoresis as under **Subheading 3.6.** (**Fig. 1**).
4. Pool fractions containing mono-, di-, or trinucleosomes separately, concentrate to more than 5 µg/mL using Microcon-30, and dialyze against TEMP buffer

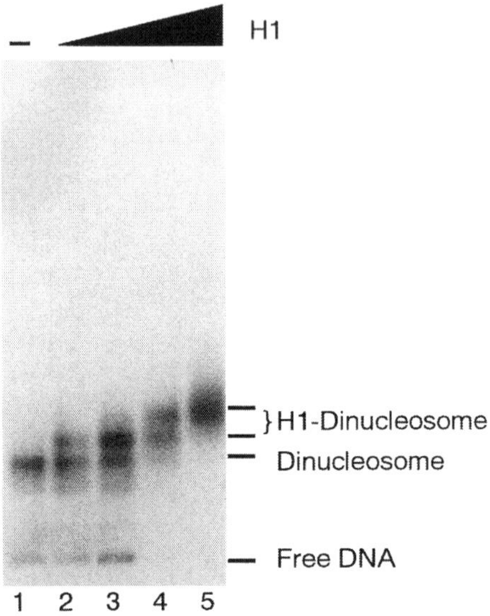

Fig. 2. Binding of histone H1 to reconstituted dinucleosomes. Reconstituted dinucleosomes were mixed with free DNA before various amounts of histone H1 were added, and analyzed by 0.7% nucleoprotein agarose gel electrophoresis. In lanes 1–5, 10 ng (DNA content) of reconstituted dinuleosome cores were mixed with 0, 1, 2, 4, and 8 ng of histone H1 from calf thymus, respectively *(27)*. The positions of free DNA, dinucleosome, and H1-dinucleosome, are indicated.

at 4°C to remove sucrose. Store samples in siliconized tubes on ice until use (*see* **Note 13**).

3.8. Binding of Linker Histones to Reconstituted Chromatin

Reconstituted short chromatin is a useful substrate to investigate binding of chromatin proteins—for example, linker histones—and sequence-specific DNA binding proteins to nucleosomal DNA *(11,21,26)*. By several criteria, the native interaction between histone H1 and the spaced dinucleosome can be recapitulated in vitro.

1. Incubate dinucleosomes (20 ng DNA content) with various amounts of purified histone H1 in 10 μL of binding buffer at 25°C for 20 min *(27)*.
2. Analyze histone H1 binding by loading the sample directly onto a 0.7% agarose nucleoprotein gel as under **Subheading 3.6.** (**Fig. 2**). The gel may be dried and subjected to autoradiography.

3.9. DNase I Footprinting Analysis

The DNase I cleavage pattern of reconstituted chromatin shows the positioning of nucleosome cores. Well-positioned nucleosome such as the 5S dinucleosome give typical 10 nucleotide cleavage ladders because the nuclease cuts preferentially where the DNA backbone is maximally exposed on the nucleosome surface.

1. End-labeled dinucleosomes (200 ng DNA content, >50,000 cpm) are adjusted to 2 m*M* $MgCl_2$ and digested with 1 µL DNase I solution at 25°C for 4 min. In a parallel experiment, naked DNA is digested with 1 µL of 10-fold-diluted DNase I. Digestion is stopped by addition of 1/10 volume of stop solution.
2. Transfer the reaction mixture to a 0.7% nucleoprotein agarose gel. After electrophoresis, wrap the wet gel with plastic and subject to autoradiography at 4°C for several hours.
3. Excise nucleoprotein complexes from the wet gel. Isolate DNA fragments, ethanol precipitate, and resuspend in 5 µL formamide loading buffer.
4. Apply samples on a 6% denaturing polyacrylamide gel. Run at 80 W for 1–2 h. After electrophoresis, put the gel on two sheets of 3MM paper and dry the gel.
5. Subject to autoradiography (**Fig. 3**).

4. Notes

1. A suitable DNA length for mononucleosome reconstitution is probably 200–270 bp ***(14)***. The 3′-end of the DNA is labeled using Klenow DNA polymerase and [α-^{32}P]-NTPs.
2. Sonication of isolated nuclei in ice-cold 0.2 m*M* EDTA, pH 7.5, 0.25 m*M* PMSF, is also a possible way to break chromatin. Insufficient sonication results in loss of most of the chromatin in the pellet of nuclear debris.
3. Yield depends on the conditions during nuclease digestion of nuclei. The length of digested chromatin DNA should be checked by 1% agarose gel electrophoresis. It should be approximately 146 bp to 2 kb.
4. 37.5/1 acrylamide/bisacrylamide, 18% (w/v) SDS/polyacrylamide separating gel containing 0.375 *M* Tris-HCl, pH 8.8, and 4% (w/v) SDS/polyacrylamide stacking gel containing 0.125 *M* Tris-HCl, pH 6.8.
5. In order to purify core histones from H1-depleted chromatin, skip this dialysis step.
6. It is not necessary to use the FPLC system, but column chromatography without an FPLC system will take a long time.
7. In the case that solubilized chromatin is applied to the hydroxyapatite column, wash out H1 first with 0.6 *M* NaCl buffer. During this washing, some H2A/H2B are lost and ruin the core histone stoichiometry. Therefore, it is necessary to elute

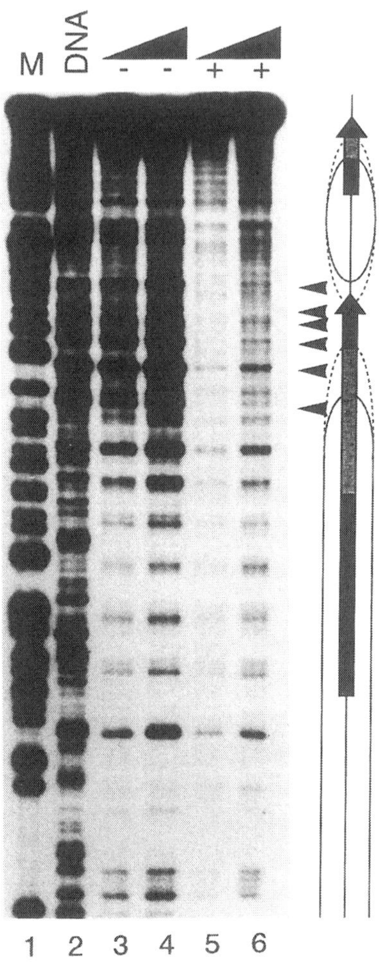

Fig. 3. DNase I footprinting of dinucleosomes. Bound or unbound histone H5 dinucleosomes were prepared and digested with DNase I. Individual complexes were isolated by 0.7% nucleoprotein agarose gel electrphoresis. DNA from these complexes was isolated and analyzed by 6% denaturing polyacrylamide gel electrophoresis. The 3′-end-radiolabeled coding strand of the 5S dinucleosome DNA is used as a template. Lane 1 contains a G-specific cleavage reaction used as markers (M). Digestion of naked DNA (lane 2), of dinucleosomes (lanes 3 and 4), and of dinucleosomes containing histone H5 (lanes 5 and 6) is shown, as indicated at the top. Filled triangles indicate increasing tamounts of DNase I (twofold) for digestion. The large vertical arrows show the location and orientation of 5S RNA gene. Solid and dotted ovals indicate the regions contacted by the nucleosome cores and nucleosomes containing H5, respectively. Arrowheads highlight selected positions protected around linker DNA after binding of histone H5 *(21)*.

H2A/H2B and H3/H4 separately, with 1 *M* NaCl and with 2 *M* NaCl, respectively ***(28)***, and adjust histone stoichiometry again for chromatin reconstitution.

8. The total volume of H1-depleted chromatin prepared by protocol A should be large (~50 mL). In this case, apply sample chromatin to a column using 50 mL super loop (Amersham Pharmacia) or after concentrating H1-depleted chromatin to 1 mg/mL using a concentrator (MWCO 30K).
9. It is possible to use biotinylated DNA fragments that are generated by PCR using 5′-unlabeled and 5′-biotinylated primers ***(29)***. Biotin-labeled dinucleosome are detected by streptavidin conjugated with alkaline phosphatase and chemiluminescent substrate, e.g., CPD-Star (Boehringer), on nylon membrane after nucleoprotein gel electrophoresis. Biotinylated nucleosome is not suitable for footprinting analysis. It is also possible to make use of DNA fragments that contain unique DNA damage sites or DNA methylation sites at specific positions. They can be generated by using single-stranded DNA and modified oligonucleotides ***(30)***.
10. When a high concentration of specific chromatin templates is necessary for a subsequent assay (e.g., in vitro transcription assay), use cold specific DNA fragments as carrier DNA.
11. The nucleosome contains 200 bp DNA (MW 130,000) and a histone octamer (MW 108,000), so that the ratio of histone to DNA (w/w) is almost equal to the molar ratio of histone octamer to nucleosome DNA.
12. Reconstitution efficiency is affected by the length and structure of both labeled and carrier DNA. It is recommended that at least three different ratios of histone to DNA (w/w), for example, 0.6/1, 0.9/1, 1.2/1, are tested in the first reconstitution to find the best condition. A ratio that is too high (>2) makes aggregates. Also, it is not recommended that either the volume is below 75 µL or the DNA concentration is below 50 µg/mL.
13. Reconstituted chromatin should be stable for a few months under these conditions.
14. It should be noted that chromatin reconstituted by this method contains a large amount of nonspecific donor chromatin.
15. Reconstituted nucleosomes are destabilized by heating the gel during electrophoresis. Do not run the gel at current in excess of 2.5 mA/cm.
16. Six gradients may be prepared simultaneously using the gradient master in a short time.

Acknowledgments

The 5S dinucleosome system was established in Dr. Alan Wolffe's laboratory, NIH. We thank Dr. A. Wolffe for valuable comments, Dr. Jeffrey Hayes for helpful discussions and useful comments on the manuscript, and Drs. Y. Ohkuma and S. Aota for critical reading of the manuscript.

References

1. Jaenisch, R. (1997) DNA methylation and imprinting: why bother? *Trends Genet.* **13,** 323–329.

2. Kass, S. U., Pruss, D., and Wolffe, A. P. (1997) How does DNA methylation repress transcription? *Trends Genet.* **13,** 444–449.
3. Ng, H. H. and Bird, A. (1999) DNA methylation and chromatin modification. *Curr. Opin. Genet. Dev.* **9,** 158–163.
4. Graessmann, M. and Graessmann, A. (1993) DNA methylation, chromatin structure and the regulation of gene expression, in *DNA Methylation: Molecular Biology and Biological Significance* (Jost, J. P. and Saluz, H. P., eds.), Birkhauser-Verlag, Basel, pp. 404–424.
5. Kass, S. U., Landsberger, N., and Wolffe, A. P. (1997) DNA methylation directs a time-dependent repression of transcription initiation. *Curr. Biol.* **7,** 157–165.
6. Wolffe, A.P. (1995) *Chromatin: Structure and Function.* Academic, San Diego, CA.
7. Nan, X., Ng, H. H., Johnson, C. A., Laherty, C. D., Turner, B. M., Eisenman, R. N., and Bird, A. (1998) Transcriptional repression by the methyl-CpG-binding protein MeCP2 involves a histone deacetylase complex. *Nature* **393,** 386–389.
8. Jones, P. L., Veenstra, G. J., Wade, P. A., Vermaak, D., Kass, S. U., Landsberger, N., Strouboulis, J., and Wolffe, A.P. (1998) Methylated DNA and MeCP2 recruit histone deacetylase to repress transcription. *Nature Genet.* **19,** 187–191.
9. Jeddeloh, J. A., Stokes, T. L., and Richards, E. J. (1999) Maintenance of genomic methylation requires a SWI2/SNF2-like protein. *Nature Genet.* **22,** 94–97.
10. Godde, J. S., Kass, S. U., Hirst, M. C., and Wolffe, A. P. (1996) Nucleosome assembly on methylated CGG triplet repeats in the fragile X mental retardation gene 1 promoter. *J. Biol. Chem.* **271,** 24,325–24,328.
11. Nightingale, K. and Wolffe, A.P. (1995) Methylation at CpG sequences does not influence histone H1 binding to a nucleosome including a *Xenopus borealis* 5 S rRNA gene. *J. Biol. Chem.* **270,** 4197–4200.
12. Campoy, F. J., Meehan, R. R., McKay, S., Nixon, J., and Bird, A. (1995) Binding of histone H1 to DNA is indifferent to methylation at CpG sequences. *J. Biol. Chem.* **270,** 26,473–26,481.
13. McArthur, M. and Thomas, J. O. (1996) A preference of histone H1 for methylated DNA. *EMBO J.* **15,** 1705–1714.
14. Rhodes, D. and Laskey, R. A. (1989) Assembly of nucleosomes and chromatin. *Methods Enzymol.* **170,** 575–585.
15. Simpson, R. T., Thoma, F., and Brubaker, J. M. (1985) Chromatin reconstituted from tandemly repeated cloned DNA fragments and core histones: a model system for study of higher order structure. *Cell* **42,** 799–808.
16. Tse, C., Sera, T., Wolffe, A. P., and Hansen, J. C. (1998) Disruption of higher-order folding by core histone acetylation dramatically enhances transcription of nucleosomal arrays by RNA polymerase III. *Mol. Cell Biol.* **18,** 4629–4638.
17. Steger, D. J., Eberharter, A., John, S., Grant, P. A., and Workman, J. L. (1998) Purified histone acetyltransferase complexes stimulate HIV-1 transcription from preassembled nucleosomal arrays. *Proc. Natl. Acad. Sci. USA* **95,** 12,924–12,929.
18. Ito, T., Bulger, M., Pazin, M. J., Kobayashi, R., and Kadonaga, J. T. (1997) ACF, an ISWI-containing and ATP-utilizing chromatin assembly and remodeling factor. *Cell* **90,** 145–155.

19. Almouzni, G. (1998) Assembly of chromatin and nuclear structures in *Xenopus* egg extracts, in *Chromatin: A Practical Approach* (Gould, H., ed.), Oxford University Press, New York.
20. Becker, P. B., Tsukiyama, T., and Wu, C. (1994) Chromatin assembly extracts from *Drosophila* embryos. *Methods Cell Biol.* **44,** 207–223.
21. Ura, K., Hayes, J. J., and Wolffe, A. P. (1995) A positive role for nucleosome mobility in the transcriptional activity of chromatin templates: restriction by linker histones. *EMBO J.* **14,** 3752–3765.
22. Ura, K., Nightingale, K., and Wolffe, A. P. (1996) Differential association of HMG1 and linker histones B4 and H1 with dinucleosomal DNA: structural transitions and transcriptional repression. *EMBO J.* **15,** 4959–4969.
23. Ura, K., Kurumizaka, H., Dimitrov, S., Almouzni, G., and Wolffe, A. P. (1997) Histone acetylation: influence on transcription, nucleosome mobility and positioning, and linker histone-dependent transcriptional repression. *EMBO J* **16,** 2096–2107.
24. Thoma, F. and Koller, T. (1977) Influence of histone H1 on chromatin structure. *Cell* **12,** 101–107.
25. Noll, H. and Noll, M. (1989) Sucrose gradient techniques and applications to nucleosome structure. *Methods Enzymol.* **170,** 55–116.
26. Owen-Hughes, T. and Workman, J.L. (1994) Experimental analysis of chromatin function in transcription control. *Crit. Rev. Eukaryot. Gene. Expr.* **4,** 403–441.
27. Hayes, J. J. and Wolffe, A. P. (1993) Preferential and asymmetric interaction of linker histones with 5S DNA in the nucleosome. *Proc. Natl. Acad. Sci. USA* **90,** 6415–6419.
28. Ura, K. and Wolffe, A. P. (1996) Reconstruction of transcriptionally active and silent chromatin. *Methods Enzymol.* **274,** 257–271.
29. Sato, M., Ura K., Hohmura, I., Tokumasu, F., Yoshimura, S., Hanaoka, F., and Takeyasu, K. (1999) Atomic force microscopy sees nucleosome positioning and histone H1-induced compaction in reconstituted chromatin. *FEBS Lett.* **452,** 267–271.
30. Ura, K., Araki, M., Saeki, H., Masutani, C., Ito, T., Iwai, S., et al. (2001) ATP-dependent chromatin remodeling facilities nucleotide excision repair of UV-induced DNA lesions in synthetic dinucleosomes. *EMBO J.* **20,** 2004–2014.

25

Genomic Imprinting in Plants

Rinke Vinkenoog, Melissa Spielman, Sally Adams, Hugh G. Dickinson, and Rod J. Scott

1. Imprinting in Plants Affects the Endosperm but Not the Embryo

Genomic imprinting, though most extensively studied in mammals, has long been known to perform an important role in seed development in flowering plants. In this chapter, an overview of what is known to date about genomic imprinting in flowering plants and how this knowledge came into being will be given.

Flowering plants (Angiosperms) are unique in that their initial development requires a double fertilization. After pollination, one sperm nucleus fertilizes the egg cell, giving rise to the embryo. A second sperm nucleus from the same pollen fuses with the two central cell nuclei of the female gametophyte. The triploid cell that results from this fertilization event will develop into a distinct tissue, the endosperm. As in most species both the central cell and the egg originate from a single meoitic product, they are genetically identical. The same holds true for the two sperm in the pollen. As a consequence, embryo and endosperm are genetically identical, their only difference being that the embryo is diploid, containing one maternal and one paternal set of chromosomes, while the endosperm is triploid, with two sets of maternal and one set of paternal chromosomes. Despite their common ancestry, embryo and endosperm develop along very different pathways.

One of the most dramatic consequences of imprinting in mammals is that embryos with a uniparental genetic contribution are inviable. Parthenogenetic mouse embryos, which are derived by the activation of an oocyte and therefore contain only a maternal genome, never develop to term *(1)*. Androgenetic embryos, derived by nuclear transfer of two male gamete nuclei, and gynoge-

From: *Methods in Molecular Biology, vol. 181: Genomic Imprinting: Methods and Protocols*
Edited by: A. Ward © Humana Press Inc., Totowa, NJ

netic embryos, derived by the fusion of two female gamete nuclei, also result in abortion *(2–4)*. The phenotypes of all three types of embryo in mice show a parent-of-origin effect. In parthenogenetic and gynogenetic mice, the embryo proper is relatively well developed, but the extraembryonic tissues are poorly developed or even absent. In androgenones, on the other hand, the extraembryonic tissues are well developed *(3–5)*. The underlying reasons for the differences between androgenetic and parthenogenetic/gynogenetic cells have been extensively studied in chimeric embryos. In chimeras between parthenogenetic and androgenetic cells, the parthenogenetic cells are confined to the embryo while androgenetic cells constitute the bulk of the extraembryonic trophoblast. Both cell types contribute to the yolk sac *(6,7)*. Androgenetic cells have a higher proliferation rate than gynogenetic or parthenogenetic cells. While the presence of androgenetic cells in chimeras increases the weight of the embryo, the presence of parthenogenetic cells decreases embryo weight by 30–50% *(8,9)*. Thus, even though paternally derived cells are located in the extraembryonic tissues, their presence leads to an increase in embryo weight.

In contrast to the situation in mammals, in flowering plants embryos with a genome inherited from either the seed parent or the pollen parent are viable in many species. Parthenogenesis has long been known to occur in, for instance, dandelion (*Taraxacum sp.*) and hawkweed (*Hieracium sp.*) *(10–12)* and in buttercup (*Ranunculus sp.*) *(13)*. In agamospermous *Taraxacum* microspecies, a defective meiosis leads to an oocyte with an unreduced number of chromosomes *(14–16)*. In some hawkweed species the megaspore is replaced by a cell of the nucellus or the chalaza *(17)*. Gynogenesis has been reported for many species, including commercially important plants such as onion (*Allium cepa*), durum wheat (*Triticum tugidum*), sugar beet (*Beta vulgaris*), and pot gerbera (*Gerbera jamesonii*) *(18–23)*. Similarly, plants with only a paternal genome contribution can be derived via the process of anther culture. By culturing anthers or pollen on appropriate media, pollen cells can develop into haploid embryoids and eventually grow into plants. In most species, androgenetic plants are derived from the vegetative cell of the pollen. This has been reported in, for example, oil seed rape (*Brassica napus*), maize (*Zea mais*), tulip (*Tulipa gesneriana*), wheat (*Triticum aestivum*), and barley (*Hordeum vulgare*) (***24–28***; reviewed in ***29***). Androgenesis via the sperm of the pollen is less frequently observed, but has been reported for henbane (*Hyoscamus niger*) and carrot (*Daucus carota*) *(30,31)*. One has to bear in mind that most flowering plants are hermaphrodite and thus act as both seed and as pollen parents. With regard to this, it can be argued whether embryos derived from the vegetative pollen cell (which is not a gamete insofar as it is never involved in fertilization) can be called true androgenones. However, the fact that plants obtained from haploid or diploid embryos derived from one parent are viable indicates that, in contrast to mammals, genomic imprinting has little or no severe impact on the development of the Angiosperm embryo.

The correct genomic constitution of the endosperm, however, is of great importance for seed development. With a few rare exceptions, restricted mainly to the Asteraceae family, apomictic plants still need pollination and fertilization of the central cell to form a functional endosperm *(16,32,33)*. Strikingly, in many species, shifting the usual balance of the two maternal genomes to one paternal genome in the endosperm results in parent-of-origin effects analogous to those described above for mice. Thus, it seems that though in the early development of the seed of flowering plants both the embryo and the endosperm are of vital importance, genomic imprinting in most species affects the development of the endosperm and not of the embryo. Where there are effects on embryo development, these are thought to be of secondary origin and as a result of imprinting effects on the endosperm.

Given the relative importance of the endosperm in genomic imprinting in plants, its function and development will be discussed briefly.

2. Function of the Endosperm

In contrast to Gymnosperms, in which the female gametophyte is a source of nutrients for the developing embryo, the ovule in flowering plants contains limited food reserves. In most Angiosperms, the endosperm has taken over this embryo-feeding role. Most probably, in dicotyledons the embryo can develop without nutrient influx from the endosperm until it has reached the globular or the heart stage *(34,35)*. From this stage on, the endosperm apparently becomes indispensable as it functions as a sink and a source that acquires resources from maternal tissues for later use by the developing or germinating embryo *(36,37)*. In this respect, the endosperm performs an analogous function to the placenta in mammals.

3. Endosperm Development

Within the Angiosperms there is a great variety in size, structure, and development of the endosperm. At its maximum size, the endosperm can constitute the bulk of the mature kernel as in, for instance, maize and other cereals. In the other extreme, the endosperm can be very small indeed, as for example in the orchids, where, after initial fertilization of the central cell nuclei little or no proliferation takes place. In general, two types of endosperm can be distinguished. The *persistent* endosperm is maintained into the mature seed and functions as a nutrient store for the seedling during and shortly after germination. This type of endosperm is found in many monocotyledonous (monocot) plant species, for instance, cereals. Many dicotyledonous (dicot)

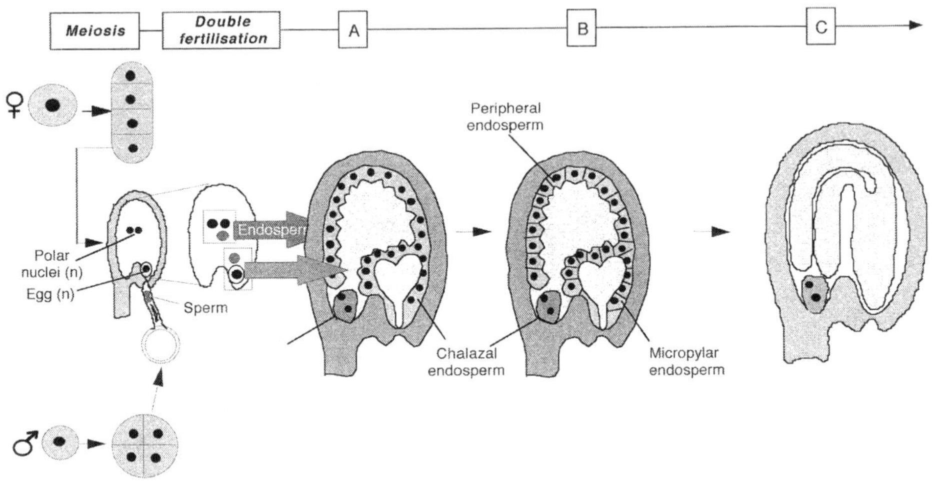

Fig. 1. Double fertilization and seed development in *Arabidopsis thaliana*. Meiosis in the female germline gives rise to an ovule containing eight genetically identical, haploid nuclei. In the male germline, a pollen is formed containing three genetically identical, haploid nuclei: two generative sperm nuclei and one vegetative nucleus. Fertilization of the egg by a sperm gives rise to a diploid embryo. Fertilization of the two polar nuclei of the central cell by the second sperm results in the formation of a triploid primary endosperm nucleus. Replication and division of this nucleus leads to the formation of a syncytial endosperm lining the inner wall of the seed coat (**A**). Three different types of endosperm can be distinguished: chalazal, micropylar, and peripheral. When the embryo is in late heart stage, the micropylar and peripheral endosperm cellularise (**B**). In a mature seed, the embryo has absorbed virtually the whole endosperm (**C**).

seeds have *transient* endosperms, which reach their maximum size before the seed reaches maturity and are then consumed by the growing embryo. In mature seeds of such plants, the endosperm is thus reduced to a few cell layers or may be even completely absent.

As an example of an transient endosperm, the growth and development of the endosperm of *Arabidopsis thaliana* will be discussed.

3.1. Endosperm Development in Arabidopsis Thaliana

After fertilization, the now triploid primary endosperm nucleus replicates and divides (**Fig. 1**). As in many other species, *Arabidopsis* endosperm development is initially free nuclear: the mitotic cycle does not include cytokinesis. Consequently, the central cell forms a syncytium, which can consist of hundreds of individual nuclei. The endosperm differentiates into three distinct types. At the micropylar pole, micropylar peripheral endosperm surrounds the embryo and the suspensor. Central peripheral endosperm constitutes the bulk of the endosperm. A third type of endosperm develops at the chalazal pole. This chalazal endosperm is easily distinguished from both other types because of the high density of its cell mass. The chalazal endosperm is located adjacent to the chalazal proliferating tissue, a maternal tissue that lies adjacent to the site of nutrient unloading from the vascular system. This location suggests that the chalazal endosperm may be involved in translocation of nutrients into the endosperm ***(38–40)***. Further indications for a nutrient-importing function of the chalazal endosperm come from seed ultrastructure studies in several species and radiolabeling of water-insoluble photoassimilates in developing soybean seeds ***(38,41–43)***. This is reminiscent of the situation in cereals, where the cells in either the chalazal and/or the micropylar endosperm develop into special transfer cells that have extensive wall ingrowths into the maternal tissue. In these haustoria the surface area available for transport is significantly increased ***(35,44–46)***.

In *Arabidopsis*, the central cell expands as the peripheral endosperm proliferates. When the embryo is approximately in late heart stage, the seed reaches its maximum size. Starting at the micropylar pole, the endosperm now cellularizes. First, anticlinal cell walls are formed. The nuclei lining the inner cell wall of the central cell replicate and divide once more and subsequently are separated by periclinal cell walls. This process is repeated by the newly formed second layer of nuclei, until the whole former central cell is filled with cellularized micropylar and peripheral endosperm. Endosperm cells, once formed, no longer replicate ***(47,48)***. Cellularization thus seems to mark the end of endosperm proliferation. The chalazal endosperm cellularizes later ***(47)***. Exceptions do occur, however: in the pea, the endosperm never cellularizes but is absorbed by the embryo in the free-nuclear state ***(49)***. From the time of endosperm cellularization onwards, the developing embryo absorbs the endosperm, and presumably assimilates the endosperm cells and its contents (**Fig. 1**).

4. Genomic Imprinting in Plants—Historical Overview

4.1. The Origin of Seed Failure in Interploidy Crosses

The first clues to the existence of genomic imprinting in plants came from the study of seed development following interploidy crosses (for a review, see also **ref.** ***50***). In some plant species, crosses between diploid and tetraploid plants give rise to viable seed. However, in most plant species, 2x × 4x and 4x × 2x crosses result in seed failure. As this was found to be the case even between diploids and their colchicine-derived autotetraploids, the realization grew that seed failure must be the result of quantitative rather than qualitative differences between the genomes contributed by either parent. In the course of the last century, different hypotheses were proposed to explain seed failure in interploidy crosses. Müntzing ***(51,52)*** suggested that the ploidy balance between maternal : endosperm : embryo tissues was critical for normal seed development. Any deviation from the observed 2:3:2 ratio was predicted to result in seed failure. Although this theory seems to have been widely accepted and referred to over a long time, as early as 1932, Watkins was able to rule out the importance of the ploidy of the maternal tissue ***(53)***. Working with diploid and autotetraploid *Primula* and *Campanula* species, he came to the conclusion that it is the endosperm:embryo relation that is important. His finding that the relative ploidy of the maternal tissue is of no importance for the success of seed development was confirmed by Howard in 1939 ***(54)***. Using diploid and autotetraploid *Brassica oleracea*, he found that seeds from a 4x × 2x cross develop to a comparable size to exceptional 2x × 2x seeds in which the mother plant had produced diploid instead of haploid embryo sacs. While the maternal tissue ploidy differs between the crosses, being 4x for the first and 2x for the latter, in both cases the embryo : endosperm ratio is 3x : 5x.

Cooper and Brink ***(44,55)*** were the first to report that the seed failure in interploidy crosses is due to failure of the endosperm itself. They proposed that embryo death is only a secondary effect of a disturbed endosperm development in interploidy crosses. Cooper and Brink were among the first to make a detailed study of the initial development of embryo and endosperm in crosses between diploid and tetraploid plants. This would eventually shed light on the impact of genomic imprinting in plant development.

4.2. Seed Development in Interploidy Crosses

In the early 1950s, Håkansson and other researchers provided detailed descriptions of the phenotypes of seeds produced in reciprocal crosses between diploid and tetraploid plants, in both monocot and dicot species ***(56–59)***. It became apparent that in many species such seeds show specific parent-of-origin effects dependent on the ploidy of the seed and the pollen parent. These characteristics of 2x × 4x and 4x × 2x crosses were reciprocal in nature. In 2x × 4x seeds, Håkansson emphasised the following peculiarities of the endosperm:

1. Mitotic irregularities were frequently observed. In both rye (*Secale cereale*, **ref. *59***) and barley (*Hordeum vulgare*, **ref. *58***), endomitosis and resulting giant nuclei were observed in the endosperm. Mitotic endosperm irregularities were also commonly seen in the offspring of interploidy crosses of the dicot *Galium mollugo* ***(60)*** and in interploidy crosses between different species within the genus *Triticum* ***(61)*** or within the genus *Avena* ***(62)***.

2. Cellularization of the syncytial endosperm is delayed. In barley and maize a slight delay in the onset of formation of cell walls was observed ***(58)***. In 2x × 4x rye seeds, however, there was no sign of cellularization, or at its best it was very much delayed, even though high numbers of endosperm nuclei were achieved.
3. In some species, abnormal meristematic activity was observed in the mycropylar region of the endosperm. Growth of this part of the endosperm resulted in enclosement of the embryo in endosperm tissue both in barley ***(58)*** and in maize ***(56)***. This is reminiscent of the situation in the dicot *Galeopsis pubescens*, where in 2x × 4x seeds the micropylar haustorium is well developed and shows extraovular outgrowth with large nuclei ***(57)***. This may indicate a tendency for enhanced development of those (basal) parts of the endosperm that are in close contact with maternal tissues and presumably play an important role in the import of nutrients into the endosperm.
4. Deposition of starch occurred late compared to 2x × 2x seeds, and usually in small quantities, for example, in barley ***(58)*** and in maize ***(56)***. This late onset of starch synthesis and deposition may be correlated with the delay in cellularization observed in these species.

In endosperm of the reciprocal 4x × 2x cross Håkansson noted the following characteristics:

1. Mitotic irregularities were rare or absent.
2. Cellularization occurred early in many species, including maize, barley, and rye. Fagerlind ***(60)*** reported in 1937 that "cell walls in a 4x × 2x cross in the dicot *Galium mollugo* were formed earlier than I ever observed in any other *Galium*-cross."
3. Abnormal meristematic activity was not observed; the volume of the endosperm remains small and mitosis stops early.
4. Starch deposition began early.

Although all interploidy crosses described above resulted in seed failure at some stage during development, the early stages of endosperm development displayed a strong indication of a parent-of-origin effect pattern. High ploidy × low ploidy crosses showed a tendency to form a small endosperm with a low cell cycle activity that cellularized early, while seed of the reciprocal cross had a tendency toward a bigger endosperm in which a longer and more active mitotic stage was correlated with a later onset of cell wall formation. As for the cause of these phenotypes, Håkansson and others still believed this had be sought in the different ploidy ratios of embryo and endosperm.

4.3. Parental Ratios Within the Endosperm Are Critical

It was not until the 1960s that the emphasis finally shifted from the ploidy ratio between different tissues toward the balance of paternal and maternal genomes within the endosperm itself. In 1962, von Wangenheim published his results on the development of endosperm resulting from interploidy crosses in the evening primrose, *Oenothera* (*63*). *Oenothera* is an exception within the Angiosperms in that its central cell has a single haploid nucleus rather than the usual two. As a consequence, after pollination and double fertilization, both embryo and endosperm have the same number of chromosomes. Thus, *Oenothera* provided a unique means of testing the importance of ploidy ratio of maternal, embryo, and endosperm to the development of the seed. Both a 4x × 2x and a 2x × 4x cross resulted in a 3*n* embryo and a 3*n* endosperm. In the former cross both "zygotes" consisted of two sets of maternal and one set of paternal and in the latter of one maternal and two paternal sets of chromosomes. Despite both reciprocal crosses having the same ploidy and the same endosperm:embryo ratio, the phenotypes of the developing seeds and endosperms were strikingly different. In line with the results of interploidy crosses in cereals described above, 2x × 4x crosses in *Oenothera* resulted in seeds with large, late-cellularizing endosperms, in whose nuclei endopolyploidy was frequently observed. In many seeds the chalazal endosperm was overproliferated. On the other hand, 4x × 2x seeds developed small, early-cellularizing endosperms with a small chalazal endosperm. Von Wangenheim, in excluding a role for the ploidy level of the maternal tissue, concluded that the different parental contributions to the endosperm were responsible for the opposite phenotypes in the reciprocal crosses. Since the paternal:maternal ratio represented the only difference between the genomes of a 4x × 2x and a 2x × 4x seed, von Wangenheim proposed that the origin of the differences had to be extrachromosomal. Extrapolating a theory proposed by Kihara and Nishiyama, he first speculated that pollen, in addition to its genome, contributed a certain unknown active compound that inhibited the formation of cell walls as well as differentiation of parts of the endosperm. Only after a series of mitoses and the accompanying growth of the endosperm syncytium would this compound have become so diluted that cellularization finally could take place. In order for this theory to be consistent, the concentration of this compound had to be related to both the ploidy of the pollen nucleus and that of the recipient central cell nucleus. However, calculations of how the concentration of this hypothetical compound would decrease in the course of endosperm development in both 2x × 4x and 4x × 2x seeds could not explain the reciprocal outcomes observed in both crosses. A second theory, which states that it is the embryo sac that carries an extrachromosomal compound that stimulates the formation of cell walls, proved even harder to defend. Von Wangenheim finally concluded that "the observed phenomena are more likely to be caused by [the presence of] an extrachromosomal compound which is capable of autoreduplication."

Von Wangenheim's observation that it was the ratio of paternal to maternal genomes in the endosperm that is critical for endosperm development was confirmed in 1966 by Nishiyama and Inomata ***(64)***. Working with diploid and autotetraploid *Brassica*, they concluded that a maternal:paternal ratio of 2:1 is required for normal development . Deviation from this ratio led to endosperm dysfunction and hence to seed failure. A more restricted theory was proposed by Sarkar and Coe ***(65)***, who from their work with maize suggested that the triploidy of the endosperm itself is of vital importance for proper endosperm development.

A definitive answer to the question of what is required for normal endosperm development was provided by the *indeterminate gametophyte (ig)* mutation in maize. Maize plants carrying this mutation are aberrant in the formation of polar nuclei. This may result in the production of female gametophytes with extra polar nuclei in the central cell. Kermicle was able to obtain a number of viable triploid plants from a cross between a diploid *ig/ig* and a tetraploid *wt* pollen donor ***(66)***. He explained this by assuming that those plants resulted from seeds that combined a triploid (1 maternal:2 paternal) embryo with a hexaploid (4 maternal:2 paternal) endosperm, in which the required 2:1 balance was restored. This hypothesis was confirmed by Lin in 1984 ***(67)***. Using diploid *ig/ig* females and both diploid and tetraploid pollen donors, he obtained diploid as well as triploid embryos, which were combined with a range of different endosperm karyotypes. Both triploid and hexaploid endosperms could yield normal seeds, but only if the maternal:paternal contribution was either 2:1 or 4:2. Endosperms with a ratio of 5:1 or 2:2 invariably failed to produce viable seeds. Strikingly, tetraploid endosperms with an aberrant constitution of 3 maternal:1 paternal genomes were also found to give rise to viable seeds. Although such seeds were smaller than seeds with a 2:1 or a 4:2 endosperm, their survival indicates that although a 2:1 dosage of maternal and paternal genomes in the endosperm is required for normal seed development, at least in certain species small deviations from this ratio are tolerated (***67***; reviewed in ***68***). In this regard it is noteworthy that whereas in maize an endosperm with one extra set of maternal genomes (3:1) is viable, the reciprocal endosperm, that is, with an extra set of paternal genomes (2:2), is not. This is reminiscent of the outcome of interploidy crosses of other species: there are several reports of interploidy crosses that fail to give viable seed when the pollen parent has the highest ploidy, even when the reciprocal cross may be successful (e.g., ***44,50,53,59,69,70***).

4.4. Parent-of-Origin Effects in *Arabidopsis thaliana*

In *Arabidopsis sp.*, Rédei analysed seed set following interploidy crosses ***(71)***. He found that 4x × 2x crosses produced a high proportion of good seed, but the reciprocal 4x × 2x cross resulted in a low percentage of viable seeds. However, the first study in which seed development following interploidy crosses was analyzed in detail was by Scott et al. *(72)*. They found that in contrast to the findings of Rédei and unlike the closely related *Brassica* species, in *Arabidopsis* crosses between diploid and tetraploid plants in either direction gave rise to a high percentage of viable seeds. This offered the opportunity to study seed development in such crosses in detail from the moment of fertilization until the seed reaches maturity.

Although viable seeds were produced in both 4x × 2x and 2x × 4x crosses, these seeds were very different from each other, as well as from balanced seeds produced by self-pollinated 2x, 4x, or 6x plants, in final size, weight, and development of the different parts of the endosperm. Seeds of a 4x × 2x cross are significantly smaller and lighter than 2x × 2x seeds, while a 2x × 4x cross gives rise to bigger and heavier mature seeds. Using Feulgen staining combined with confocal microscopy, the development of embryo and endosperm in the different crosses was analyzed. In 4x × 2x crosses, both endosperm mitosis and embryo differentiation were found to be slower than in balanced crosses. Endosperm cellularization began slightly early compared to balanced crosses. As a result of this, cellularization led to a small endosperm with a limited number of large cells. The chalazal endosperm remained small and binucleate and often began to collapse after 5 DAP (days after pollination), when the embryo is still only at the heart stage. Seeds with a paternal excess showed complementary phenotypes to those with a maternal excess, that is, endosperm hyperplasia. In 2x × 4x crosses, embryos developed at about the same rate as in balanced crosses. The central peripheral endosperm, however, underwent accelerated mitosis, giving rise to a large number of peripheral endosperm nuclei. The endosperm also cellularized late, so that when cytokinesis finally occured there were many, small endosperm cells. The chalazal endosperm was often enlarged and vacuolated. Chalazal nodules, peripheral endosperm protoplasts that developed characteristics of the chalazal endosperm, could also be very large.

Although crosses between diploid and tetraploid plants yield viable seed, interploidy crosses between diploid and hexaploid nearly always resulted in seed failure. For the first few days, embryogenesis in 2x × 6x crosses proceeded at about the same rate as in 2x × 4x crosses, but never passed the heart stage. Hyperplasia of the endosperm was very dramatic. The central peripheral endosperm divided rapidly without cellularizing. Both micropylar and chalazal endosperm, as well as chalazal nodules, became hugely overgrown and vacuolate, eventually engulfing the embryo and filling the seed. Seeds from 6x × 2x crosses had a similar but more extreme hypoplasic phenotype than those from 4x × 2x crosses. The peripheral endosperm cellularized early, and embryos

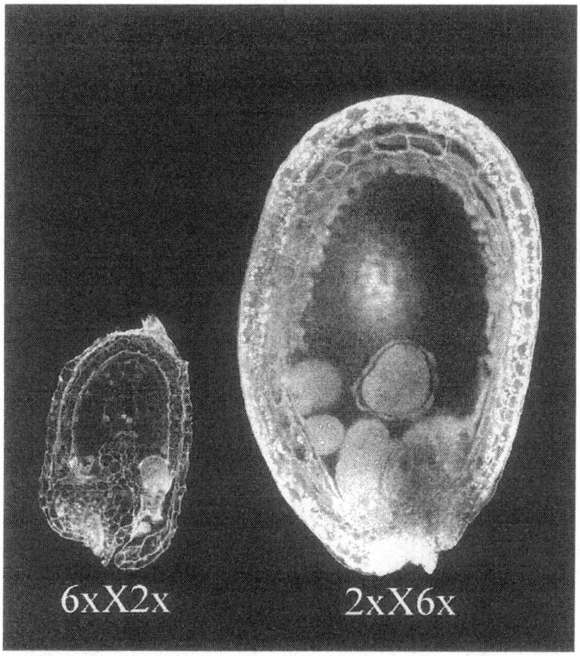

Fig. 2. The effect of imbalanced crosses on seed development in *Arabidopsis thaliana*. Left, a seed from a 6x × 2x cross at DAP 5. Characteristic for this extreme maternal excess phenotype are the absence of chalazal endosperm and a small seed with few peripheral endosperm nuclei. The peripheral endosperm has cellularized prematurely, resulting in a few, large cells. Right, a seed from a 2x × 6x cross at DAP 5, with a large, uncellularized peripheral endosperm containing many nuclei and a massive chalazal endosperm with a number of chalazal nodules. In both crosses the embryo is at the globular stage. Seeds from both crosses fail to germinate.

aborted by the globular to heart stage. No distinct micropylar peripheral endosperm was seen, and the chalazal endosperm disappeared by 5 DAP (**Fig. 2**).

The results from the interploidy crosses in *Arabidopsis* are very much in line with the earlier observations on initial seed development in other species, discussed above ***(57–59)***. Seeds with double the number of paternal relative to maternal genomes show accelerated mitosis and delayed cellularization of the endosperm, have a well-developed chalazal endosperm, and are abnormally large at maturity. In contrast, those containing a double dose of maternal genomes exhibit reduced endosperm mitosis and precocious cellularization. They develop small chalazal endosperms and are abnormally small at maturity. In all cases, embryo development appears roughly normal (though slow when there is a maternal excess).

As others working predominantly on maize ***(50,67,73–78)*** had previously, Scott et al. concluded that the observed parent-of-origin effects were best explained by assuming that endosperm development requires the activity of imprinted genes.

4.5. Parental Conflict and Imprinting in Flowering Plants

The results of the interploidy crosses in *Arabidopsis* and other species are in accordance with the parental conflict theory of Haig and Westoby ***(50,75)***. This model predicts that maternally and paternally derived alleles will be selected to have opposite effects on endosperm (and, hence, ultimately embryo) growth. Adding paternal genomes to the seed is expected to provide extra doses of the uniparentally expressed alleles that increase seed size, while extra maternal genomes are predicted to provide an excess of alleles that limit seed size. The *Arabidopsis* results provide support for the parental conflict theory, as *Arabidopsis* seeds with double the normal dose of paternal genomes produce large endosperms and embryos, whereas those containing a double dose of maternal genomes have the opposite effect (**Fig. 3**).

There is an apparent contradiction in the assumed presence of genomic imprinting in *Arabidopsis thaliana*. As this species is an inbreeding plant that in the wild reproduces almost solely by self-fertilization, it unites mother and father in a single individual ***(79)***. As in this case one can hardly speak of a "parental conflict," it seems there should be no need for uniparentally expressed

Fig. 3. *(opposite page)* Model of the effect of parental genome dosage on seed development in flowering plants. (**A**) In this model it is assumed that some loci affecting endosperm growth are imprintable, while some are nonimprintable. Each of these types can be subdivided into loci that promote endosperm growth (*white*) and those that inhibit (*black*). Here we represent only the imprintable loci (vertical lines on upstream regions are sites for imprint associated methylation). A crucial aspect of the model is that imprintable growth-promoting genes are maternally inactivated, while imprintable growth inhibiting genes are paternally inactivated. (**B**) Each "germline" nucleus contains both classes of loci. Germ cells in flowering plants are derived from somatic cells late in development. There is no sequestered germline, and inactivation via imprinting is presumed to occur during gametogenesis (the polar nuclei as well as the egg are considered as female gametes here) In this model, polar nuclei transmit inactivated growth-promoting genes (*white*) as well as active growth-inhibiting genes (*black*), while sperm transmit the reciprocal. Both polar nuclei and sperm also carry potentially active growth-promoting and -inhibiting genes that are not imprintable (not shown) Double fertilization yields a triploid endosperm containing two maternal genomes and one paternal genome (as well as a diploid embryo). (**C**) Crosses between individuals of different ploidies generate unbalanced endosperm, which results in abnormal seed development. In the model, this is explained in terms of the

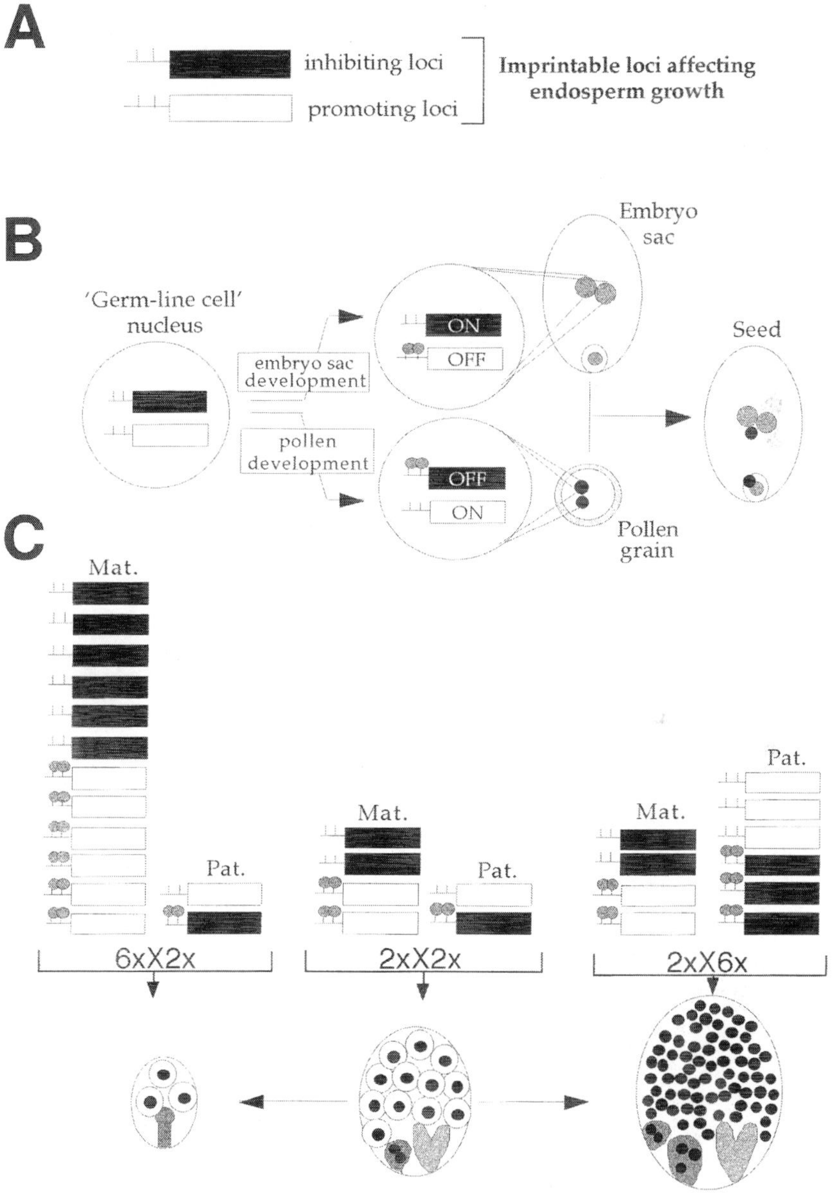

Fig. 3. *(continued)* relative numbers of active imprintable growth-promoting and inhibiting genes in the endosperm. Paternal excess generates endosperms with a high active *positive*:*negative* ratio (>1:2) and consequently increased vigor because sperm genomes contribute active growth-promoting genes but only inactive growth-inhibiting genes. Maternal excess has the reciprocal effect.

genes. However, it is probable that *A. thaliana*, like other inbreeding plants, evolved from outcrossers *(80)*. Therefore it is suggested that *A. thaliana* retains a parental imprinting system inherited from outcrossing ancestors, which has partially broken down as inbreeding has become predominant, allowing development of viable seeds with a limited degree of maternal or paternal excess *(72)*.

The advantage of using *Arabidopsis* in studies of imprinting and parent-of-origin effects lies in the fact that interploidy crosses in this species do lead to viable seeds. It demonstrates that, by influencing the development of the transient endosperm, eventually the embryo and thus the whole seed are affected by altering the ratio of maternal to paternal genomes in the endosperm. In other words, extra paternal genomes, which are expected to contribute active copies of maternally imprinted genes, stimulate endosperm growth and development, leading to a bigger endosperm with most probably a larger capacity for storing nutrients. As a result, the embryo in such a seed will reach a larger size at maturity.

5. Mechanisms of Imprinting in Flowering Plants: The Role of Methylation

Little is known about the parental imprinting mechanism in plants, although there is evidence that, as in mammals, DNA methylation is involved. In maize endosperm, imprinted *zein* genes are expressed only when inherited from the seed parent, and these loci are methylated at fewer sites on maternally then paternally derived chromosomes *(77)*. Differential methylation also corresponds with parent-specific expression of the *R* locus *(76,81)*.

Recently, Adams et al. *(82)* analyzed seed development in crosses in which the genome of one or either parent had undergone extensive demethylation. Transgenic *Arabidopsis* lines expressing the *Methyltransferase I* antisense gene (*METI* a/s) have only 15% of the methylation level of a wild-type plant *(83)*. If DNA methylation is indeed essential to the imprinting mechanism in *Arabidopsis*, and if the antisense transgene prevents imprinting-specific methylation, one would predict that hypomethylated plants produce gametes in which imprinted alleles have lost most or all of their silencing. Indeed, in crosses using *METI* a/s plants as one of the parents, both seed weight and the development of the endosperm showed that hypomethylation closely phenocopies the effect of interploidy crosses. *METI* a/s × 2x seeds had a strong paternal excess phenotype with high seed weight, many endosperm nuclei, delayed endosperm cellularization, and overgrown chalazal endosperm, although both parents were diploid, and the seed was nourished by a hypomethylated mother that suffers a variety of defects in vegetative and floral development *(83)*. This behavior is consistent with a model in which hypomethylation of the maternal genome in *METI* a/s plants has prevented silencing of endosperm-promoting genes that would normally be expressed only from the paternal genome ***(50,72,75)***. Meanwhile, the wild-type paternal genome contributes its normal complement of silenced endosperm-inhibiting genes and active endosperm-promoting genes. The net effect, according to the model, is that endosperm has an excess of imprinted genes that behave as if they were inherited from the father, thus phenocopying an excess of paternal genomes.

As predicted, seed resulting from a 2x × *METI* a/s cross phenocopied the maternal excess obtained in a 4x × 2x cross. Hypomethylated pollen gave rise to small seeds with fewer peripheral endosperm nuclei and a small chalazal endosperm. Cellularization in such endosperms was early compared to wild-type seeds. This phenotype can be explained by assuming that in the pollen donor, the activity of the *METI* a/s gene has erased genomic imprints from, or prevented silencing of, paternally imprinted genes, thereby reactivating these genes. As paternally imprinted genes are predicted to have an endosperm-inhibiting effect, the endosperm genome after fertilization will have an extra set of active, endosperm-inhibiting genes. In other words, demethylation of pollen genomes "maternalizes" such genomes, whereas demethylation in female gametes leads to "paternalization" (**Fig. 4**). Crosses with demethylated plants that were hemizygous for the METI a/s gene (hemi-Met) demonstrated that the determining factor is the demethylated status of the gamete genome and not the presence of the transgene. Seed development in offspring of crosses with a hemi-Met as a parent showed the expected, above-described phenotypes, independently of the presence or absence of the *METI* a/s transgene in the seed.

It is not known in which way methylation may lead to gene silencing in flowering plants. In a number of plant species, transgene silencing has been found to be correlated with a change in both methylation level and in chromatin structure. In *Petunia hybrida*, the maize *A1* transgene is expressed in some lines but silent in others. In *A1*-expressing lines the transgene is hypomethylated and sensitive to DNase I and nuclease S7 digestion. In lines where the transgene is silent, however, the locus is hypermethylated and significantly less sensitive to digestion, suggesting a more condensed chromatin structure on the transgenic locus ***(84)***. Likewise, in *Arabidopsis*, transgene silencing of the *HPT* gene is correlated with both increased methylation and an increased resistance to DNase I and micrococcal nuclease, again indicating a change in chromatin structure ***(85,86)***. These data suggest that, as in animals (reviewed in **ref.** *87*), in plants methylation, chromatin structure and gene silencing may well be linked. This need not be always the case, however. Activation of the pea *rbcS* gene involves a change in chromatin structure, but not in methylation status ***(88)***.

Recently, an *Arabidopsis* gene has been cloned that plays an important role in the maintenance of transcriptional gene silencing ***(89)***. A mutation in this

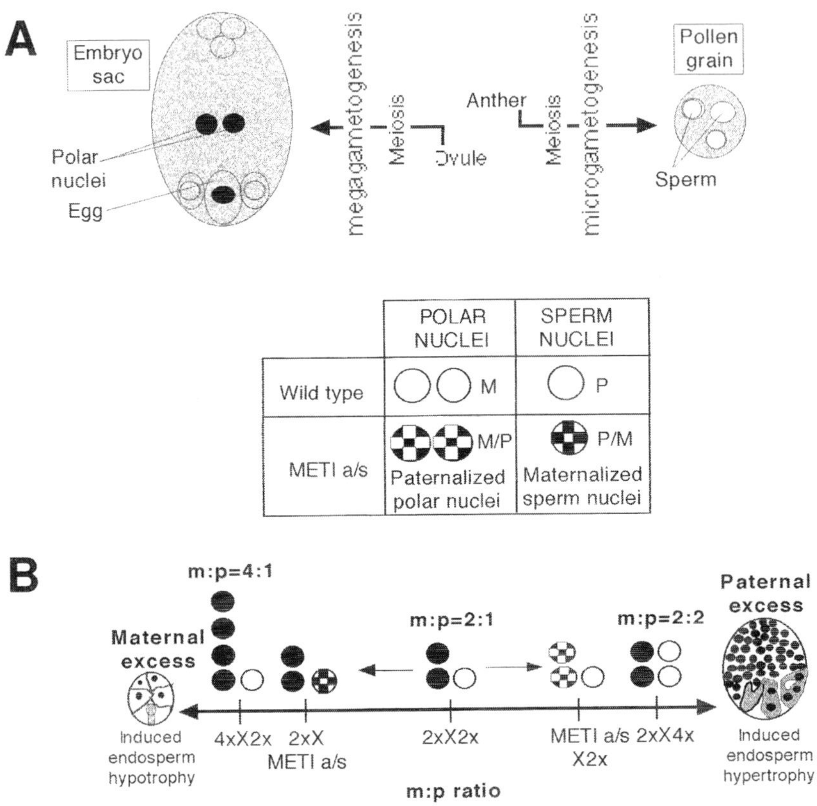

Fig. 4. Model of the effect of global DNA hypomethylation on parental imprinting in *Arabidopsis thaliana.* **(A)** Normally, endosperms contain two copies of the maternal genome, contributed by the polar nuclei, and one copy of the paternal genome, contributed by the sperm. In the maternally inherited genomes, endosperm-promoting genes are expected to be imprinted and hence silenced. In paternally inherited genomes, these genes will be expressed, but endosperm-inhibiting genes are expected to be imprinted. When maternal genomes are contributed by a *MET I* a/s parent, the endosperm-promoting genes are expected to be largely derepressed, producing a "paternalized" genome. Similarly, a *MET I* a/s pollen parent is expected to contribute a "maternalized" genome. **(B)** Interploidy crosses (e.g., 4x × 2x or 2x × 4x) result in seeds with extra maternal or paternal genomes, and therefore extra doses of active maternal or paternal alleles of imprinted loci. Maternal or paternal excess has dramatic and complementary effects on seed development. A diploid *MET I* a/s parent does not contribute extra genomes but appears to contribute extra doses of active endosperm-promoting or -inhibiting genes, resulting in phenotypes similar to those produced by parental genomic imbalance.

Morpheus Molec ule (MOM) gene leads to reactivation of transcriptionally silent transgenic loci. Interestingly, the methylation level of at least one of these loci does not change: even after nine generations, the reactivated *HPT* transgenic locus remains hypermethylated. As in addition no change in methylation status of a 180-bp CEN repeat could be detected, it is likely that in a *mom* mutant background reactivation of silenced genes and methylation patterns are inherited independently.

What is the function of *MOM* in gene silencing? The MOM protein may act downstream of methylation. It may for instance be involved in the link between methylation and the actual transcriptional silencing. Alternatively, it is also possible that *MOM* regulates gene expression in a separate, methylation-independent way. It will be interesting to determine what effect the *mom* mutation has on the chromatin structure of affected genes, and to what extent *MOM* regulates the expression of imprinted genes.

6. Imprints in Arabidopsis Are Not Essential for Development

The seed development of plants with a hypomethylated parent suggests the importance of methylation and the activity of the *Methyltransferase I* gene in *Arabidopsis* in establishing imprinting-associated methylation. As inbreeding of demethylated plants yields viable seeds, one can also conclude that genomic imprinting is not a prerequisite for seed development in this species. Apparently, removal of all or most genomic imprints from both parental genomes does not prevent embryo or endosperm development. Jaenisch ***(90)*** proposed that removal of imprints or of imprinted genes themselves should have few developmental consequences, as they exist in "'paired sets' of genes involved in the same pathway" (e.g., of growth promoters and inhibitors, as predicted by Haig and colleagues). This has been difficult to test in mammals as embryos with reduced methylation die during gestation ***(91)***.

In contrast to mammals, in flowering plants one would expect to observe consequences for seed development if imprints are erased in both parents. It is the endosperm genome that is mainly subject to imprinting, and as described earlier, this genome consists of two maternally inherited and one paternally inherited genome copies. As a consequence of this, paternally imprinted genes are expressed from two copies of the maternally inherited genome. Maternally imprinted genes, on the other hand, are expressed from only one allele, inherited via the pollen donor. Thus, in the endosperm genome the bias of two sets of maternally expressed, endosperm-inhibiting genes to one of paternally expressed, endosperm-promoting genes is the normal, "balanced" situation that results in a wild-type endosperm. Overall removal of imprints in both parents would shift this bias from a 2:1 ratio to a 3:3 ratio: all three copies of the endosperm genome are now allowing expression from formerly imprinted genes. This is predicted to mimic the effect of a 2x × 4x cross, in which the endosperm genome consists of two maternal and two paternal copies of the genome. In other words, a "paternal excess" phenotype with a large endosperm containing many peripheral endosperm nuclei that cellularizes late and forms a large chalazal endosperm would be expected.

What is seen, however, is that in *Arabidopsis*, seeds resulting from a *METI* a/s × *METI* a/s cross more closely resemble wild-type seeds than those produced by crosses between one *METI* a/s and one wild-type plant. They do, however, contain small chalazal endosperms and weigh less than wild-type seeds, both features indicating a maternal excess. One aspect of the phenotype, however, is consistent with this prediction: the number of peripheral endosperm nuclei in *METI* a/s × *METI* a/s crosses is higher than in 2x × 2x and about sixfold greater than in 4x × 2x crosses ***(72,82)***.

It is not known why self-fertilization of hypomethylated plants leads to seeds that show a combination of maternal and paternal excess phenotypes, but several factors may contribute. First, demethylation is not complete in *METI* a/s plants ***(83)***, perhaps in part because of other methyltransferases in the *Arabidopsis* genome that are not affected by the *METI* a/s transgene ***(92,93)***. Partial demethylation could affect gamete genomes or individual sequences unequally. In addition, some genes may even become hypermethylated in a *METI* a/s background, like the (nonimprinted) *SUPERMAN* locus ***(94)***. Finally, due to the complicated regulation of imprinted genes, global DNA hypomethylation in mouse can repress as well as activate imprinted alleles ***(95,96)***. It is conceivable that the same occurs in plants, although it seems more likely that the overwhelming effect of hypomethylation is to activate normally silent imprinted alleles.

7. Imprinted Genes in Flowering Plants

In contrast to the situation in mammals, where several well-characterized examples of imprinted genes are known, in flowering plants to date only a small number of genes are known or even suspected to be imprinted. In maize, four loci show a strong parent of origin effect and hence are likely to be imprinted. These are the *R* gene, which encodes a transcription factor active in the regulation of anthocyanin biosynthesis in the aleuron layer of the endosperm, the *delta zein-regulator (dzr)*, a storage protein regulator, a *zein* gene, and an *alpha tubulin* gene ***(76–78,97***; reviewed in ***98,99)***. None of these genes seems to be involved in seed development.

Recently, three mutants have been isolated in *Arabidopsis thaliana* that shed more light on the backgrounds of genomic imprinting in this species. All three genes, *MEDEA/FIS1/EMB173 (Fertilization Independent Seed-1), FIS2 (Fertilization Independent Seed 2)* and *FIE (Fertilization Independent Endosperm)/FIS3* ***(100–106)*** are involved in the regulation of the development of the endosperm. *FIS2* contains a putative zinc-finger motif and three putative nuclear localization signals, suggesting that the FIS2 protein might be a transcription factor ***(106)***.

7.1. A role for MEDEA, FIS2, and FIE in Endosperm Development

Some mutant alleles of *FIE*, *MEA*, and *FIS2* confer a degree of autonomous endosperm development; in other words, the central cell proliferates and develops into an endosperm-like structure even in the absence of pollination and fertilization. Without being stimulated by fertilization, the central cell nucleus starts replicating and dividing and the central cell develops into a premature endosperm in the absence of embryo development. In addition to endosperm development, both the seed coat and the silique wall, which constitute maternal tissue, proliferate. Autonomous endosperm development is limited: differentiation into either chalazal or micropylar endosperm does not take place. In an unpollinated mutant *fie* ovule, endosperm development arrests in the free nuclear stage ***(103)***. In contrast, the autonomous endosperm in a mutant *mea* or *fis2* ovule will cellularize ***(101,103)***. One of the functions of *MEA*, *FIS2*, and *FIE* must be to control the fertilization-dependent block of endosperm development in the female gametophyte. Mutant seeds cannot be rescued by fertilization with pollen carrying the wild-type allele. One of the possible explanations would be to assume that the genes are paternally imprinted, in which case the wild-type allele carried by the pollen would be silenced.

A closer look at what happens after fertilization of a mutant *mea* or *fie* ovule by wild-type pollen provides an insight into the processes regulating seed development. Pollination of an ovule carrying the putative loss-of-function *mea-3* allele with wild-type pollen leads to a seed in which the embryo arrests at the heart stage. In such seeds, the peripheral endosperm overproliferates without cellularization, resulting in endosperm with approximately 150% the number of nuclei produced by a wild-type seed ***(105)***. Fertilization of a *fie* mutant ovule with wild-type pollen leads to a comparable phenotype. A *fie-1/FIE* heterozygote as seed parent pollinated by a wild-type pollen donor produces viable and shriveled seeds in a 1:1 ratio. In the plump seeds, which carry a wild-type maternal *FIE* allele, embryos reach maturity, and endosperm development appears to be identical to that in wild-type seeds ***(101,102,107)***. In shriveled seeds, inferred to be the result of the fertilization of a *fie* mutant ovule, the embryo becomes vacuolate and does not develop past the late heart–early torpedo stage, the endosperm fails to cellularize, and the chalazal endosperm undergoes massive overproliferation. Both the *fie* and *mea* phenotypes are reminiscent of the lethal paternal excess phenotype obtained in a 2x × 6x cross in *Arabidopsis* ***(72)***. If *MEA* and *FIE* are assumed to be involved in inhibiting proliferation of the central cell and endosperm, this would explain why these undergo extra divisions in both unpollinated and pollinated *fie-1* and *mea-3* mutants. According to this model, the seed phenotypes of pollinated *fie-1* and *mea-3* mutants on one hand and wild-type 2x × 6x seeds on the other all result from failure to inhibit endosperm proliferation. In the case of *mea-3* and *fie-1*, the loss of wild-type MEA and FIE protein causes derepression of gene activity promoting endosperm growth. In the case of the interploidy cross, an overdose of paternal genomes contributes to the seed extra active alleles of genes that promote endosperm growth.

The analogy between the *fie* and *mea* mutants and the (lethal) paternal excess in a 2x × 6x seed can be extended. If either mutant leads to a lethal "paternalizing" effect on seed development, then introducing factors that "maternalize" seed development are predicted to shift back the paternal excess phenotype, possibly to such an extent that seed viability could be restored. In the case of *MEA*, it has been found that *mea* mutant ovules can be rescued by pollination, provided that the resulting seed is homozygous for a loss-of-function mutant allele of the *Decrease in DNA-Methylation 1* gene (*DDM1*) ***(108)***. In homozygous *ddm1* mutants, the overall cytosine methylation has been reduced by 70% ***(109)***. However, *ddm1* mutations do not affect methyltransferase activity *(110)* and *DDM1* has recently been found to be a member of the SWI2/SNF2 family of chromatin remodeling proteins ***(111,112)***. It is possible that hypomethylation or chromatin remodeling or both have activated the silenced paternal copy of the *MEA* gene ***(108)***. Alternatively, it is also possible that it is the altering of the methylation status of the sperm genome itself that rescues the maternal *mea* mutant, possibly by activating genes that function downstream of *MEA*.

Ovules carrying a *fie-1* mutant allele can be rescued by hypomethylated pollen from a plant expressing the *METI* a/s transgene ***(107)***. Interestingly, rescue of *fie* seeds was only obtained by using demethylated pollen carrying the wild-type *FIE* allele. It was concluded that an active wild-type paternal *FIE* allele is needed for the *METI* a/s-mediated rescue. Most probably, *FIE* is a paternally imprinted gene that is silent in wild-type pollen. In a hypomethylated background of the pollen genome, the normally silenced *FIE* allele apparently becomes reactivated. The rescue of a maternal fie mutant by such pollen can be explained in two ways. It is possible that *FIE* is crucial for endosperm development. The paternally inherited, active copy of the *FIE* gene, probably in concert with the overall "maternalizing" of the paternal genome, inhibits endosperm development to such an extent that the lethal paternal overexcess phenotype caused by the maternal *fie* mutation is attenuated to a state in which the seed can survive. It is also possible, however, that the maternalization of the sperm genome is due entirely to the overall demethylation of the pollen genome, which is predicted to lead to removal of imprints from endosperm-inhibiting genes. Indeed, as described earlier, demethylation of the genome of a pollen parent has been shown to have such an effect ***(82)***. An active *FIE* allele may be needed for proper embryo development. In this case, the failure of demethylated *fie* pollen to rescue a mutant *fie* ovule would not be due to an endosperm, but to an embryo defect. With regard to this, it is interesting to note that while *mea/mea* and *fis2/fis2* homozygous plants do occur sporadically, despite extensive searching *fie/fie* plants have not been recorded ***(103,107)***.

Rescued *fie-1* mutant seeds showed many features of paternal genomic excess, including large size and weight, and overproliferation and delayed cellularization of endosperm ***(107)***. Vielle-Calzada et al. ***(108)*** also reported that rescued *mea/MEA* seeds were enlarged, containing large embryos and sometimes excess persistent and partially cellularized endosperm. These phenotypes can be interpreted as indicating that reactivation of paternally silenced alleles—which in effect maternalizes the sperm genome—can compensate partially but not completely for the paternalizing effect of maternal *fie-1* or *mea-3* mutations. The activity of wild-type *FIE* and *MEA* will probably inhibit (over)expression of endosperm-promoting genes, which would otherwise result in a lethal paternal excess phenotype. This endosperm-inhibiting activity might come too late to fully restore the mutant phenotype caused by the loss of *FIE* or *MEA* function in the female gametophyte, though. It is also possible, however, that the partial rescue is due to a gene dosage effect of the Polycomb proteins (e.g., ***113***). The single copy of activated *FIE* or *MEA* from the paternally inherited genome might not be as effective as the two copies normally transmitted by the diploid central cell.

7.2. Uniparental Expression in the Endosperm

Interestingly, *mea*, *fis2*, and *fie* are all maternal effect mutations: the aberrant phenotype is observed only when the mutant allele is inherited via the mother. A heterozygote seed in which the mutant allele is inherited via the father is indistinguishable from a homozygote wild-type seed. This could be explained by assuming that these three genes are subject to paternal imprinting and hence are expressed only from the maternal copy of the genome. There are several indications that for *MEA* as well as *FIS2*, in the endosperm genome the paternally inherited allele is silenced throughout the initial stages of seed development. Expression of β-glucoronidase (GUS) driven by either the *FIE* or the *FIS2* promoter could only be detected in the endosperm when the construct was inherited via the seed parent and not when it was inherited via the pollen parent (Luo and Chaudhury; reviewed in **ref.** ***93***).

Vielle-Calzada et al. ***(108)*** analyzed the expression of *MEA* in embryo and endosperm using *in situ* hybridization. Using a *MEA*-specific probe, they showed the presence of two nuclear dots in the polar nuclei both before and after fertilization. Nuclear dots have been observed in mammals and insects. In *Drosophila*, they are indicative of the presence of nascent transcripts of actively expressed genes ***(114)***. The presence of only two nuclear dots after fertilization suggests that the paternal *MEA* allele remains silent after fertilization and that the observed dots correspond to the two maternal *MEA* alleles. No dots could be observed in the nucleus of the egg ***(108)***. Using parent plants with distinguishable *MEA* alleles, the parent specific expression of the *MEA* gene was analyzed. With reverse transcriptase polymerase chain reaction (RT-PCR) analysis on cDNA isolated from complete siliques at 54 h after pollination, when the embryos are in the midglobular stage, only expression of the maternally inherited allele could be detected. The authors conclude that in the early stages of seed development the paternally inherited *MEA* alleles are silent in both embryo and endosperm.

Silencing of the paternal allele in the endosperm, but not the embryo, was confirmed by Kinoshita et al. ***(115)***. They managed to dissect *Arabidopsis* seeds under a stereomicroscope into embryo and endosperm plus seed coat fractions at 4, 6, 7, and 8 DAP, corresponding to the heart, torpedo, walking stick, and early maturation stages of embryo development. Pure endosperm fractions were obtained from seeds of 7 DAP. Using ecotypes that have distinguishable *MEA* alleles, the expression of both parental alleles in reciprocal crosses was analysed by RT-PCR. Expression of the paternal allele in whole-seed samples could be detected in all stages under study. Paternal and maternal *MEA* mRNA could be detected in embryos at 6, 7, and 8 DAP. In contrast, only expression from the maternal allele could be detected in endosperm plus seed coat material at 6 and 7 DAP, and in isolated endosperm at 7 DAP. From these observations, it can be concluded that during these stages the paternally inherited *MEA* allele is specifically silenced in the endosperm, but not in the embryo.

The *MEA* gene thus provides the first example of a plant gene where an observed parent-of-origin effect on seed morphogenesis can be linked to and explained by genomic imprinting at the molecular level. If the expression pattern of the *MEA* gene is representative of other paternally imprinted genes in flowering plants, it at least partially explains why the effects of interploidy and interspecific crosses are observed only in the endosperm and not in the embryo. Apparently the uniparental silencing of imprinted genes, at least during some stages of seed development, takes place only in the endosperm and not in the embryo.

Interestingly, Vielle-Calzada et al. ***(116)*** recently published data that suggest that the whole paternal genome in both embryo and endosperm may be silenced during the first 3 or 4 d after fertilization in *Arabidopsis thaliana.* By screening a library of enhancer detector and gene trap lines expressing the *GUS* gene, they identified 19 transposants that show *GUS* expression in the developing seed, either in embryo, endosperm, or both. In reciprocal crosses between these transposants and wild-type plants, it became apparent that when the transposants were used as seed donors, *GUS* expression was detected from very early stages of seed development onwards. In contrast, when the transposants were used as pollen donors to fertilize wild-type ovules, *GUS* expression could only be detected in the seeds from up to 80 h after pollination, when the embryo is in the globular stage. Having shown that in at least one of these genes the absence of expression from the paternal allele is not related to transgene silencing, the authors concluded that in all genes under study, the paternal allele is silenced. As the genes tested are distributed throughout the genome and represent a wide variety of functions, they proposed that most, if not all, of the paternal genome is silenced during early seed development in both embryo and endosperm. This overall silencing early during seed development may explain why Vielle-Calzada et al. ***(108)*** could not detect expression of the *FIE* gene in the embryo at 54 h after pollination, whereas Kinoshita et al. ***(115)*** detected embryonic expression from 4 DAP onwards.

The reason for this delayed transcriptional activation of the paternally inherited genome is not clear, nor is the mechanism behind it. The authors suggest that the silencing of the paternal genome probably occurs during sperm cell differentiation and may be related to either a tight packaging of the DNA into heterochromatin, or an alternative methylation level of sperm DNA. As a consequence of the overall silencing of the paternal genome in the initial stages of seed development, one would have to assume that at 3 or 4 DAP the barriers preventing expression from the paternal genome are removed. An exception would have to be made for the paternally imprinted genes in the endosperm genome. Alternatively, these would have to become silenced immediately after the activation of the paternal genome.

8. The Role of Polycomb Group Proteins in Genomic Imprinting

Both the *FIE* and the *MEA* gene encode Polycomb group proteins. In *Drosophila* the best-understood function of Polycomb proteins is to maintain transcriptional repression of homeotic genes through many rounds of cell division by forming complexes that modulate chromatin configuration or prevent access of transcription factors ***(117)***. Both *MEA* and *FIE* are expressed in flowers before fertilization and in developing siliques afterwards ***(102,104,105)***. The MEA protein contains a SET domain (so called because it was initially found in the *Drosophila* genes *S*uppressor of variegation, *E*nhancer of Zeste, and *T*rithorax), and shows homology to the SET domain polycomb protein Enhancer of Zeste [E(z)] in *Drosophila*. The FIE protein contains several WD40 motifs, and shows highest homology to the extra sex combs (esc) protein in *Drosophila* and the Embryonic ectoderm development (Eed) protein in mice and humans ***(118,119)***. The WD40 motif is thought to promote protein–protein interactions. The presence of multiple WD motifs allows the protein to be bind to multiple other proteins at the same time ***(120,121)***. WD polycomb proteins have been reported to be active in repressing expression of insect and mammalian genes during embryo development. In *Drosophila* ***(122,123)***, mouse ***(118)***, and human ***(119)*** it has been reported that WD polycomb group proteins interact with a SET domain polycomb group protein.

In *Drosophila* and in mammals, polycomb group proteins function to maintain a repressed state of homeotic genes. Polycomb group proteins function in complexes that bind to chromatin and downregulate gene expression through epigenetic silencing ***(117,124,125)***. The polycomb group proteins are thought not to initially repress expression, but to maintain a repressed state of already silenced target genes ***(117)***. The protein complexes formed by polycomb group proteins and other proteins interact with Polycomb-response elements (PRE) in the DNA. Silencing of the gene containing the PRE and presumably other genes in the vicinity as well is thought to take place though packaging the DNA in a higher level of condensation ***(117)***.

Given the observation that loss of function mutations in *FIE*, *MEA*, or *FIS2* gives rise to highly comparable phenotypes, it is possible that the proteins of these genes also form a complex. The absence of any one of them may lead to disruption or inactivation of the whole complex. In *Drosophila*, the WD40 polycomb protein esc is thought to form a complex that comprises other polycomb group proteins plus the zinc-finger transcription factor Hunchback. Interaction of this complex with the transcription machinery at the *Ultrabithorax* locus leads to silencing of the *Ubx* gene, most likely by packaging the DNA into a condensed chromatin form ***(126)***. As the WD polycomb group protein FIE is an *Arabidopsis* homolog of esc, it is not unlikely that it will perform a similar function in the plant's genome. In this case, FIE may form a complex with FIS2, a putative zinc-finger transcription factor, and the SET-polycomb protein MEA. In early *Arabidopsis* seed development, this complex would then negatively regulate the expression of endosperm-promoting genes from the central cell and endosperm genome ***(93,102)***.

9. Repression of Endosperm Proliferation Involves both *FIE* and DNA Methylation

In plants that are heterozygous for the loss of function *fie-1* allele, proliferation of the central cell without pollination is seen in half of the ovules ***(101,102)***. This phenotype has been interpreted as showing that wild-type *FIE* represses endosperm development before fertilization. However, much of the endosperm developmental programme in *fie-1* mutants does not take place. For example, there is no regional specification of micropylar and chalazal endosperm, and cellularization of the peripheral endosperm does not take place. Therefore there is a block to complete endosperm development in unpollinated *fie-1* mutants. Aside from a wild-type *FIE* allele, several components of normal seed development are missing from unpollinated *fie-1* ovules: pollination and fertilization themselves, and therefore gene expression they might trigger; and a paternally transmitted genome, which is not equivalent to maternal genomes because of genomic imprinting.

Recently it was shown that demethylating *fie-1/FIE* heterozygotes using the *METI* a/s construct allowed autonomous endosperm to develop much further than previously reported ***(107)***. In half of the ovules of a hemizygous *METI* a/s, unpollinated *fie-1/FIE* plant—presumably those carrying wild-type *FIE* alleles—the central cells did not proliferate. This indicates that hypomethylation on its own does not promote fertilization-independent seed development. The other half of the ovules did show autonomous endosperm development. Those ovules fell into two classes. In type 1, the central cell usually underwent more rounds of mitosis than in normally methylated *fie-1* mutants, and the endosperm cellularized, but there was still little or no regional differentiation. In the type 2 seedlike structures, autonomous endosperm development went much further. These seeds resembled sexually produced seeds, containing more peripheral endosperm nuclei, which cellularized later than in type 1, and large chalazal and micropylar endosperms. In neither of the types were embryos found. The presence of two types of autonomous endosperm may be caused by the presence or absence of the *METI* a/s transgene in the ovule, or it may reflect a less direct effect of the transgene in the mother plant; for example, hypomethylation might affect different DNA sequences in different embryo sacs.

In which way does hypomethylation relieve the partial endosperm block in *fie-1* mutants? One explanation would be that some genes necessary for full endosperm development are maternally imprinted, and that this is mediated by means of methylation. Demethylation would lead to activation of these genes, which would in effect supply the missing paternal genome. As demethylation of the maternal genome alone does not lead to autonomous endosperm development, clearly not all components of this pathway are regulated by methylation. In normally methylated plants, fertilization of a *fie-1* ovule leads to a strong paternal excess. This, in concert with *FIE* encoding a polycomb group protein, suggests that *FIE* itself represses endosperm-promoting genes in the maternal gametophyte. So, it is possible that there are two (possibly overlapping) pathways of controlling endosperm-promoting genes in female gametophytes: (1) some genes are maternally imprinted by means of methylation; (2) another set of genes is controlled by the expression of the maternal *FIE* gene. Alternatively, *FIE* and DNA methylation could participate in repressive complexes at the same loci. The *fie-1* mutation alone may not be sufficient to completely release the expression of all these genes (**Fig. 5**).

10. The FIS Complex Confers Sexual Identity to Gamete Genomes in *Arabidopsis*

Earlier (**Subheading 7.1.**) we described how pollination of a mutant *fie* ovule with wild-type pollen results in seed abortion. The phenotype of the abortive seeds closely resembles that of a 2x × 6x or a *METI* a/s × 4x seed: the embryo does not develop past the late heart–early torpedo stage, the chalazal endosperm undergoes massive overproliferation, and the endosperm fails to cellularize *(72,107)*. One explanation for this observation is that seed abortion in a *fie* × 2x cross results directly from massive paternal excess in the endosperm. Since the abortive endosperms resemble those generated by a 2x × 6x cross, we speculate that the m:p ratio must be close to 2:3 (**Fig. 6**).

Consistent with this proposition is the fact that the lethal phenotype encountered in *fie* × 2x crosses can be rescued by providing the *fie* mutant ovule with demethylated instead of wild-type pollen, provided this carries a wild-type *FIE* allele. The reactivation of paternally imprinted, endosperm-inhibiting genes in such pollen is predicted to restore the lethal 2m:3p ratio to a viable 3m:3p ratio, giving rise to seed developing with a viable paternal excess phenotype, which resemble those generated by a 2x × 4x cross (**Fig. 6**).

Under **Subheadings 7.1.** and **8.**, we argued that *FIE* as well as *FIS1/MEDEA* and *FIS2* may be involved in the regulation of the expression of imprinted genes. These proteins, possibly in a complex with other proteins (from now on referred to as the FIS complex), are predicted to prevent autonomous endosperm development in the absence of fertilization and inhibit endosperm-promoting genes after fertilization. Another way to interpret the *fie* autonomous and fertilized phenotype is to propose that *FIE* participates in establishing or maintaining the gender of the female gametes (polar nuclei).

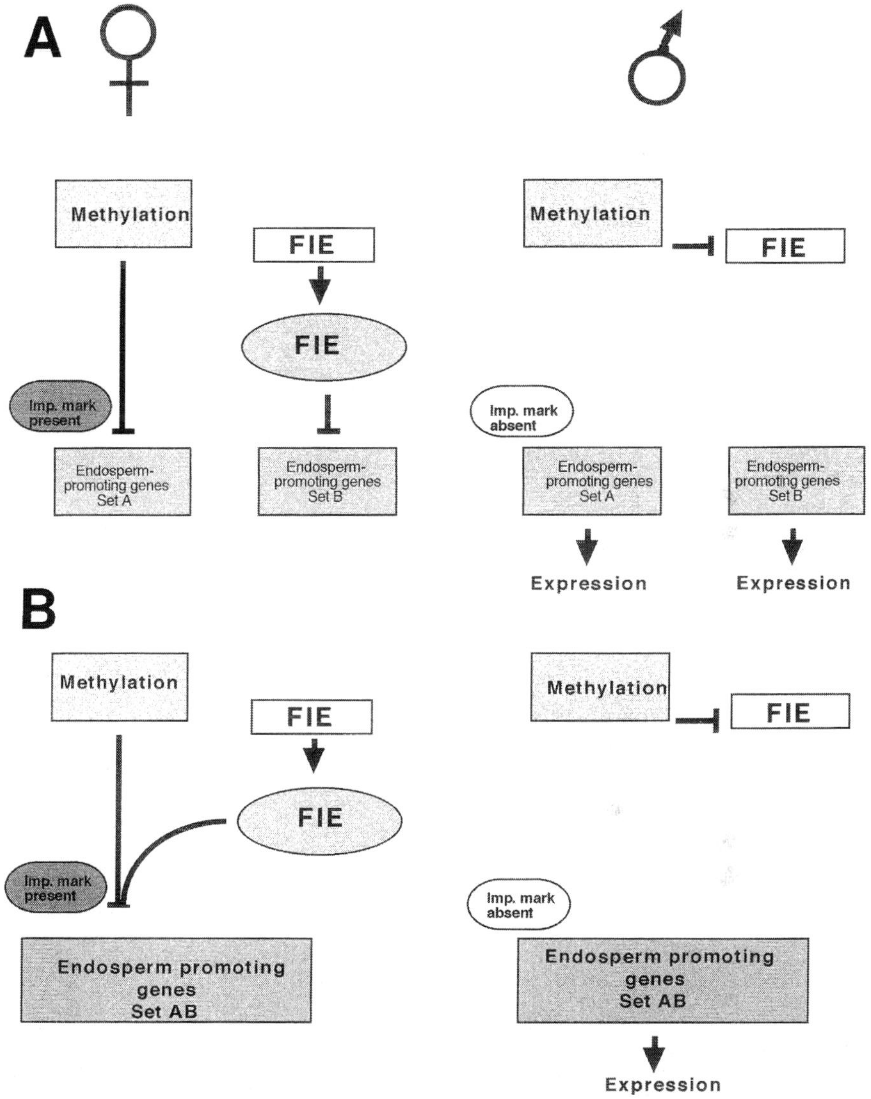

Fig. 5. Hypothetical models for the role of *FIE* in the regulation of endosperm development in the ovule. **(A)** The FIE protein represses expression of a set of endosperm-promoting genes (Set B). Other endosperm-promoting genes are down-regulated by methylation (Set A). In the pollen genome, both sets of endosperm-promoting genes are expressed. As a (hypothetical) imprinting mark on the genes of Set A is absent in the male germline, the genes are no target for methylation and hence can be expressed. The *FIE* gene is paternally imprinted (probably involving methylation) In the absence of active FIE protein, the endosperm-promoting genes in Set B can be expressed as well. **(B)** Methylation and the FIE protein are components of the imprinting mechanism that represses expression from the endosperm promoting genes (represented as Set AB) Possibly, the FIE protein (presumably in a complex with FIS1, FIS 2, and other proteins) maintains transcriptional repression of genes that are uniparentally methylated.

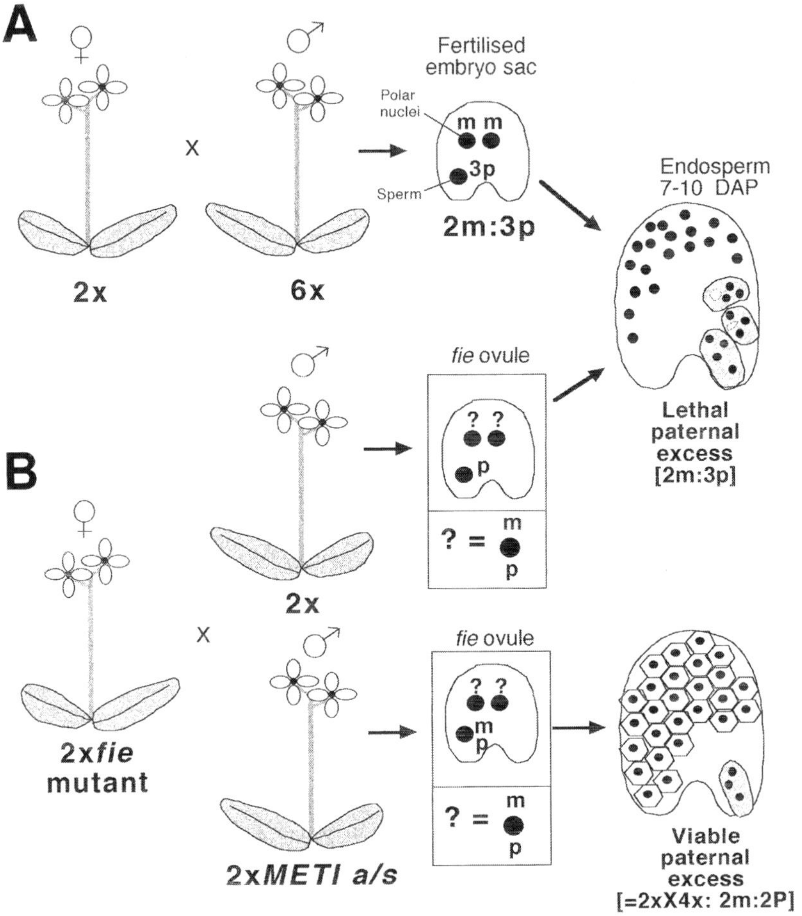

Fig. 6. Fertilized *fie* mutant ovules abort with a paternal excess phenotype. **(A)** A seed resulting from a 2x × 6x cross expresses two sets of maternally inherited endosperm-inhibiting genes (2m) for every three sets of paternally inherited endosperm-promoting genes (2m:3p). This leads to the development of a lethal paternal excess phenotype in the endosperm: massive, overgrown chalazal endosperm, many peripheral endosperm nuclei, and no cellularization. **(B)** In a *fie* mutant ovule, the imprinting status of the polar nuclei is not known. Fertilization with wild-type pollen (and, thus, adding one set of active endosperm-promoting genes to the endosperm), however, phenocopies a 2x × 6x cross (2m:3p) and results in a lethal paternal excess phenotype. Fertilization of a *fie* mutant ovule with demethylated pollen (expressing both a set of endosperm-promoting and a set of endosperm-inhibiting genes from which the imprint has been lifted) phenocopies a 2x × 4x cross (2m:2p) and results in a mild (viable) paternal excess phenotype. It is therefore likely that the genomes of the polar nuclei in a *fie* mutant ovule are not imprinted and express both the endosperm-inhibiting and the endosperm-promoting genes (1m:1p).

Gametes are morphologically and biochemically different in ways that reflect their different roles as vehicles in the fertilization process. However, nuclear transfer experiments in mammals ***(2–4)*** and the outcomes of interploidy crosses in flowering plants ***(57,58,72)*** reveal that gamete genomes in these organisms are nonequivalent. The basis of this genomic nonequivalence appears to be due to genomic imprinting. Therefore, in the absence of imprinting, would the gamete genomes of flowering plants, or indeed mammals, have a gender of their own? Or perhaps more correctly, a memory of the gender of the individual that produced them? As discussed earlier (**Subheading 1.**), viable parthenogenetic, gynogenetic, and androgenetic embryos can be formed in flowering plants. This is in accordance with the hypothesis that the embryo genome is not subject to imprinting. Inactivation of the FIS complex and thus removal or prevention of maintenance of imprints in the female gametophyte could be interpreted as additionally erasing imprinting from the polar nuclei. The formation of autonomous endosperm in a *fie* mutant suggests that maternally imprinted genes have been reactivated, thereby generating a genetic 2m:2p ratio in the central cell. In other words, by removing or preventing the application of imprints the genome has returned to, or remained in, the same neutral or hermaphrodite state as the progenitor somatic cells (i.e., 1m:1p). Therefore, gametic gender can be seen as a consequence of imprinting, and whether a gamete is "female" or "male" as depending on which class of genes is imprinted: female gametes have the set of endosperm-promoting genes imprinted, while production of male gametes implies imprinting of the endosperm-inhibiting genes.

Logically, therefore, in the absence of gender, any combination of gamete genomes could in principle lead to a viable embryo. In organisms in which genomic imprinting of the genome is thought to be absent (viz., species in which development of the offspring is not dependent on nutrient influx from a maternal tissue), it should therefore be possible to obtain viable uniparental embryos. In contrast to the situation in mammals, where androgenetic and gynogenetic embryos abort with opposite phenotypes, nonimprinted uniparental embryos are not only predicted to be viable but also to show no parent-of-origin effects in their development.

To date, natural and induced uniparental embryos have been reported for a number of species representing several vertebrate and invertebrate taxa. Uninseminated turkey eggs can develop parthenogenetically into viable turkeys ***(127)***. Naturally occurring parthenogenesis in reptiles has been recorded for several lizard species ***(128–130)***. In the amphibian species *Xenopus laevis*, both gynogenetic and androgenetic individuals can be experimentally induced and grown into adulthood ***(131–133)***. The first unisexual vertebrate discovered was a fresh-water fish, the Amazon molly (*Poecilia formosa*) ***(134)***. The Amazon molly is an "all female" species that reproduces solely by gynogenesis. During reproduction, the entire female genome is transmitted to the next generation. Sperm of related, sexual species is needed to activate embryogenesis in the ova of the gynogenetic female, but the paternal genome is incorporated nor expressed ***(135)***.

Androgenetic and gynogenetic embryos can be induced in many species of fish, including commercially interesting species such as salmon and trout and the much-studied zebrafish ***(136–139)***. In zebrafish, the development of induced haploid and diploid andro- and gynogenetic embryos has been compared. Development of haploid zebrafish embryos arrests in an early stage. Haploid androgenotes are indistinguishable in appearance from haploid gynogenotes, indicating that the failure to develop into adults is not due to parent-of-origin effects. Diploid andro- and gynogenotes, obtained by inhibiting the first mitotic division of the haploid embryo, complete embryogenesis and develop into adults ***(136,139,140)***. Although parent-of-origin transgene methylation has been observed in zebrafish ***(141)***, it appears that even if endogenous genes are imprinted, this has no detectable effects on the development of uniparental organisms.

In insects, parthenogenesis is found in many species, for example, aphids and grasshoppers ***(142–145)***. Natural androgenesis has only been reported for the Sicilian stick insect ***(146,147)***, but both androgenesis and gynogenesis (leading to viable adults) can be induced in *Drosophila* ***(148,149)***. All tested members of the order Hymenoptera, which comprises bees, wasps, and ants, are haplodiploid: the sexes differ in ploidy. Whereas the females are diploid and develop from fertilized eggs, male bees, wasps, and ants are haploid and arise gynogenetically from unfertilized eggs ***(150,151)***. In wasps, it is the diploid status itself and not the contribution of the male genome that determines the female gender. Infection of the parthenogenesis-inducing *Wolbachia* bacterium disrupts meiosis in female hosts and leads to the formation of diploid eggs, which develop into female wasps ***(152)***. Haplodiploidy occurs also in other insect orders and in Arthropods other than insects, such as ticks and mites ***(151)***.

Genomic imprinting has been observed in insects. In fact, the term "imprinting" in an epigenetic context was first used to describe the maternally inherited control of sex-specific elimination of X chromosomes in the fly *Sciara coprophila* ***(153,154)***. As in zebrafish, the occurrence of androgenetic, gynogenetic, and parthenogenetic insects and their development into adulthood suggests that, unlike in mammals and flowering plants, genomic imprints do not play a pivotal role in early development of insect embryos.

The widespread occurrence of natural uniparental embryogenesis and the possibilty to induce androgenesis and/or gynogenesis in species representing four of the five vertebrate classes and many invertebrate taxa can be explained by assuming that in these taxa the gamete genomes do not have a gender. Any combination of gamete genomes (or, at least in some species, any haploid gamete genome) can result in the formation and development of a viable embryo. Where the development of gynogenetic or parthenogenetic and androgenetic embryos has been studied within a species, no parent-of-origin effects have been found, again indicating an apparent absence of differences between the male and female gamete genomes. In the gametes of mammals and in the endosperm genome of flowering plants, it is highly likely that it is the establishments of genomic imprints that determines the gender of the gamete genome. With genomic imprints present and sets of genes uniparentally silenced, gamete genomes of different parental origin have become complementary and hence only a combination of maternal and paternal genomes can be successful. Because of their suggested role in maintaining genomic imprints in the polar nuclei, *FIS1*, *FIS2*, and *FIE* in *Arabidopsis* may thus give the central cell genome her maternal identity. One can therefore speculate about the existence of genes that impose and/or maintain maleness onto the sperm genomes. Paternally imprinted genes are predicted to inhibit endosperm development. At the moment, we can only speculate about the nature of such genes. They may be genes involved in import of sugars in the endosperm, or cell cycle-inhibiting genes controlling the rate of proliferation in the endosperm. It is possible that the expression of such imprinted genes is controlled by paternally expressed upstream regulatory genes (possibly polycomb genes). Analogous to *MEA*, *FIS2*, and *FIE* on the female side, the activity of such regulatory genes would lead to parent-of-origin-specific silencing of genes and thereby impose a male gender onto the genome.

Plants carrying loss of function mutations in *FIS1/MEDEA* have been accused of being bad mothers and hence the gene was named after Medea, who sacrificed her children to revenge their father's infidelity ***(104)***. By first imposing a female identity on her genome, we think that *FIS1* is more like Medea's illustrious predecessor, Eve.

11. Possible Applications of Genomic Imprinting in Plants

It will be obvious that large parts of the knowledge of genomic imprinting in plants are still lacking. We are just beginning to identify some of the imprinted genes and determine their function in endosperm development in Angiosperms. Though the importance of DNA methylation for the establishment and/or maintenance of genomic imprints is becoming clear now, the way in which alleles are marked for parent-of-origin-specific methylation is not known, nor is the exact role of methylation. Nonetheless, it has already become clear that the development of embryo and endosperm and the involvement of genomic imprinting is an area of research that is not only scientifically highly interesting, but also offers several opportunities at the level of bioengineering.

First, as has been shown by both interploidy crosses and crosses with demethylated parents in *Arabidopsis*, differential genomic imprints in the parental genomes influence endosperm size and development and thereby eventually embryo and seed size. Many of the world's most important crop species, such as wheat, rye, rice, barley, and maize, have a persistent endosperm. In contrast to *Arabidopsis*, in such plants the endosperm is still present in the mature seed and can make up to 90% of the seed weight. In cereals, where the mature seed is the crop product, endosperm therefore constitutes the bulk of the harvest. Opportunities to influence (increase) endosperm size could thus lead to an enhanced yield.

As described above, changing the methylation level of one of the parental gamete genomes would be one of the ways to achieve this. When more imprinted genes, and more components of the imprinting machinery in flowering plants, are characterized, additional and more sophisticated ways to influence the expression of imprinted genes are likely to become possible. For instance, transgenic plants overexpressing one or more maternally imprinted, endosperm-promoting genes could be generated. When such plants are used as pollen donors, this will make the pollen even more vigorous concerning the promotion of endosperm growth. Depending on the nature of silencing of maternally imprinted genes, it may even be possible to have such genes expressed from the maternally inherited copy of the endosperm genome as well. If the mark that eventually leads to imprinting and silencing in the maternal copy is located in the promoter sequence, then expression of the gene from a different promoter should be possible. In the mouse *Igf2r* gene, the information necessary for both *de novo* methylation of the imprinted paternal allele and an allele-discrimination signal are located on a 113-bp sequence in intron 2 ***(155)***. Deletion or mutation of these signals abolishes *de novo* methylation of a normally methylated site in the imprinted allele, as well as differential methylation of the parental alleles in the embryo. Acquiring comparable data on the sequences involved in the regulation of imprinting of plant genes will make the manipulation of parental imprinting more likely.

Influencing the extent of genomic imprinting in plants will also be very useful in the establishment of interspecific crosses. Many related plant species have developed interspecific cross barriers that appear to be of an epigenetic rather than a genetic character.

In 1942, Stephens noted that a cross between 4x Asiatic cotton and 2x American cotton managed to produce viable seeds despite their being of a different ploidy ***(156)***. As this was in conflict with the then-accepted required

2:3:2 ratio for maternal, endosperm, and embryo tissue, he suggested that within each species, a single set of chromosomes has a certain strength that may differ from that in related species, whether of the same or of a different ploidy. More examples of interspecific interploidy crosses in different species followed, with similar results ***(157)***. Howard, by comparing the seed weight following different interspecific and interploidy crosses, was able to calculate that the relative "seed strengths" of the diploid *Nasturtium officinale* and its allotetraploid *N. uniseriatum* were 1 and 1.41 instead of the expected 1 and 2 ***(158)***. Perhaps the most striking examples were found in the genus *Solanum* ***(159,160)***. Seed set in the interploidy cross between the two tetraploid species *S. acaule* and *S. tuberosum* invariably fails due to endosperm breakdown. However, when an autotetraploid of *S. acaule* was used, the cross could be made, notwithstanding the fact that the endosperm now consisted of 8 maternal:2 paternal sets of chromosomes. Johnston et al. ***(159)*** accounted for these and similar observations with their endosperm balance number (EBN) hypothesis. According to this theory, in the case where there is a difference in "genome strength" between the parents, an effective ploidy ratio—which is not necessarily the same as the absolute ploidy ratio—has to be reached within the endosperm. The endosperm balance number hypothesis has been shown to be of highest value in the genus *Solanum*. Johnston and Hanneman ***(161)*** assigned the diploid species *S. chacoense* a random EBN of 2. When crossing this species and its autotetraploid (EBN 4) into other *Solanum* species, only those crosses that gave rise to an EBN balance of 2:1 were successful. In this way, a number of 2x and 4x *Solanum* species could be assigned an endosperm balance number of 1, 2, or 4 ***(68,161)***.

The EBN can be interpreted in terms of genomic imprinting. Species with a high EBN are likely to be highly imprinted species, that is, species in which uniparental imprinting has a strong effect on the development of the seed. This can be explained either by assuming that a large number of genes in such a species is imprinted, or because imprinted genes have a particularly strong effect on seed development ***(162)***. If differences in EBN indeed can be accounted for by differences in gamete genome imprinting, then changing the level of imprinting should allow interspecific crosses between species with a different EBN without the need for polyploidization of one of these species.

A third and probably most promising implication of engineering the level of genomic imprinting in flowering plants is the possible creation of apomictic plants. Many crop plants are produced as F1 hybrids between inbred strains. Because of the heterosis effect, F1 hybrids are more vigorous then either parent. The obvious drawback is that each F1 generation has to be produced by cross-fertilization of the two parent strains. Introducing apomixis into the desired F1 hybrid strain would allow this strain with its desired genetic background to be propagated into future generations without meiotic recombination and segregation. One of the problems involved in trying to obtain apomictic plants is the dependency on double fertilization. For the successful sexual reproduction of a plant, both a functional embryo and a functional endosperm are needed. In an ideal, true apomict, seeds would develop without pollination, implying that both embryo and endosperm development would have to be initiated and carried on without fertilization. Most natural apomicts have successfully overcome the fertilization-dependent barrier of embryo development. A $2n$ embryo, derived from either the fusion of two haploid maternal gametes or from an aberrant $2n$ maternal gamete, is formed and develops within the seed. However, with a few exceptions, in all these seeds the endosperm is still dependent on fertilization of the polar nuclei by a sperm, and in most cases the requirement of a 2:1 maternal:paternal ratio remains ***(32,50,163)***. Screening for mutants that allow both endosperm and embryo development in the absence of fertilization has not yet been successful.

In *Arabidopsis thaliana* it is now possible to obtain seeds in which the endosperm develops autonomously in the absence of fertilization to such an extent that at least morphologically it cannot be distinguished from a sexual endosperm ***(107)***. Combining the mutant *fie-1* allele and hypomethylation leads to the development of such an endosperm in mutant ovules. If these, purely maternally derived endosperms are functional, that is, if they are capable of importing and storing enough nutrients and expressing the genes needed for breakdown and conversion of those nutrients when embryo growth and development demands it, then in principle they could support the development of an (apomictic) embryo. If so, the autonomous endosperm mutants may be the ideal background for mutagenizing and screening for autonomous embryo development in *Arabidopsis*. If this proves to be successful, then characterization of the genes and pathways involved in autonomous development of either tissue may open the way to the introduction of apomixis in other (crop) species.

References

1. Kaufman, M. H., Barton, S. C., and Surani, M. A. (1977) Normal postimplantation development of mouse parthenogenetic embryos to the forelimb bud stage. *Nature* **265,** 53–55.
2. Barton, S. C., Surani, M. A., and Norris, M. L. (1984) Role of paternal and maternal genomes in mouse development. *Nature* **311,** 374–376.
3. Surani, M. A. H., Barton, S. C., and Norris, M. L. (1984) Development of reconstituted mouse eggs suggests imprinting of the genome during gametogenesis. *Nature* **308,** 548–550.
4. McGrath, J. and Solter, D. (1984) Completion of mouse embryogenesis requires both the maternal and paternal genomes. *Cell* **37,** 179–183.

5. Surani, M. A. H., Barton, S. C., and Norris, M. L. (1986) Nuclear transplantation in the mouse: heritable differences between maternal and paternal genomes after activation of the embryonic genome. *Cell* **45,** 127–136.
6. Thomson, J. A. and Solter, D. (1988) The developmental fate of androgenetic, parthenogenetic and gynogenetic cells in chimeric gastrulating mouse embryos. *Genes Dev.* **2,** 1344–1351.
7. Surani, M. A. H., Barton, S. C., and Norris, M. L. (1987) Influence of parental chromosomes on spatial specificity in androgenetic vs. parthenogenetic chimaeras in the mouse. *Nature* **326,** 395–397.
8. Nagy, A., Paldi, A., Dezso, L., Varga, L., and Magyar, A. (1987) Prenatal fate of parthenogenetic cells in mouse aggregation chimeras. *Development* **101,** 67–71.
9. Surani, M. A., Kothary, R., Allen, N. D., Singh, P. B., Fundele, R., Ferguson-Smith, A. C., and Barton, S. C. (1990) Genome imprinting and development in the mouse. *Development* **113**(Suppl.), 89–98.
10. Schwere, S. (1896) Zur Entwicklungsgeschichte der Frucht von *Taraxacum officinale* Web.; ein Beitrag zur Embryologie der Compositen, "*Flora*" Bd. **82,** pp. 32–66, Taf. II-V, Marburg.
11. Raunkiær, C. (1903) Kimdannelse uden Befrugtning hos Mælkebøtte (*Taraxacum*). *Botanisk Tidsskrift* **25,** 110–140.
12. Murbeck, S. (1904) Parthenogenese bei den Gattungen *Taraxacum* und *Hieracium*. *Bot. Not.* **6,** 285–296.
13. Rutishauser, A. (1954) Entwicklungserregung der Eizelle bei pseudogamen Arten der Gattung *Ranunculus*. (in German, with English, Italian, and French summary) *Bull. Schweiz. Akad. Med. Wissensch.* **10,** 491–512.
14. Rutishauser, A. (1967) Fortpflanzungsmodus und Meiose apomiktischer Blütenpflanzen. *Protoplasmatologia* **6**(3), 3–243.
15. Richards, A. J. (1970) Eutriploid facultative agamospermy in *Taraxacum*. *New Phytol.* **69,** 761–774.
16. Richards, A. J. (1973) The origin of *Taraxacum* agamospecies. *Bot. J. Linn. Soc.* **66,** 189–211.
17. Pogan, E. and Wcislo, H. (1995) Embryological analysis of *Hieracium pilosella* L from Poland. *Acta Biol. Cracov. Ser. Bot.* **37,** 53–61.
18. Bruun, L. (1991) Histological and semiquantitative approaches to in vitro cellular-responses of ovule, embryo and endosperm in sugar beet, *Beta vulgaris* L. *Sex. Plant Reprod.* **4,** 64–72.
19. Ferrant, V. and Bouharmont, J. (1994) Origin of gynogenetic embryos of *Beta vulgaris* L. *Sex. Plant Reprod.* **7,** 12–16.
20. Hansen, A. L., Gertz, A., and Joersbo, M. (1995) Short-duration colchicine treatment for in vitro chromosome doubling during ovule culture of *Beta vulgaris* L. *Plant Breeding* **114,** 515–519.
21. Miyoshi, K. and Asakura, N. (1996) Callus induction, regeneration of haploid plants and chromosome doubling in ovule cultures of pot gerbera (*Gerbera jamesonii*) *Plant Cell Rep.* **16,** 1–5.

22. Mdarhri-Alaoui, M., Saidi, N., Chlyah, A., and Chlyah, H. (1998) Green haploid plant formation in durum wheat through in vitro gynogenesis. *C.R. Acad. Sci. Ser. III: Life Sci.* **321,** 25–30.
23. Michalik, B., Adamus, A., and Nowak, E. (2000) Gynogenesis in Polish onion cultivars. *J. Plant Physiol.* **156,** 211–216.
24. Clapham, D. (1971) In vitro development of callus from the pollen of *Lolium* and *Hordeum*. *Z. Pflanzenzüchtg.* **65,** 285–292.
25. Ouyang, K. H., Hu, H., Chuang, C. C., and Tseng, C.-C. (1973) Induction of pollen plants from anthers of *Triticum aestivum* L. cultured in vitro. *Sci. Sinic.* **16,** 79–95.
26. Pretova, A., de Ruijter, N. C. A., van Lammeren, A. A. M., and Schel, J. H. N. (1993) Structural observations during androgenic microspore culture of the 4C1 genotype of *Zea mays* L. *Euphytica* **65,** 61–69.
27. van den Bulk, R. W., de Vries van Hulten, H. P. J., Custers, J. B. M. and Dons, J. J. M. (1994) Induction of embryogenesis in isolated microspores of tulip. *Plant Sci.* **104,** 101–111.
28. Binarova, P., Hause, G., Genklova, V., Cordewener, J. H. G., and Campagne, M. M. V. (1997) A short severe heat shock is required to induce embryogenesis in late bicellular pollen of *Brassica napus* L. *Sex. Plant Reprod.* **10,** 200–208.
29. Atanassov, A., Zagorska, N, Boyadjiev, P., and Djilianov, D. (1995) In-vitro production of haploid plants. *World J. Microbiol. Biotechnol.* **11,** 400–408.
30. Raghavan, V. (1976) Role of the generative cell in androgenesis in henbane. *Science* **191,** 388–389.
31. Tyukavin, G. B., Shmykova, N. A., and Monakhova, M. A. (1999) Cytological study of embryogenesis in cultured carrot anthers. *Russian J. Plant Physiol.* **46,** 767–773.
32. Richards, A. J. (1986) *Plant Breeding Sytems.* George Allen & Unwin, London.
33. Sterk, A. A., Hommels, C. H., Jenniskens, M. J. P. J., Neuteboom, J. H., den Nijs, J. C. M., Oosterveld, P., and Segal, S. (1987) *Paardebloemen-planten zonder vader.* Koninklijke Nederlandse Natuurhistorische Vereniging, Utrecht, The Netherlands.
34. Schulz, P. and Jensen, W. A. (1974) *Capsella* embryogenesis: the development of the free nuclear endosperm. *Protoplasma* **80,** 183–205.
35. Bhatnagar, S. P. and Sawhney, V. (1981) Endosperm—its morphology, ultrastructure, and histochemistry. *Int. Rev. Cytol.* **73,** 55–102.
36. Lopes, M. A. and Larkins, B. A. (1993) Endosperm origin, development, and function. *Plant Cell* **5,** 1383–1399.
37. Berger, F. (1999) Endosperm development. *Curr. Opin. Plant Biol.* **2,** 28–32.
38. Schulz, P. and Jensen, W. A. (1971) *Capsella* embryogenesis: the chalazal proliferating tissue. *J. Cell Sci.* **8,** 201–227.
39. Mansfield, S. G. and Briarty, L. G. (1990) Development of the free-nuclear endosperm in *Arabidopsis thaliana* (L.) *Arabidopsis Inf. Serv.* **27.**
40. Mansfield, S. G. (1994) Endosperm development, in *Arabidopsis, An Atlas of Morphology and Development.* (Bowman, J., ed.), Springer-Verlag, Berlin-Heidelberg-New York, pp. 385–397.

41. Håkansson, A. (1956) Seed development of *Brassica oleracea* and *B. rapa* after certain reciprocal pollinations. *Hereditas* **42,** 373–396.
42. Mansfield, S. G., Briarty, L. G., and Erni, S. (1991) Early embryogenesis in *Arabidopsis thaliana.* I. The mature embryo sac. *Can. J. Bot.* **69,** 447–460.
43. Chamberlin, M. A., Horner, H. T., and Palmer, R. G. (1993) Nutrition of the ovule, embryo sac, and young embryo in soybean: an anatomical and autoradiographic study. *Can. J. Bot.* **71,** 1152–1168.
44. Brink, R. A. and Cooper, D. C. (1947) The endosperm in seed development. *Bot. Rev.* **13,** 423–541.
45. Maheswari, P. (1950) *An Introduction to the Embryology of the Angiosperms.* McGraw-Hill, New York.
46. Vijayaraghavan, M. R. and Prabhakar, K. (1984) The endosperm, in *Embryology of Angiosperms* (Johri, B. M., ed.), Springer-Verlag, Berlin, pp. 319–376.
47. Mansfield, S. G. and Briarty, L. G. (1990b) Endosperm cellularization in *Arabidopsis thaliana* L. *Arabidopsis Inf. Serv.* **27.**
48. Brown, R. C., Lemmon, B. E., Nguyen, H., and Olsen, O.-A. (1999) Development of endosperm in *Arabidopsis thaliana. Sex. Plant Reprod.* **12,** 32–42.
49. Marinos, N. G. (1970) Embryogenesis of the pea (*Pisum sativum*) I. The cytological environment of the developing embryo. *Protoplasma* **70,** 261–279.
50. Haig, D. and Westoby, M. (1991) Genomic imprinting in endosperm: its effect on seed development in crosses between species, and between different ploidies of the same species, and its implications for the evolution of apomixis. *Phil. Trans. R. Soc. Lond. B* **333,** 1–13.
51. Müntzing, A. (1930) Über Chromosomenvermehrung in *Galeopsis*—Kreuzungen und ihre phylogenetische Bedeutung. *Hereditas* (Lund) **14,** 153–172.
52. Müntzing, A. (1933) Hybrid incompatibility and the origin of polyploidy. *Hereditas* (Lund) **18,** 33–55.
53. Watkins, A. E. (1932) Hybrid sterility and incompatibility. *J. Genet.* **25,** 125–162.
54. Howard, H. W. (1939) The size of seeds in diploid and autotetraploid *Brassica oleracea* L. *J. Genet.* **38,** 325–340.
55. Cooper, D. C. and Brink, R. A. (1945) Seed collapse following matings between diploid and tetraploid races of *Lycopersicon pimpinellifolium. Genetics* **30,** 376–401.
56. Cooper, D. C. (1951) Caryopsis development following matings between diploid and tetraploid strains in *Zea mays. Am. J. Bot.* **38,** 702–708.
57. Håkansson, A. (1952) Seed development after 2x, 4x crosses in *Galeopsis pubescens. Hereditas* **38,** 425–448.
58. Håkansson, A. (1953) Endosperm formation after 2x, 4x crosses in certain cereals, especially in *Hordeum vulgare. Hereditas* **39,** 57–64.
59. Håkansson, A. and Ellerström, S. (1950) Seed development after reciprocal crosses between diploid and tetraploid rye. *Hereditas* **36,** 256–296.
60. Fagerlind, F. (1937) Embryologische, zytologische und bestäubungsexperimentelle Studien in der Familie *Rubiaceae* nebst Bemerkungen über einige Polyploiditätsprobleme. *Acta Horti Bergiana* **11,** 195–470.

61. Boyes, J. W. and Thompson, W. P. (1937) The development of the endosperm and embryo in reciprocal interspecific crosses in cereals. *J. Genet.* **34.**
62. Kihara, H. and Nishiyama, I. (1932) Different compatibility in reciprocal crosses of *Avena* with special reference to tetraploid hybrids between hexaploid and diploid species. *Jpn. J. Bot.* **6.**
63. von Wangenheim, K.-H. (1962) Zur Ursache der Abortion von Samenanlagen in Diploid-Polyploid-Kreuzungen. II. Unterschiedliche Differenzierung von Endospermen mit gleichem Genom. *Z. Vererbungslehre* **93,** 319–334.
64. Nishiyama, I. and Inomata, N. (1966) Embryological studies on cross-incompatibility between 2x and 4x in *Brassica. Jpn. J. Genet.* **41,** 27–42.
65. Sarkar, K. R. and Coe, E. H., Jr. (1971) Anomalous fertilisation in diploid-tetraploid crosses in maize. *Crop Sci.* **11,** 539–542.
66. Kermicle, J. L. (1971) Pleiotropic effects on seed development of *the indeterminate gametophyte* gene in maize. *Science* **166,** 1422–1424.
67. Lin, B.-Y. (1984) Ploidy barrier to endosperm development in maize. *Genetics* **107,** 103–115.
68. Ehlenfeldt, M. K. and Ortiz, R. (1995) Evidence on the nature and origins of endosperm dosage requirements in *Solanum* and other Angiosperm genera. *Sex. Plant Reprod.* **8,** 189–196.
69. Thompson, W. P. (1930) Causes of difference in success of reciprocal interspecific crosses. *Am. Naturalist* **64,** 407–421.
70. Woodell, S. R. J. and Valentine, D. H. (1961) Studies in British primulas. IX. Seed incompatibility in diploid-autotetraploid crosses. *New Phytol.* **60,** 282–294.
71. Rédei, G. (1964) Crossing experiments with polyploids. *Arabidopsis Electronic Information Service* **1.**
72. Scott, R. J., Spielman, M., Bailey, J., and Dickinson, H. G. (1998) Parent-of-origin effects on seed development in *Arabidopsis thaliana. Development* **125,** 3329–3341.
73. Kermicle, J. L. (1970) Dependence of the R-mottled aleurone phenotype in maize on mode of sexual transmission. *Genetics* **66,** 69–85.
74. Lin, B.-Y. (1982) Association of endosperm reduction with parental imprinting in maize. *Genetics* **100,** 475–486.
75. Haig, D., and Westoby, M. (1989) Parent-specific gene expression and the triploid endosperm. *Am. Nat.* **134,** 147–155.
76. Kermicle, J. L. and Alleman, M. (1990) Gametic imprinting in maize in relation to the angiosperm life cycle. *Development* **113**(Suppl.), 9–14.
77. Lund, G., Ciceri, P., and Viotti, A. (1995) Maternal-specific demethylation and expression of specific alleles of zein genes in the endosperm of *Zea mays* L. *Plant J.* **8,** 571–581.
78. Lund, G., Messing, J., and Viotti, A. (1995) Endosperm-specific demethylation and activation of specific alleles of alpha-tubulin genes of *Zea mays* L. *Mol. Gen. Genet.* **246,** 716–722.

79. Abbot, R. J. and Gomes, M. F. (1989) Population genetic structure and outcrossing rate in *Arabidopsis thaliana* (L.) Heynh. *Heredity* **62,** 411–418.
80. Stebbins, G.L. (1974) *Flowering Plants: Evolution Above the Species Level. London, Edward Arnold.*
81. Finnegan, E. J., Peacock, W. J., and Dennis, E. S. (1998) Reduced DNA methylation in plants. *Annu. Rev. Plant Physiol. Plant Mol. Biol.* **49,** 223–247.
82. Adams, S., Vinkenoog, R., Spielman, M., Dickinson, H. G., and Scott, R. J. (2000) Parental imprinting in *Arabidopsis* requires DNA methylation. *Development* **127,** 2493–2502.
83. Finnegan, E. J., Peacock, W. J., and Dennis, E. S. (1996) Reduced DNA methylation in *Arabidopsis thaliana* results in abnormal plant development. *Proc. Natl. Acad. Sci. USA* **93,** 8449–8454.
84. van Blokland, R., ten Lohuis, M. and Meyer, P. (1997) Condensation of chromatin in transcriptional regions of an inactivated plant transgene: evidence for an active role of transcription in gene silencing. *Mol. Gen. Genet.* **257,** 1–13.
85. Assaad, F. F. and Signer, E. R. (1992) Somatic and germinal recombination of a direct repeat in *Arabidopsis*. *Genetics* **132,** 553–566.
86. Ye, F. and Signer, E. R. (1996) RIGS (repeat-induced gene silencing) in *Arabidopsis* is transcriptional and alters chromatin configuration. *Proc. Natl. Acad. Sci. USA* **93,** 10,881–10,886.
87. Razin, A. (1998) CpG methylation, chromatin structure and gene silencing—a three way connection. *EMBO J.* **17,** 4905–4908.
88. Görz, A., Schafer, W., Hirasawa, E., and Kahl, G. (1988) Constitutive and light-induced Dnase I hypersensitive sites in the rbcS-genes of pea (*Pisum sativum*). *Plant Mol. Biol.* **11,** 561–573.
89. Amedeo, P., Habu, Y., Afsar, K., Mittelsten Sceid, O., and Paszkowski, J. (2000) Disruption of the plant gene *MOM* releases transcriptional silencing of methylated genes. *Nature* **405,** 203–206.
90. Jaenisch, R. (1997) DNA methylation and imprinting: why bother? *Trends Genet.* **13,** 323–329.
91. Li, E., Bestor, T. H., and Jaenisch, R. (1992) Targeted mutation of the DNA methyltranferase gene results in embryonic lethality. *Cell* **69,** 915–926.
92. Genger, R. K., Kovac, K. A., Dennis, E. S., Peacock, W. J., and Finnegan, E. J. (1999) Multiple DNA methyltransferase genes in *Arabidopsis thaliana*. *Plant Mol. Biol.* **41,** 269–278.
93. Finnegan, E. J., Peacock, W. J., and Dennis, E. S. (2000) DNA methylation, a key regulator of plant development and other processes. *Curr. Opin. Genet. Dev.* **10,** 217–223.
94. Jacobsen, S. E. and Meyerowitz, E. M. (1997) Hypermethylated *SUPERMAN* epigenetic alleles in *Arabidopsis*. *Science* **277,** 1100–1103.
95. Li, E., Beard, C., and Jaenisch, R. (1993) Role for DNA methylation in genomic imprinting. *Nature* **366,** 362–365.

96. Tilghman, S. M. (1999) The sins of the fathers and mothers: genomic imprinting in mammalian development. *Cell* **96,** 185–193.
97. Chaudhuri, S. and Messing, J. (1994) Allele-specific parental imprinting of *dzr1*, a posttranscriptional regulator of zein accumulation. *Proc. Natl. Acad. Sci. USA* **91,** 4867–4871.
98. Matzke, M. and Matzke, A. J. M. (1993) Genomic imprinting in plants: parental effects and trans-inactivation phenomena. *Annu. Rev. Plant Physiol. Plant Mol. Biol.* **44,** 53–76.
99. Martienssen, R. (1998) Chromosomal imprinting in plants. *Curr. Opin. Genet. Dev.* **8,** 240–244.
100. Castle, L. A., Errampalli, D., Atherton, T. L., Franzmann, L. H., Yoon, E. S., and Meinke, D. W. (1993) Genetic and molecular characterization of embryonic mutants identified following seed transformation in *Arabidopsis*. *Mol. Gen. Genet.* **241,** 504–514.
101. Ohad, N., Margossian, L., Hsu, Y.-C., Williams, C., Repetti, P., and Fischer, R. L. (1996) A mutation that allows endosperm development without fertilization. *Proc. Natl. Acad. Sci. USA* **93,** 5319–5324.
102. Ohad, N., Yadegari, R., Margossian, L., Hannon, M., Michaeli, D., Harada, J. J., Goldberg, R. B., and Fischer, R. L. (1999) Mutations in *FIE*, a WD Polycomb group gene, allow endosperm development without fertilization. *Plant Cell* **11,** 407–415.
103. Chaudhury, A. M., Ming, L., Miller, C., Craig, S., Dennis, E. S., and Peacock, W. J. (1997) Fertilization-independent seed development in *Arabidopsis thaliana*. *Proc. Natl. Acad. Sci. USA* **94,** 4223–4228.
104. Grossniklaus, U., Vielle-Calzada, J.-P., Hoeppner, M. A., and Gagliano, W. (1998) Maternal control of embryogenesis by *MEDEA*, a Polycomb group gene in *Arabidopsis*. *Science* **280,** 446–450.
105. Kiyosue, T., Ohad, N., Yadegari, R., Hannon, M., Dinneny, J., Wells, D., Katz, A., Margossian, L., Harada, J. J., Goldberg, R. B., and Fischer, R. L. (1999) Control of fertilization-independent endosperm development by the *MEDEA* polycomb gene in Arabidopsis. *Proc. Natl. Acad. Sci. USA* **96,** 4186–4191.
106. Luo, M., Bilodeau, P., Koltunow, A., Dennis, E. S., Peacock, W. J., and Chaudhury, A. (1999) Genes controlling fertilization-independent seed development in *Arabidopsis thaliana*. *Proc. Natl. Acad. Sci. USA* **96,** 296–301.
107. Vinkenoog, R., Spielman, M., Adams, S., Fischer, R. L., Dickinson, H. G., and Scott, R. J. (2000) Hypomethylation promotes autonomous endosperm development and rescues post-fertilisation lethality in fie-mutants. *Plant Cell* **12,** 2271–2282.
108. Vielle-Calzada, J.-P., Thomas, J., Spillane, C., Coluccio, A., Hoeppner, M. A., and Grossniklaus, U. (1999) Maintenance of genomic imprinting at the *Arabidopsis medea* locus requires zygotic DDM1 activity. *Genes Dev.* **13,** 2971–2982.
109. Vongs, A., Kakutani, T., Martienssen, R. A., and Richards, E. J. (1993) *Arabidopsis thaliana* DNA methylation mutants. *Science* **260,** 1926–1928.

110. Kakutani, T., Jeddeloh, J. A., and Richards, E. J. (1995) Characterisation of an *Arabidopsis thaliana* DNA hypomethylation mutant. *Nucleic Acids Res.* **23,** 130–137.
111. Jeddeloh, J. A., Stokes, T. L., and Richards, E. J. (1999) Maintenance of genomic methylation requires a SWI2/SNF2-like protein. *Nature Genet.* **22,** 94–97.
112. Peterson, C. L. and Workman, J. L. (2000) Promoter targeting and chromatin remodeling by the SWI/SNF complex. *Curr. Opin. Genet. Dev.* **10,** 187–192.
113. Campbell, R. B., Sinclair, D. A. R., Couling, M., and Brock, H. W. (1995) Genetic interactions and dosage effects of polycomb group genes of *Drosophila. Mol. Gen. Genet.* **246,** 291–300.
114. Shermoen, A. W. and O'Farrell, P. H. (1991) Progression of the cell cycle through mitosis leads to abortion of nascent transcripts. *Cell* **67,** 303–310.
115. Kinoshita, T., Yadegari, R., Harada, J. J., Goldberg, R. B., and Fischer, R. L. (1999) Imprinting of the MEDEA Polycomb gene in the *Arabidopsis* endosperm. *Plant Cell* **11,** 1945–1952.
116. Vielle-Calzada, J-P., Baskar, R., and Grossniklaus, U. (2000) Delayed activation of the paternal genome during seed development. *Nature* **404,** 91–94.
117. Pirotta, V. (1997) PcG complexes and chromatin silencing. *Curr. Opin. Genet. Dev.* **7,** 249–258.
118. van Lohuizen, M., Tijms, M., Voncken. J. W., Schumacher, A., Magnuson, T., and Wientjens, E. (1998) Interaction of mouse Polycomb-Group (Pc-G) proteins Enx1 and Enx2 with Eed: indication for separate Pc-G complexes. *Mol. Cell Biol.* **18,** 3572–3579.
119. Sewalt, R. G. A. B., van der Vlag, J., Gunster, M. J., Hamer, K. M., den Blaauwen, J. L., Satijn, D. P. E., Hendrix, T., van Driel, R., and Otte, A. P. (1998) Characterization of interactions between the mammalian Polycomb-Group proteins Enx1/EZH2 and EED suggests the existence of different mammalian Polycomb-Group protein complexes. *Mol. Cell Biol.* **18,** 3586–3595.
120. Sondek, J., Boh, A., Lambright, D. G., Hamm, H. E., and Sigler, P. B. (1996) Crystal structure of a G_A protein βγ dimer at 2.1 Å resolution. *Nature* **379,** 369–374.
121. Ng, J., Li, R. H., Morgan, K., and Simon, J. (1997) Evolutionary conservation and predicted structure of the *Drosophila* extra sex combs repressor protein. *Mol. Cell Biol.* **17,** 6663–6672.
122. Jenuwein, T., Laible, G., Dorn, R., and Reuter, G. (1998) SET domain proteins modulate chromatin domains in eu- and heterochromatin. *Cell. Mol. Life Sci.* **54,** 80–93.
123. Jones, C. A., Ng, J., Peterson, A. J., Morgan, K., Simon, J., and Jones, R. S. (1998) The *Drosophila* esc and E(z) proteins are direct partners in Polycomb group-mediated repression. *Mol. Cell Biol.* **18,** 2825–2834.
124. Ma, H. (1997) Polycomb in plants. *Trends Genet.* **13,** 167.
125. Preuss, D. (1999) Chromatin silencing and *Arabidopsis* development: a role for Polycomb proteins. *Plant Cell* **11,** 765–767.

126. Gutjahr, T., Frei, E., Spicer, C., Baumgartner, S., White, R. A. H., and Noll, M. (1995) The Polycomb-group gene, *extra sex combs*, encodes a nuclear member of the WD-40 repeat family. *EMBO J.* **14,** 4296–4306.
127. Cassar, G., Mohammed, M., John, T. M., Gadzinski, P., and Etches, R. J. (1998) Differentiating between parthenogenetic and positive development embryos in turkeys by molecular sexing. *Poultry Sci.* **77,** 1463–1468.
128. Moritz, C., McCallum, H., Donnellan, S., and Roberts, J. D. (1991) Parasite loads in parthenogenetic and sexual lizards (*Heteronotia binoei*): support for the Red Queen hypothesis. *Proc. R. Soc. Lond. B* **244,** 145–149.
129. MacGulloch, R. D., Murphy, R. W., Kupriyanova, L. A., and Darevski, I. S. (1997) The Caucasian rock lizard *Lacerta rostombekovi*: a monoclonal parthenogenetic vertebrate. *Biochem. Syst. Ecol.* **25,** 33–37.
130. Price, A. H. (1992) Comparative behavior in lizards of the genus *Cnemidophorus* (Teiidae) with comments on the evolution of parthenogenesis in reptiles. *Copeia* **2,** 323–331.
131. Krotoski, D. M., Reinschmidt, D. C., and Tompkins, R. (1985) Developmental mutants isolated from wild-caught *Xenopus laevis* by gynogenesis and inbreeding. *J. Exp. Zool.* **233,** 443–449.
132. Yu, H. J., Shi, C. P., and Liang, J. H. (1986) Production of tetraploid and androgenetic diploid adults of *Xenopus laevis* by suppression of first cleavage. *Kexue Tongbao* **31,** 720.
133. Tompkins, R. and Reinschmidt, D. (1991) Experimentally induced homozygosity in *Xenopus laevis*. *Methods Cell Biol.* **36,** 35–44.
134. Hubbs, C. L. and Hubbs, L. C. (1932) Apparent pathenogenesis in nature, in a form of fish of hybrid origin. *Science* **76,** 628–630.
135. Vrijenhoek, R. (1994) Unisexual fish: model systems for studying ecology and evolution. *Annu. Rev. Ecol. Syst.* **25,** 71–96.
136. Streisinger, G., Walker, C., Dower, N., Knauber, D., and Singer, F. (1981) Production of clones of homozygous diploid zebra fish (*Brachyodanio rerio*). *Nature* **291,** 293–296.
137. May, B., Henley, K. J., Krueger, C. C., and Gloss, S. P. (1988) Androgenesis as a mechanism for chromosome set manipulation in brook trout (*Salvelinus fontinalis*). *Aquaculture* **75,** 57–70.
138. Ihssen, P. E., McKay, L. R., McMillan, I., and Phillips, R. B. (1990) Ploidy manipulation and gynogenesis in fishes: cytogenetic and fisheries applications. *Trans. Am. Fish. Soc.* **119,** 698–717.
139. Corley-Smith, G. E., Lin, C. J., and Brandhorst, B. P. (1996) Production of androgenetic zebrafish (*Danio rerio*). *Genetics* **142,** 1265–1276.
140. Hörstgen-Schwark, G. (1993) Production of homozygous diploid zebra fish (*Brachydanio rerio*). *Aquaculture* **112,** 25–37.
141. Martin, C. C. and McGowan, R. (1995) Parent-of-origin specific effects on the methylation and expression of a transgene in the zebrafish, *Danio rero*. *Genet. Res. Camb.* **65,** 21–28.

142. Drosopoulos, S. (1976) Triploid pseudogamous biotype of the leaf-hopper *Muellerianella fairmairei*. *Nature* **263,** 499–500.
143. Drosopoulos, S. (1978) Laboratory synthesis of a pseudogamous triploid species of the genus *Muellerianella* (Homoptera, Delphacidae). *Evolution* **32,** 916–920.
144. Normark, B. B. (2000) Molecular systematics and evolution of the aphid family Lachnidae. *Mol. Phylogen. Evol.* **14,** 131–140.
145. Gautam, D. C., Crema, R., and Pagliai, A. M. B. (1993) Cytogenetic mechanisms in Aphids. *Bollettino di Zoologia* **60,** 233–244.
146. Tinti, F. and Scali, V. (1993) Chromosomal evidence of hemiclonal and all-paternal offspring production in *Bacillus rossius grandii benazzii* (Insecta, Phasmatodea). *Chromosoma* **102,** 403–414.
147. Tinti, F. and Scali, V. (1995) Allozymic and cytological evidence for hemiclonal, all paternal, and mosaic offspring of the hybridogenetic stick insect *Bacillus rossius grandii grandii*. *J. Exp. Zool.* **273,** 149–159.
148. Komma, D. J. and Endow, S. A. (1995) Haploidy and androgenesis in *Drosophila*. *Proc. Natl. Acad. Sci. USA* **92,** 11,884–11,888.
149. Presgraves, D. C. (2000) A genetic test of the mechanism of Wolbachia-induced cytoplasmic incompatibility in *Drosophila*. *Genetics* **154,** 771–776.
150. Cook, J. M. (1993) Sex determination in the Hymenoptera: a review of models and evidence. *Heredity* **71,** 421–435.
151. Beukeboom, L. W. (1995) Sex determination in Hymenoptera: a need for genetic and molecular studies. *BioEssays* **17,** 813–817.
152. Huigens, M. E., Luck, R. F., Klaassen, R. H. G., Maas, M. F. P. M., Timmermans, M. J. T. N., and Stouthamer, R. (2000) Infectious parthenogenesis. *Nature* **405,** 178–179.
153. Crouse, H. V. (1960) The controlling element in sex chromosome behaviour in *Sciara*. *Genetics* **45,** 1425–1443.
154. Crouse, H. V. (1966) An inducible change in state on the chromosome inheritance and the problem of "imprinting" in *Sciara* (Sciaridae, Diptera). *Chromosoma* **18,** 230–253.
155. Birger, Y., Shemer, R., Perk, J., and Razin, A. (1999) The imprinting box of the mouse *Igf2r* gene. *Nature* **397,** 84–88.
156. Stephens, S. G. (1942) Colchicine-produced polyploids in Gossypium: an autotetraploid asiatic cotton and certain of its hybrids with diploid species. *J. Genet.* **44,** 272–295.
157. von Wangenheim, K.-H. (1957) Untersuchungen über den Zusammenhang zwischen Chromosomenzahl und Kreuzbarkeit bei *Solanum*-Arten. *Zeitschrift für indukt. Abstammungs- und Vererbungslehre* **88,** 21–37.
158. Howard, H. W. (1947) Seed size in crosses between diploid and autotetraploid *Nasturtium officinale* and *allotetraploid N. uniseriatum*. *J. Genet.* **48,** 111–118.
159. Johnston, S. A., den Nijs, T. P. M., Peloquin, S. J., and Hanneman, R. E. (1980) The significance of genic balance to endosperm development in interspecific crosses. *Theor. Appl. Genet.* **57,** 5–9.

160. Ortiz, R. and Ehlenfeldt, M. K. (1992) The importance of endosperm balance number in potato breeding and the evolution of tuber-bearing *Solanum* species. *Euphytica* **60,** 105–113.
161. Johnston, S. A. and Hanneman, R. E. (1982) Manipulations of endosperm balance number overcome crossing barriers between diploid *Solanum* species. *Science* **217,** 446–448.
162. Quarin, C. L. (1999) Effect of pollen source and pollen ploidy on endosperm formation and seed set in pseudogamous apomictic *Paspalum notatum*. *Sex. Plant Reprod.* **11,** 331–335.
163. van Dijk, P. and J. van Damme (2000) Apomixis technology and the paradox of sex. *Trends Plant Sci.* **5,** 81–84.

Index

X, Y, Z